T0250392

# BIOLOGY AND ECOLOGY OF EDIBLE MARINE BIVALVE MOLLUSCS

*Biology and Ecology of Marine Life*

# BIOLOGY AND ECOLOGY OF EDIBLE MARINE BIVALVE MOLLUSCS

**Ramasamy Santhanam, PhD**

Apple Academic Press Inc.
3333 Mistwell Crescent
Oakville, ON L6L 0A2 Canada

Apple Academic Press Inc.
9 Spinnaker Way
Waretown, NJ 08758 USA

© 2018 by Apple Academic Press, Inc.

First issued in paperback 2021

*Exclusive worldwide distribution by CRC Press, a member of Taylor & Francis Group*
No claim to original U.S. Government works

ISBN 13: 978-1-77-463064-8 (pbk)
ISBN 13: 978-1-77-188626-0 (hbk)

All rights reserved. No part of this work may be reprinted or reproduced or utilized in any form or by any electric, mechanical or other means, now known or hereafter invented, including photocopying and recording, or in any information storage or retrieval system, without permission in writing from the publisher or its distributor, except in the case of brief excerpts or quotations for use in reviews or critical articles.

This book contains information obtained from authentic and highly regarded sources. Reprinted material is quoted with permission and sources are indicated. Copyright for individual articles remains with the authors as indicated. A wide variety of references are listed. Reasonable efforts have been made to publish reliable data and information, but the authors, editors, and the publisher cannot assume responsibility for the validity of all materials or the consequences of their use. The authors, editors, and the publisher have attempted to trace the copyright holders of all material reproduced in this publication and apologize to copyright holders if permission to publish in this form has not been obtained. If any copyright material has not been acknowledged, please write and let us know so we may rectify in any future reprint.

**Trademark Notice:** Registered trademark of products or corporate names are used only for explanation and identification without intent to infringe.

---

**Library and Archives Canada Cataloguing in Publication**

Santhanam, Ramasamy, 1946-, author
Biology and ecology of edible marine bivalve molluscs / Ramasamy Santhanam, PhD.
Includes bibliographical references and index.
Issued in print and electronic formats.
ISBN 978-1-77188-626-0 (hardcover).--ISBN 978-1-315-11153-7 (PDF)
1. Bivalves. I. Title.

| QL430.6.S27 2018 | 594'.4 | C2018-900507-6 | C2018-900508-4 |

---

**Library of Congress Cataloging-in-Publication Data**

Names: Santhanam, Ramasamy, 1946- author.
Title: Biology and ecology of edible marine bivalve molluscs / Ramasamy Santhanam, PhD.
Description: Toronto ; New Jersey : Apple Academic Press, 2018. | Series: Biology and ecology of marine life | Includes bibliographical references and index.
Identifiers: LCCN 2018002159 (print) | LCCN 2018003541 (ebook) | ISBN 9781315111537 (ebook) | ISBN 9781771886260 (hardcover : alk. paper) Subjects: LCSH: Bivalves. Classification: LCC QL430.6 (ebook) | LCC QL430.6 .S27 2018 (print) | DDC 594/.4--dc23
LC record available at https://lccn.loc.gov/2018002159

---

Apple Academic Press also publishes its books in a variety of electronic formats. Some content that appears in print may not be available in electronic format. For information about Apple Academic Press products, visit our website at **www.appleacademicpress.com** and the CRC Press website at **www.crc-press.com**

# ABOUT THE AUTHOR

**Ramasamy Santhanam, PhD**

Dr. Ramasamy Santhanam is the former Dean of Fisheries College and Research Institute, Tamil Nadu Veterinary and Animal Sciences University, India. His fields of specialization are marine biology and fisheries environment. Presently, he is serving as a resource person for various universities of India. He has also served as an expert for the "Environment Management Capacity Building," a World Bank-aided project of the Department of Ocean Development, India. He has been a member of the American Fisheries Society, United States; World Aquaculture Society, United States; Global Fisheries Ecosystem Management Network (GFEMN), United States; and the International Union for Conservation of Nature's (IUCN) Commission on Ecosystem Management, Switzerland. To his credit, Dr. Santhanam has 22 books on Marine Biology/Fisheries Science and 70 research papers.

# BIOLOGY AND ECOLOGY OF MARINE LIFE BOOK SERIES

**Series Author:**
Ramasamy Santhanam, PhD
Former Dean, Fisheries College and Research Institute,
Tamil Nadu Veterinary and Animal Sciences University,
Thoothukudi 628008, India
E-mail: rsanthaanamin@yahoo.co.in

**Books in the Series:**

- Biology and Culture of Portunid Crabs of the World Seas
- Biology and Ecology of Edible Marine Bivalve Molluscs
- Biology and Ecology of Edible Marine Gastropod Molluscs
- Biology and Ecology of Venomous Marine Snails
- Biology and Ecology of Venomous Stingrays
- Biology and Ecology of Toxic Pufferfish

# CONTENTS

# LIST OF ABBREVIATIONS

| | |
|---|---|
| ACE | angiotensin I-converting enzyme |
| DPPH | 2,2-diphenyl-1-picrylhydrazyl |
| EAA | essential amino acids |
| FW | fresh weight |
| GAGs | glycosaminoglycans |
| MeOH | methanol |
| MICs | minimum inhibitory concentrations |
| MUFA | monounsaturated fatty acids |
| NEE | nonessential amino acids |
| NW | net weight |
| PBS | phosphate buffer saline |
| PSP | paralytic shellfish poisoning |
| PUFA | polyunsaturated fatty acids |
| SFA | saturated fatty acids |
| SSO | seaside organism disease |
| TFA | total fatty acids |

# PREFACE

Marine bivalve molluscs play an important role in marine ecosystems by serving as habitat and prey for a variety of sea life. This diverse group of species, estimated at about 15,000, inhabits virtually the entire world's oceans, from the balmy tropics to the subzero Arctic, and from the deep ocean to sandy and rocky shorelines. Among the marine bivalves, a total of about 500 species comprising oysters, pearl oysters, mussels, and clams have long been a part of the diet of coastal human populations. These bivalve molluscs represented almost 10% of the total world fishery production. The NOAA estimated the 2011 economic value of commercial bivalve mollusc harvesting at about $1 billion annually in the United States, and the weight of the harvest was estimated at about 75 million tons (Anon., https://oceanservice.noaa.gov/facts/bivalve.html). These bivalve molluscs also represent 26% in volume and 14% in value of the total world aquaculture production. World bivalve molluscs production (capture + aquaculture) has increased substantially in the last 50 years, going from nearly 1 million tons in 1950 to about 14.6 million tons in 2010. China is by far the leading producer of bivalve molluscs, with 10.35 million tons in 2010, representing 70.8% of the global molluscan shellfish production and 80% of the global bivalve mollusc aquaculture production. Other major bivalve producers are Japan, the United States, the Republic of Korea, Thailand, France, and Spain (Botana, 2014).

Many species of marine bivalves are also commercially important. The pearl oysters form the main source of commercially available pearls because the calcareous concretions produced by most other species have no luster. Crushed shells are added as a calcareous supplement to the diet of laying poultry.

Throughout history, brightly colored scallop shells have been used for jewelry and decoration. Several species of edible marine bivalves have been reported to possess valuable bioactive compounds that are the potential sources for the development of novel drugs. Bioceramic powder extracted from the clam shell *Anadara antiquata* has been reported to be potentially useful in the rehabilitation and reposition of human fractured bones.

Edible marine bivalve molluscan species like Atlantic thorny oyster, *Spondylus americanus*; Thorny oyster, *Spondylus* sp.; bear paw clam, *Hippopus hippopus*; China clam, *Hippopus porcellanus*; blue clam, boring clam, *Tridacna crocea*; fluted giant clam, *Tridacna squamosa*; Gigas aka "giant" clam; *Tridacna gigas*, maxima clam, *Tridacna maxima*; and Southern giant clam, *Tridacna derasa,* are of great use in marine aquaria owing to their beautiful coloration.

As the majority of bivalve species exhibit an extremely limited mobility or are completely sessile as adults, they represent the contamination of their habitat ideally. As these bivalves accumulate certain toxic heavy metals from wastes, they may be used as bio-indicators or bioremediators of marine pollution.

Although several books are available on molluscs, a comprehensive book on the biology and ecology of edible marine bivalve molluscs is still wanting. Keeping this in consideration, this attempt has now been made. In this book, aspects, namely, (1) biology of edible marine bivalves; (2) profile (habitat, distribution, diagnostic features, biology, etc.) of about 300 species, (3) aquaculture values, (4) nutritional values, (5) pharmaceutical values, and (6) diseases and parasites have been covered in an easy-to-read style with neat illustrations. It is hoped that the present publication will be of great use for students and researchers of disciplines such as fisheries science, marine biology, aquatic biology, and fisheries and zoology, as well as serving as a standard reference book for the libraries of all universities and research centers.

I am highly indebted to Dr. K. Venkataramanujam, former Dean, Fisheries College and Research Institute, TANUVAS, Thoothukudi, India; and Dr. S. Ramesh, Professor, Ratnam Institute of Pharmacy, Nellore, Andhra Pradesh, India, for their valuable suggestions. I sincerely thank all my international friends who were very kind enough to collect and send images of certain species of edible marine bivalve molluscs for the present purpose. I also thank Mrs. Albin Panimalar for her help in photography and secretarial assistance.

Suggestions from the users are welcome.

**—Ramasamy Santhanam**

# CHAPTER 1

# INTRODUCTION

## CONTENTS

## ABSTRACT

The characteristics of the class bivalvia along with the human uses/ commercial importance of edible marine bivalves are given in this chapter.

The class Bivalvia (Lamellibranchiata or Pelecypoda) of the phylum Mollusca comprises more than 15,000 species of commercially important true oysters, pearl oysters, mussels, scallops, clams, etc. These molluscs are characterized by a shell that is divided from front to back into left and right valves. The valves are connected to one another at a hinge. In most molluscan species, the respiratory gills have become modified into organs of filtration called ctenidia. In keeping with a largely sedentary and deposit-feeding or suspension-feeding lifestyle, the bivalves have lost the head and the radular rasping organ which are the characteristic features of most molluscs.

**Human uses/commercial importance of edible marine bivalves:** Besides their value in the different fields of scientific research, edible marine bivalves play an important role in human consumption and other uses, such as:

1. nutritional values,
2. commercial fisheries,
3. aquaculture production,
4. pearl production,
5. aquarium uses,
6. pharmaceutical applications,
7. ornamental values, and
8. environmental values.

**Commercial fisheries:** The total marine catch of molluscs is twice that of crustaceans, and the great majority of this is bivalve. About million metric tons of edible bivalves are harvested throughout the world each year, and it is about 10% of the total world fishery production. The worldwide scallop fishery is a billion-dollar industry. The largest wild scallop fishery in the world is based on the scallop species Atlantic sea scallop, *Placopecten magellanicus*, found in the northeastern United States and eastern Canada.

**Nutritional values:** Almost all bivalves are edible and fall into the main categories of oysters, mussels, scallops, and clams. The most important edible species oysters are representatives of the genus *Crassostrea*, notably *Crassostrea gigas* in the western Pacific, *Crassostrea virginica* in North America, and *Crassostrea angulata* in Portugal. In Mexico, clams and cockles are usually eaten raw on the half-shell, or in cocktails or salads. Sometimes, the red clams, *Megapitaria aurantiaca*, and the black clam, *Megapitaria squalida*, are shucked, chopped, prepared with other ingredients, and broiled in their shells. The *Atlantic rangia*, cross-barred Venus, and other small clams are used for soups and cooked dishes with rice or spaghetti. Oysters are eaten in cocktails or on the half-shell. In addition, a tiny quantity (0.1% of landings) is smoked and canned. Shells for export are mainly the mother of pearl or nacre for cosmetics, clam shell for buttons, and abalone shell for jewelry. In Mexico, processed molluscan shells are used as poultry feed, and as building material, handcrafts, and jewelry.

**Seafood:** It is reported that 0.7 million tons of scallops are consumed throughout the world every year. These are followed by oysters (0.17 million tons) and mussels (0.14 million tons).

i. **Edible oysters:** From ancient times, the Romans and Greeks used the oysters as a main dish in their meals. The Red Indians were great oyster eaters. The maximum amount of oyster meats are collected from the two genera, namely, *Ostrea* and *Crassostrea*. In Japan, the United States, and Europe, over billion pounds of oyster meat are sold each year. In Europe, *Ostrea edulis* and *C. angulata* are very popular among oyster eaters. The Virginia oyster, *C. virginica*, is harvested from southern Canada to Mexico. Oysters are sold in the markets alive or after freezing.

ii. **Edible clams:** The edible clams are harvested throughout the world and its meat is a source of food since prehistoric times. The United States, Japan, Malaysia, and Europe consume a large proportion of clams. Major portion of the edible clams are harvested in Japan and the United States. The quahogs (Veneridae), razor clams (Solenidae), and soft-shelled clam (Myacidae) are in great demand in the markets of the United States. The quahog clam, *Mercenaria mercenaria*, is harvested in the Gulf of Mexico and in the Gulf of St. Lawrence. The largest clam fishery in the world is based upon the surf clam, *Spisula solidissima*, which is found in the United States. The cockles are harvested in northwest Europe and Malaysia. The most favorite American clam is soft-shelled clam,

*Mya arenaria* which is used extensively for the nature of its sweet meat. In Malaysia, the ark shells (*Anadara* sp.) are used extensively as food.

iii. **Edible scallops:** The sweet, delicate meat of the scallops holds a high rank to the people of shell-fish consumers, especially to the gourmets. The people of the United States, Japan, and southern Australia consume a large amount of scallop meats. The deep sea scallop, *P. magellanicus*, is very popular in the east coast of the United States. The scallops are also extensively harvested in Japan and Australia. Scallop meat is valued as a food in many western as well as eastern countries (China and Japan). In some countries, such as the United States and Canada, only the well-developed adductor muscle is used as meat, while in other countries apart from the adductor muscle, other parts of the scallop are also used. Scallop fishery is an industry found in many countries across the world with an annual value running into billions of dollars. Fortunately, the industry has been well managed in many countries.

iv. **Edible sea mussels and other edible molluscs:** The edible sea mussels mainly include *Mytilus* and *Perna* (Mytilidae). The European edible mussels, *Mytilus edulis*, though not popular in the United States but in France, has gained vast popularity. In Chile, the mussel fishing has acquired the front rank among the South American countries.

Marine bivalves continue to be an important part of the diet of people living on the coastal areas. Traditionally, shellfish gathering was the work of women and girls, who harvested a variety of bivalve species including clams, *Paphies* spp., *Dosinia anus*, and *Austrovenus stutchburyi*; oysters, *Saccostrea cucullata* and *Ostrea chilensis*; mussels, *Perna canaliculus* and *M. edulis*; and scallop, *Pecten novaezelandiae*. These species are still commonly eaten today.

**Toxin accumulation:** Several species of edible marine bivalves have been reported to accumulate toxin, and this is largely due to their eating on toxic, red-tide causing dinoflagellate *Alexandrium catanella* that may persist for months in their tissues. The butter clam, *Saxidomus giganteus* (Veneridae), is especially vulnerable to paralytic shellfish poisoning (PSP) and great caution should be used before eating them (RaLonde, 1996). Sea otters and seabirds are able to detect such toxin-tainted clams and avoid them, though humans cannot. The US Food and Drug Administration considers that anything above 80 µg of toxin is not safe for human consumption. Further, discharge of industries effluents containing toxic heavy metals often poses problems for the habitats of edible marine bivalves. Therefore, caution should be exercised while eating marine bivalves. It is always advisable to eat edible bivalve molluscs after proper depuration/treatment methods.

**Aquaculture values:** A number of edible bivalve species are raised commercially. The European mussel *M. edulis* has been introduced into the northern Pacific, and now, aquaculture practices flourish widely in Japan and China. Several species of scallops belonging to the genera *Pecten, Placopecten,* and *Amusium* are also being cultivated world-wide. Among clams, *M. arenaria* and *M. mercenaria* are cultivated in the North Atlantic; and *Venerupis japonica* and *Tapes philippinarum* in the Pacific.

**Commercial farming:** Bivalve molluscs represented almost 10% of the total world fishery production, but 26% in volume and 14% in value of the total world aquaculture production. World bivalve molluscs production (capture + aquaculture) has increased substantially in the last 50 years, going from nearly 1 million tons in 1950 to about 14.6 million tons in 2010. China is by far the leading producer of bivalve molluscs, with 10.35 million tons in 2010, representing 70.8% of the global molluscan shellfish production and 80% of the global bivalve mollusc aquaculture production. All of the Chinese bivalve production is cultured. Other major bivalve producers in 2010 were Japan (819,131 t), the United States (676,755 t), the Republic of Korea (418,608 t), Thailand (285,625 t), France (216,811 t), and Spain (206,003 t). By species, the bivalve mollusc production by aquaculture in 2010 consisted of 38.0% clams, cockles, and ark shells, 35.0% oysters, 14.0% mussels, and 13.0% scallops and pectens (Karuna-sagar, nd).

**Pearl production:** Bivalves of the genera *Pinctada* and *Pteria* have been collected in many tropical seas for their natural pearls. In many countries, most notably Japan, pearl oyster fisheries have been well developed.

**Pearl producing oysters (*Pinctada and Pteria*):** All members of the genus *Pinctada* are known for the production of large pearls of commer-cial value, and attempts have been made to harvest pearls commercially from many of the species. However, the common species that are currently of significant commercial interest are *Pinctada maxima, Pinctada margaritifera, Pinctada fucata, Pinctada radiate,* and *Pinctada albina.* The different species produce different types of pearls. Tahitian pearls come from the black-lip oyster, *P. margaritifera*; South Sea pearls from the white-lip oyster, *P. maxima*; and Akoya cultured pearls from the Japanese pearl oyster, *P. fucata.* The quality of pearls from *P. margaritifera* is the highest quality out of all the pearl oysters. Japan and Australia are the largest producers of cultured pearls. The species *P. fucata* and *Pteria penguin* are mainly used for this purpose in Japan and *P. maxima* in Australia.

Natural pearls

**Half-pearls:** The other important pearl-producing species of the genus *Pteria* is *P. penguin*, penguin-wing oyster. This species is used for the production of "half-pearls" or "blister pearls") for the jewelry industry. In this process, hemispherical pieces of plastic or shell are glued to the inner shell of the living wing oyster. Subsequently, the mantle tissue of the wing oyster coats the hemispherical piece with nacre and the resulting pearl halves are then cut from the shell. The Pacific wing-oyster *Pteria sterna* naturally produces gray, pink, golden, green, and purplish pearls.

**Non-nacreous natural pearls:** An important by-product of the worldwide scallop fishery is the extremely rare scallop pearl, a non-nacreous pearl, with a unique three-dimensional effect that produces a metallic sheen. The shapes of such pearls are drop-shaped, oval, button, round, off-round, or baroque. Colors are white to deep royal purple with varying shades of orange or pink (Scarratt and Hann, nd). Species producing non-nacreous natural pearls include the scallops *Nodipecten nodosus* and *P. magellanicus* and the giant clam, *Tridacna gigas*. Since the resurgence in demand for natural pearls, beginning from the 1990s, scallop pearls have come to the forefront as the latest natural pearl, commanding premium prices.

**Black pearls:** Black pearls are produced by the American stiff pen shell, *Atrina rigida*. Though these pearls are suitable for jewelry, they are extremely rare (Sturman et al., 2014). Other species producing black pearls include *Atrina vexillum, Atrina fragilis, Atrina pectinata, Atrina maura, Pinna bicolor, Pinna muricata, Pinna rudis*, and *Pinna rugosa*.

Non-nacreous pen pearls (pen pearls)

**Pearls in other marine bivalves:** Other marine bivalve molluscs from which pearls have been obtained are *Modiolus, Mytilus, Malleus, Pinna nigra, Pinna squamosa, Pinna nobilis, Placuna placenta, T. gigas, Venus margarifia*, and *O. edulis*. Other genera which do not possess any mother-of-pearl layer, but in which sometimes porcellanous pearls are found are *Spondylus, Pecten, Anomia, Cytherea, Lutraria, Tellina, Mya, Hippopus, Solen, Arca*, and *Glycymeris* (Alagarswami, 1965).

  *Solen sicarius*: This species seems to produce a very small but perfectly round pearl that measures just a hair over 1 mm (Holm, nd).

*Periglypta listeri*: A single living animal of this species was found with a series of loose pearls lining the center of the inside shell of this species (Bieler et al., 2004).

*Mercenaria* spp.: Free and attached pearls have been reported from several other venerids including *Mercenaria* spp. (Bieler et al., 2004).

*Gafrarium tumidum*: A shell pearl collected from this species was porcellanopus, whitish, and somewhat spherical except for the place that was in contact with the shell which become a little rough and flattened. The spherical surface was smooth and slightly glossy (Alagarswami, 1965).

**Traditional uses of marine bivalve shells:** Humans have valued marine mollusc shells since prehistoric times. Shells have been used for currency, jewelry, ornaments, tools, horns, games, medicine, and as magical or religious symbols.

**Scrapers and cutters:** Mussel shells were used for cutting hair and for scraping flax leaves to expose the fibers below the outer green layer. Tuatua shells were used for scaling fish.

**Bowls and containers:** *Dosinia* species and scallop shells were used to hold pigments for tattooing and in the preparation of ashtrays.

**Ornamental values:** The outer shell of the windowpane oyster, *Placuna*, is used, primarily in the Philippines, in the manufacture of lampshades, trays, mats, and bowls. In developing countries, many kinds of bivalve shells are used in the manufacture of jewelry and ornaments. Scallop shells valued as a collector's item: The decorative, brightly colored fan-shaped shells of some scallops are highly valued by shell collectors. The regular and pleasing geometric shape of the scallop shell has inspired artists as a decorative motif and has been used as a motif on the edge of draperies, clothing, and also in jewelry. Shell-craft industries in Southeast Asia still use thousands of tons of shells annually for mother-of-pearl products. The ornamental shell trade, which primarily includes shells exploited for their decorative or rareness value, is also substantial (Venkatesan, 2010).

**Medicinal values:** Several species of edible bivalves have been reported to be of great use in the production of pharmaceutical compounds with activities such as anticancer, antiviral, antioxidant, anti-inflammatory, etc. (vide Chapter 6).

**Industrial applications:** Crushed shells of several edible marine bivalves are used as poultry food additives, as loose paving in gardens, and for decoration. Shell lime prepared from shells is used for white washing on the walls of buildings. The shells of the giant clams (*Tridacna* spp.) are used in the manufacture of fashion accessories and for decorative purposes.

**Aquarium uses:** Edible marine bivalve molluscan species, like Atlantic Thorny oysters, *Spondylus americanus* and *Spondylus* sp.; Bear paw clam, *Hippopus hippopus*, China clam, *Hippopus porcellanus*, Boring clam, *Tridacna crocea*, Fluted giant clam, *Tridacna squamosa*, "Giant" clam, *T. gigas*, Maxima clam, *Tridacna maxima*, and Southern giant clam, *Tridacna derasa*, are of great use in marine aquaria owing to their beautiful coloration.

**Food and food additives:** Carotenoids from nonedible portions of several species of edible bivalves could be extracted for use as food and food additives. For example, the carotenoid amarouciaxanthin A has been isolated from the species, *Paphia euglypta*.

**Environmental values**: Several edible marine bivalve species are presently used in environmental surveillance program. Notable species include *Mytilus* spp., *C. virginica*, and *Ostrea equestris*. In the aforesaid programs, the mollusc tissues are periodically analyzed for heavy metals, radionuclides, halogenated hydrocarbons, and petroleum hydrocarbons, and useful data on baseline levels of these substances are collected. Mussel watch program are adopted by several countries to collect information on coastal ecosystem pollution worldwide.

**Edible marine bivalves as ideal bioindicators, biomonitors, or bioremediators of heavy metal pollution:** As the majority of edible bivalve species exhibit an extremely limited mobility or are completely sessile as adults, they represent the contamination of their habitats. As these bivalves accumulate certain toxic heavy metals from the waste, they may be used as indicators of pollution. For example, the untreated shells of *Anadara inaequivalvis* have been used as low-cost, natural biosorbents for the successful removal of the heavy metal ions such as (Cu(II) and Pb(II) ion) from industrial effluents and thus offer vast scope for treating waste waters (Bozbaş and Boz, 2016). *Anadara tuberculosa* has been reported to remove the particulate matter of semi-intensive shrimp farms effluents (Miranda et al., 2009). The black clam, *Chione fluctifraga*, can be used to bioremediate discharge effluents produced by shrimp aquaculture (Martínez-Córdova et al., 2011). The edible marine bivalve species capable of removing toxic heavy metals are given below.

| Species | Heavy metal |
|---|---|
| *Ruditapes philippinarum* | Mercury |
| *Saccostrea cucullata* | Mercury |
| *Mytilus galloprovincialis* | Cadmium |
| *Crenomytilus grayanus* | Cadmium |
| *Laternula elliptica* | Cadmium |
| *Mytilus edulis* | Cadmium |
| *Mytilus galloprovincialis* | Chromium |
| *Mya arenaria* | Thallium |
| *Mytilus edulis* | Thallium |
| *Mytilus galloprovincialis* | Lead |
| *Perna viridis* | Lead |

**Source:** Chiarelli and Roccheri (2014).

## KEYWORDS

- **edible marine bivalves**
- **human and commercial uses**
- **pearls**
- **nutritional values**
- **pharmaceutical applications**
- **ornamental values**
- **environmental values**

# CHAPTER 2

# BIOLOGY AND ECOLOGY OF EDIBLE MARINE BIVALVE MOLLUSCS

## CONTENTS

**ABSTRACT**

A brief account of the different aspects of the biology and ecology of edible marine bivalve molluscs is given in this chapter.

Marine bivalves are characterized by a laterally compressed body with an external shell of two halves which are hinged dorsally. Most bivalves are marine and there are no terrestrial forms. Fertilization is usually external, followed by trochophore and veliger stages larvae and metamorphosis to adult form.

## 2.1 EDIBLE AND COMMERCIALLY IMPORTANT MARINE BIVALVES

Among the marine bivalve molluscs, the following groups are considered to be most important:

1. true (edible oysters) (Ostreidae);
2. pearl oysters (Pteriidae);
3. mussels (Mytilidae);
4. scallops (Pectinidae); and
5. clams, such as ark clams (Arcidae), venus clams (Veneridae), giant clams (Tridacnidae), bean clams (Donacidae), saltwater clams (Mesodesmatidae), and (razor shells (Solenidae).

## 2.2 ECOLOGY OF EDIBLE MARINE BIVALVE MOLLUSCS

### 2.2.1 DISTRIBUTION

The marine bivalves, a highly successful class of invertebrates, are found all over the world seas and oceans. They inhabit in the tropics, as well as temperate and boreal waters. A number of species can survive and even flourish in extreme conditions.

### 2.2.2 HABITAT

Most of the marine bivalves are infaunal and live buried in sediment on the seabed. A large number of bivalve species are found in the intertidal and sublittoral zones of the oceans. A sandy sea beach may often have a very large number of bivalves that are living beneath the surface of the sand. In general, edible marine bivalves typically have one of the three habitat lifestyles, namely, (1) endobenthic burrowing (bivalves that live within the sediments); (2) epibenthic cemented (bivalves that live on the surface of sediments), and (3) epibenthic free-living or attached with byssal thread. Most clams have an endobenthic burrowing lifestyle. Depending on the species, the preferred benthic substrate may be sand, mud, or a mixture, and burrowing depth may be from a few centimeters to meters. Many bivalves use their large foot to bury into the sediment on the ocean floor. They then extend a long siphon up to the surface to suck water in for filtering and breathing. In many bivalve larvae or juveniles, a special gland, the byssal gland, can produce organic threads used for temporary attachment. In some groups, such as mussels, byssal threads permanently anchor the adults. A few groups of bivalves, such as oysters, are cemented permanently to the substrate.

## 2.3 BIOLOGY OF MARINE BIVALVE MOLLUSCS

### 2.3.1 EXTERNAL ANATOMY

The most prominent feature of bivalves is that the two valves of the shell may or may not be equal and may or may not completely enclose the inner soft parts. These valves have a variety of shapes and colors depending on species. The outer layer of the valves is periostacum, a brown leathery layer which is often missing through abrasion or weathering in older animals. Bivalves do not have obvious head or tail regions. The umbo or hinge area, where the valves are joined together, is the dorsal part of the animal. The region opposite is the ventral margin. In species with obvious siphons (clams), the foot is in the anterior–ventral position and the siphons are in the posterior area. In oysters, the anterior area is at the hinge, and in scallops, it is where the mouth and rudimentary foot are located.

### 2.3.2 INTERNAL ANATOMY

**Mantle:** The soft parts of the animal are covered by a mantle, which is composed of two thin sheaths of tissue. These two halves of the mantle are attached to the shell from the hinge ventral to the pallial line but are free at their edges. The mantle edge often has tentacles. In clams, the tentacles are at the tips of the siphon. In scallops, the mantle edge not only has tentacles but also numerous light sensitive organs—eyes. The main function of the mantle is to secrete the shell but it also has other purposes such as controlling the inflow of water into the body chamber and respiration.

**Adductor muscle(s):** In clams and mussels, the two adductor muscles are located near the anterior and posterior margins of the shell valves. The large, single muscle is centrally located in oysters and scallops. The muscle(s) close the valves and act in opposition to the ligament and resilium.

**Gills or ctenidia:** These are large leaf-like organs and are used for respiration and for filtering food from the water. Two pairs of gills are located on each side of the body.

**Foot:** At the base of the visceral mass is the foot. In species such as clams, it is a well-developed organ that is used to burrow into the substrate and anchor the animals. In scallops and mussels, it is much reduced and may have little function in adults. In oysters it is vestigial. Midway along the foot is the opening from the byssal gland through which the animal secretes a thread-like substance called "byssus" with which it can attach itself to a substrate. This is important in species such as mussels and some scallops.

**Digestive system:** A short esophagus leads from the mouth to the stomach, which is a hollow, chambered sac with several openings. The stomach is completely surrounded by the digestive diverticulum, a dark mass of tissue (liver). Stomach leads to the much-curled intestine that extends into the foot in clams and into the gonad in scallops, ending in the rectum and eventually the anus.

**Circulatory system:** Bivalves have a simple circulatory system. The heart lies in a transparent sac, the pericardium. It consists of two irregular shaped auricles and a ventricle. Anterior and posterior aorta lead from the ventricle and carry blood to all parts of the body. The venous system is a series of thin-walled sinuses through which blood returns to the heart.

**Nervous system:** In bivalves, the nervous system consists of three pairs of ganglia (cerebral, pedal, and visceral ganglia) with connectives.

**Urogenital system:** Sexes of bivalves may be separate (dioecious) or hermaphroditic (monoecious). In some species such as scallops, the sexes can be readily distinguished by eye when the gonad is full as the male gonad is white and the female is red, even in hermaphroditic species. Color of the full gonad may distinguish the sexes in some species such as mussels. Protandry and sex reversal may occur in bivalves. In some species, there is a preponderance of males in smaller animals indicating that either males develop sexually before females or that some animals develop as males first and then change to females as they become larger.

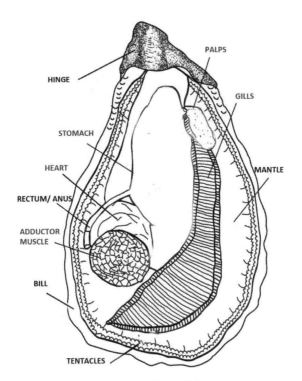

Internal anatomy of an edible oyster

### 2.3.3   FOOD AND FEEDING

Bivalves are filter feeders and feed primarily on phytoplankton and detritus. Water is drawn into the animal through the inhalant opening or the inhalant siphon, through the gills and then is returned to the surrounding water through the exhalant opening or siphon. The gills collect plankton and bind it to mucous.

### 2.3.4    PEARL FORMATION IN PEARL OYSTERS

The outer epithelial cells of the mantle of the pearl oyster secrete nacreous matter around foreign particles when they cause irritation to the soft body of the oyster. The foreign matter may be a sand grain or organic or inorganic debris. The mantle forms a pouch like epithelial sac around the intruding particle and there is a secretion of nacre in concentric layers resulting in course of time into a pearl.

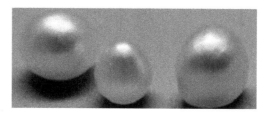

Natural pearls

### 2.3.5    LIFE HISTORY

#### 2.3.5.1    GONADAL DEVELOPMENT AND SPAWNING

In most bivalves, sexual maturity largely depends on size rather than age, and size at sexual maturity depends on species and geographic distribution. The period of spawning in natural populations differs with species and geographic location. Spawning may be initiated by several environmental factors including temperature, water currents, etc. In temperate areas, spawning is usually confined to a particular time of the year. Many bivalves undergo mass spawning and the period of spawning may be brief. Limited spawning occurs over a protracted period with one or two major pulses during this time. In some species, there may be more than one distinct spawning in a year. In hermaphroditic species, spawning is timed so that either the male or female part of the gonad spawns first.

#### 2.3.5.2    EMBRYONIC AND LARVAL DEVELOPMENT

In marine bivalves, fertilization is usually external, but in brooding species, it occurs in the mantle pallial cavity. In external fertilization, both the eggs and sperm are released into the water and the embryos develop in the water column. The fertilized egg develops into a motile trochophore in 24–36 h. The early larval stage is the straight hinge, "D" or Prodissoconch I stage. These larvae continue to swim, feed, and grow and within a week the umbones, which are protuberances of the shell near the hinge, develop. As larvae continue to grow, the umbones become more prominent and the larvae are now in the umbone or Prodissoconch II stage. Duration of the larval stage varies with species and environmental factors such as temperature but it can be 18–30 days. When these larvae approach maturity, a foot develops and gill rudiments become evident. Between periods of swimming activity, larvae settle and use the foot to crawl on a substrate. When a suitable substrate is found the larva is ready to metamorphose and begin its benthic existence. The larva is now ready to metamorphose.

### 2.3.5.3 LIFE SPAN (YEARS)

| | |
|---|---|
| *Margaritifera margaritifera* | 116 |
| *Panopea generosa* | 120 |
| *Crenomytilus grayanus* | 150 |
| *Arctica islandica* | 220 |

## 2.3.6 MORTALITIES

Too high temperatures or prolonged periods of cold temperatures along with sudden changes in temperature can be lethal to bivalves. Severe extremes in salinities, particularly low salinities can also cause extensive mortalities. Heavy siltation can smother and kill juveniles and adults. Pollution, particularly industrial pollution, can cause mortalities in juvenile and adult bivalves. Bivalves in the larval, juvenile, and adult stages are often preyed upon by a wide variety of animals that can cause severe mortalities. These animals are also hosts to parasites that can cause mortalities, particularly in the adult stage. The major cause of mortalities in bivalves, particularly of larvae and juveniles is disease.

## KEYWORDS

- **commercially important marine bivalves**
- **ecology**
- **external and internal anatomy**
- **life history**
- **life span**
- **mortality**

# CHAPTER 3

# PROFILE OF EDIBLE MARINE BIVALVE MOLLUSCS

## CONTENTS

## ABSTRACT

The profiles of the different species of edible bivalve molluscs with aspects, such as common name, distribution, habitat, description, biology, fisheries/aquaculture, and counties where the different species are eaten, are given in this chapter.

## 3.1   OYSTERS AND THEIR ALLIES (BIVALVIA: PTERIOIDA)

### 3.1.1   TRUE OYSTERS (OSTREIDAE)

The family Ostreidae comprises true oysters which are the food oysters of commerce worldwide. Oysters and scallops (Pectinidae) are related in that both groups have a central adductor muscle with the characteristic central shell scar. In the Ostreidae, the central adductor muscle is much larger and is not bounded by ridges. Radial ribbing when present is subdivided and is more irregular as is also the shell shape. Both oviparous (egg bearing) and larviparous (larvae bearing) species are found in this family. The larviparous species, of which *Ostrea edulisis* typical, show a life history of alternating sex changes in the same individual, whereas the oviparous species, for example, *Crassostrea virginica* or *Crassotrea gigas*, are essentially hermaphroditic and may produce either predominantly male or predominantly female gametes.

***Alectryonella plicatula*** (Gmelin, 1791)

**Common name(s):** Fingerprint oyster, pleated oyster

**Distribution:** Indo-West Pacific: Hong Kong, eastern Hainan, Nansha, Malaysia, and Singapore

**Habitat:** Intertidal and shallow subtidal waters; cemented on various hard substrates or objects; depth range 0–5 m

**Description:** Valve margins of this species are strongly plicate. Chomata are present all around the internal shell margins. These chomata form two to four rows of numerous pustules in right valve only. Maximum length is 12 cm.

**Biology:** Spawning has been reported to occur in this species when water temperature and chlorophyll *a* levels are highest and salinity lowest.

**Fisheries/Aquaculture:** This edible species is extensively cultivated in China. This species accounts for 50–60% of national aquaculture production in this country.

**Countries where eaten:** China and other areas of its occurrence

*Crassostrea angulata* (Lamarck, 1818)

**Common name(s):** Portuguese oyster, cupped oysters

**Distribution:** Northwest Pacific and Northeast Atlantic

**Habitat:** Coastal areas and estuaries; depth range 0–7 m

**Description:** Shell outline is variable with usually spatulate to oval individuals. Attachment area of left valve depends on the substratum. Both valves are concave with left more deeply cupped and hence a deep umbonal cavity. Both valves have dichotomous radial ribs from the umbo. Tops of the ribs are well-rounded and radial ribs with growth squamae are recognizable on both valves. The older part of the right valve, that is, the dorsal surface, is usually eroded. Chomata are absent. Adductor muscle scar is reniform. External coloration of both valves is variable from white and light purple with deep purple lines radiating from the umbo. Internally, the shell is white with patches of chalky deposits and hollow chambers. The muscle scar is colorless with occasional purple growth lines. Shell size is up to 20 cm long and 6 cm height.

**Biology:** Not reported

**Fisheries/Aquaculture:** This species is commercially harvested, eaten, and extensively cultivated in France, Spain, and Portugal.

**Countries where eaten:** France and Portugal

***Crassostrea ariakensis*** (Fujita, 1913)

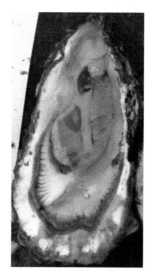

**Common name(s):** Suminoe oyster

**Distribution:** Throughout the western Pacific

**Habitat:** Intertidal hard grounds and substrate, as well as muddy creeks of warm estuaries

**Description:** This species resembles other oysters in having unequal valves and an irregular shape. The right (lower) valve is thinner, flatter, and smaller than the left. Both valves are covered with concentric growth layers (lamellae) on the outer surface, but with fewer and stronger ridges on the left (upper) side. Coloration of the lamellae on the outer surface varies from gray and yellowish brown to purple, while the inner surface of the valves is smooth and grayish white, with purple on the edges. The muscle scar on the inner surface of the valves is large and purplish. It reaches shell heights of 76 mm within 2 years postsettlement.

**Biology:** In this species, spawning occurs from July until September at a bottom water temperature of 22–26°C. Spawning is also related to water temperature and salinity.

**Fisheries/Aquaculture:** This species is commercially harvested and culti-vated in India, China, and USA (limited to a few bays in Washington, Oregon, and California).

**Countries where eaten:** This species is eaten in China and USA. Consumer acceptability tests indicate that this oyster is equal or superior to Pacific oysters in appearance, taste, and texture.

***Crassostrea belcheri*** (Sowerby, 1871)

**Common name(s):** Belcher's cupped oyster, lugubrious cupped oyster

**Distribution:** Indo-Pacific: China and Philippines.

**Habitat:** Low tide zone of estuaries and backwaters; depth range 0–5 m;

**Description:** Shell is oblong and its margins are narrow anteriorly and broad posteriorly. Lower valve is thick and upper opercular valve is thin and flat. Inner valves are white and glossy. Heart-shaped and pearly white muscle scar is present.

## Biology

**Food and feeding:** It is a suspension-feeder and feeds on diatoms (predominantly species of *Coscinodiscus*, *Thalassiosira*, *Biddulphia*, *Cocconeis*, *Achnanthes*, *Diploneis*, and *Synedra*), and detritus.

**Reproduction:** Adult oysters of this species are dioecious. Spawning occurs around the year when the oysters are kept in sea water at 18.2–30.05 ppt salinity and 28.5–30.7°C of water temperature.

**Fisheries/Aquaculture:** It is commercially exploited. It is also a potential species for aquaculture in Southeast Asia

**Countries where eaten:** Southeast Asian countries

***Crassostrea bilineata*** (Röding, 1798) (=*Crassostrea iredalei*)

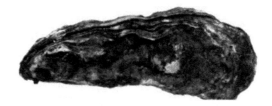

**Common name(s):** Philippine cupped oyster, slipper-cupped oyster, black oyster

**Distribution:** Western Central Pacific: Fiji, Philippines, and Tonga

**Habitat:** Brackish water areas with somewhat reduced salinity within intertidal and shallow subtidal waters; depth range 0–300 m; attached to hard objects or growing in bunches on various soft bottoms

**Description:** Shell is usually poorly sculptured and is very variable in shape. It is generally higher than long, roughly rounded, oblique triangular, or elongate ovate in outline. Left (lower) valve is rather thick but light weighted, more convex and larger than right (upper) valve. Surface of left valve is somewhat lamellate, with a few shallow to indistinct radial furrows. Right valve is flattish, concentrically lamellate, or nearly smooth. A moderately small umbonal cavity is present under the hinge of left valve. Adductor muscle scar is large and kidney-shaped. Chomata are completely absent from internal margins. Color outside of shell is dirty white and is often flushed with pale grayish brown. Right valve is with a few darker purplish gray radial bands in early stages of growth. Interior of valves are whitish and shiny and is often with irregular areas of chalky white, deep purple-brown on posterior adductor scar. Maximum shell length and height are 9 and 15 cm, respectively.

**Biology:** Not reported

**Fisheries/Aquaculture:** This species is commercially harvested and extensively cultivated in Philippines and Malaysia.

**Countries where eaten:** Southeast Asia: Brunei Darussalam, Cambodia, Indonesia, Laos, Malaysia, Philippines, Singapore, Thailand, and Vietnam

*Crassostrea columbiensis* (Hanley, 1846)

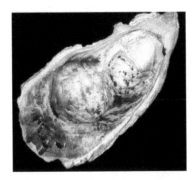

**Common name(s):** Mangrove oyster, Cortez cupped oyster

**Distribution:** Baja California to Chile

**Habitat:** Mangrove swamps; adhered to rocks or mangroves by the entire surface of lower valve

**Description:** Shape of its white shell varies from oblong to suborbicular and valves are of equal length. Shell has a wavy purple margin that is smooth inside. Adductor scar is kidney-shaped and purple in color. Upper valve may be rayed with purple or yellow on outside. Size is 75 mm in length.

**Biology:** Not reported

**Fisheries/Aquaculture:** It is extensively cultivated in Mexico and Panama

**Countries where eaten:** This species is used as food especially in Peru.

*Crassostrea corteziensis* (Hertlein, 1951)

Image not available

**Common name(s):** Cortez oyster, mangrove oyster, pleasure oyster

**Distribution:** Eastern Central Pacific and Northeast Atlantic: Mexico, Panama, and United Kingdom

**Habitat:** Brackish waters; intertidal zone; mangroves, estuaries, and coastal lagoons; associated with mangrove roots; eurythermal and euryhaline; depth range 1–2 m.

**Description:** Shell is elongated and is somewhat triangular with a wide ligamentary area. Surface shows faint radial furrows. Within shell coloration is white and margin is smooth. Maximum shell length is 25 cm.

**Biology:** Filtration and clearance rates of this species have been reported to increase with increasing temperature and decrease with increasing salinity, with the highest values at 29°C and 20 ppt. Ammonium excretion and, to lesser extent, oxygen consumption match with the variations in the feeding rate. Higher filtrations and scope for growth of oysters occur at 29°C in brackish water (20 ppt) rather than in marine-water conditions. This information has been found to be very useful for aquaculture management of this species and for establishing suitable sites to enhance their cultivation and maximize the growth of this species.

**Fisheries/Aquaculture:** This species is commercially harvested and has importance in aquaculture in Gulf of California.

**Countries where eaten**: In areas of its occurrence

***Crassostrea cuttackensis*** (Newton & Smith, 1912) (=*Crassostrea gryphoides*)

**Common name(s):** Giant oyster

**Distribution:** India

**Habitat:** Rocky and muddy areas; estuaries, backwaters, and creeks

**Description:** Shell of this species is stout, bulky, elongated, and irregular in shape. It is narrow in the anterior margin and broader in the posterior margin. Upper valve is thin, flat, and opercular. No denticles are seen on the margin. Interior of valves is white and glossy and muscle scar is more or less heart-shaped and pearly white. Shape, size, and even the impression of muscle scar change due to the environmental condition. Size: length 130–290 mm; width 56–145 mm; height 24–83 mm.

**Biology:** The false limpet Siphonaria siphon is found associated with this species especially in Sundarbans, India.

**Fisheries/Aquaculture:** This species is commercially exploited and cultivated in India and Sri Lanka.

**Countries where eaten**: India and Sri Lanka

***Crassostrea gigas*** (Thunberg, 1793)

**Common name(s):** Pacific oyster, Japanese oyster, Miyagi oyster

**Distribution:** Native to the Pacific coast of Asia; North America, Australia, Europe, and New Zealand.

**Habitat:** Estuaries; intertidal and subtidal zones; hard or rocky surfaces in shallow or sheltered waters up to 40 m deep; muddy or sandy areas; shells of other animals

**Description:** Shell is solid and inequivalve. It is extremely rough, extensively fluted, and laminated. Left (lower) valve is deeply cupped, Right valve is moderately concave. Shape of the shell varies with the environment. Color is usually whitish with many purple streaks and spots radiating away from the umbo. Interior of the shell is white, with a single muscle scar that is sometimes dark, but never purple or black. Size of shell is from 8 to 40 cm long.

**Biology**

**Ecology:** Optimal salinity range for this species is between 20 and 25‰. It also has a broad temperature tolerance, with a range of 1.8–35°C.

**Reproduction:** These oysters are protandrous hermaphrodites, most commonly maturing first as males. It has high fecundity with 8–15-cm length females producing between 50 and 200 million eggs in a single spawning. Larvae are planktotrophic. This may take 2–3 weeks, depending on water temperature, salinity, and food supply, during which time they can be dispersed over a wide area by water currents. Growth rate is very rapid and market size is attained in 18–30 months.

**Fisheries/Aquaculture:** This species is commercially harvested and cultivated in China, Japan, Taiwan, USA, France; eaten in Japan, China, Canada, Australia, Hawaii and USA, British Isles, Morocco, and France. It accounts for about 15% of the total world production.

**Countries where eaten:** It is eaten in Japan, China, Canada, Australia, Hawaii and USA, British Isles, Morocco, and France.

*Crassostrea madrasensis* (Preston, 1916)

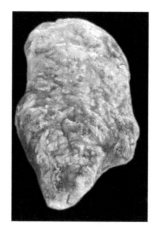

**Common name(s):** Indian backwater oyster, Indian oyster

**Distribution:** Western Indian Ocean: Pakistan to Bangladesh

**Habitat:** Brackish water; thick beds in estuaries, backwaters, creeks, ports, and harbors; intertidal zone to a depth of about 4 m

**Description:** Shell is straight and irregularly shaped and is covered by numerous foliaceous laminae. Left valve is deep and right one is slightly concave. Hinge is narrow and elongated. Adductor scar is subcentral, reniform, and dark purple in color. Inner surface of valves is white, glossy, and smooth. Purplish black coloration is seen on the inner margin of the valves. Female oyster has a more flabby appearance than a fully spawned male. Shell height is between 3.5 and 8.0 cm.

**Biology**

**Ecology:** The species has a wide range of tolerance to variations in salinity of waters in the environment.

**Food:** It feeds on diatoms, zooplankton, and detritus.

**Life span:** Its maximum reported age is 4 years.

**Reproduction:** The oyster is normally unisexual, but change of sex has been observed in a transitional hermaphroditic phase. Two spawning peaks, namely, one major (postmonsoon) in October and another minor (monsoon) in May have been observed in this species. Sex ratio of female to male is 1:1.1.

**Fisheries/Aquaculture:** This species is commercially exploited and cultivated in India.

**Countries where eaten:** India

***Crassostrea mangle*** (Amaral & Simone, 2014)

Image not available

**Common name(s):** Not designated as it is a new species

**Distribution:** Brazil

**Habitat:** Associated with the mangroves

**Description:** It is fixed in roots of mangroves and shell shape is cupped. Right valve is concave, and operculum-shaped; and left valve is larger in edge than right valve. Muscle impression is not colored and adductor muscle is more central in posterodorsal region, occupying 1/2 of total size of animal. External color of shell is varying from brown to reddish-brown; and internal is white and opalescent. Maximum size is 12 cm.

**Biology:** It is a new species.

**Fisheries/Aquaculture:** Not reported

**Countries where eaten:** Brazil

***Crassostrea nippona*** (Seki, 1934)

**Common name(s):** Iwagaki oyster, wild rock oyster, summer oyster

**Distribution:** Northwest Pacific: China and Japan

**Habitat:** Intertidal and subtidal on rocks; depth range, 0–10 m

**Description:** This species is moderately big and its shape varies from oval to round. Left shell is flat or slightly concave while the right is concave. Right shell has wavy growth scales and growth lines. It has a small umbonal cavity without chomata. Adductor muscle is reniform. Dorsal border is concave. Growth line is dark brown and its basic part is light brown. Left shell is light violet and part of the inner side is milky white. Muscle scar is polished white with violet yellow and has growth line.

**Biology:** Not reported

**Fisheries/Aquaculture:** It is farmed along the coast of the Japan Sea. These oysters are harvested during the spring and summer (May–August).

**Countries where eaten:** Japan; it is usually consumed raw and tends to be very expensive.

*Crassostrea rhizophorae* (Guilding, 1828)

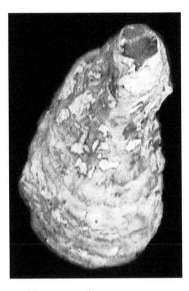

**Common name(s):** Pacific cupped oyster; mangrove oyster, mangrove-cupped oyster

**Distribution:** Atlantic Ocean: from Caribbean to Brazil and United Kingdom

**Habitat:** Brackish water; intertidal or shallow-subtidal; attached to roots of *Rhizophora mangle*, rocks or other oyster shells; intertidal or shallow-subtidal species; depth range 2–41 m

**Description:** Shell is lightweight and deep-cupped inequivalve. Left valve (attached) is larger than right. Shell shape and outline are variable. Inner margin is smooth. Presidium is transversally striated. Externally, it is dirty light gray, and internally, whitish or light gray splotched with bluish purple. It has a total length of 12 cm.

**Biology:** The gonadic stage analysis indicated that the reproduction period of this species is continuous all year, without any regressive phase.

**Fisheries/Aquaculture:** It is heavily exploited and is extensively culti-vated in Cuba and Venezuela.

**Countries where eaten:** Caribbean and Brazil. It is consumed raw, fried, grilled or boiled. It is also canned industrially.

## *Crassostrea rivularis* (Gould, 1861)

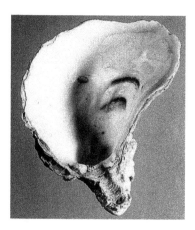

**Common name(s):** Jinjiang oyster, Suminoe oyster

**Distribution:** Western Pacific: Taiwan and China

**Habitat:** Estuaries; intertidal to subtidal; depth range 0–5 m.

**Description:** This species is characterized by large, roughly round, flat, thick shell valves with a shallow shell cavity. Left valve is thick and slightly concave and right one is about the same size or slightly larger. Aductor muscle scar is oblong and white in color. Inner surface of valves is white and bright. Maximum size: Shell length 9.5 cm; width 5 cm; and height 16 cm.

**Biology**

**Ecology:** It is mostly found in rivers and estuaries with salinities of 10–25 ppt.

**Biological indicator:** This species has also been reported to be a reliable biological indicator of cadmium pollution (Chaohua et al., 1998).

**Fisheries/Aquaculture:** It is commercially exploited and has been cultured in China for centuries.

**Countries where eaten:** China. Two forms of this species are recognized in China, that is, based on meat color. One is referred to as "red meat" and the other as "white meat." The two oysters are similar in shell morphology and sympatric. Oyster farmers, however, prefer the white oyster because of its flavor, fast growth, and high market price.

*Crassostrea sikamea* (Amemiya, 1928)

**Common name(s):** Kumamoto oyster

**Distribution:** Northwest Pacific: from Japan to South Korea, China, and Taiwan

**Habitat:** Intertidal areas, particularly on hard substrates; muddy flat as well as in estuary in tidal flats

**Description:** Shells are small, elongated or ovate, and irregularly shaped. Some oysters may be round and horseshoe shaped as a result of crowding. Morphologically, its left valve is deeply cupped and right valve is flat. Shells are relatively smooth and rarely have pronounced layers of concentric plates on the right valve. Radiant ribs are weak or absent and right shells have dark stripes radiating from the umbo to the shell edge. External coloration ranges from gray to brown. Internal coloration is chalky white with areas of dull purple or brown. Maximum length is 6.0 cm.

**Biology**

**Food and feeding:** It is a suspension feeder mainly depending on phytoplankton and detritus materials.

**Reproduction:** This species is slower in growth. In its native range, mature eggs are produced in early winter, while in the US Pacific, it comes in late summer through early winter. It is also known to have a one-way gametic incompatibility barrier, which prohibits the sperm of this species to fertilize other species.

**Fisheries/Aquaculture:** Not reported

**Countries where eaten:** Japan, China, and Taiwan

*Crassostrea tulipa* (Lamarck, 1819) (*=Crassostrea gasar*)

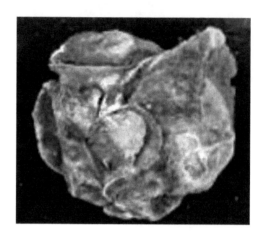

**Common name(s):** West African mangrove oyster, eastern Oyster

**Distribution:** Western Indian Ocean and Atlantic Ocean: India and from Mauritania to Angola and Brazil.

**Habitat:** Brackish water; intertidal zones of rivers and estuaries; coastal rocks and stones; grows on the bark of the stilt sections of mangrove trees

**Description:** Shell is fixed in roots of mangrove submerse and in rock and its shape is cupped or oval. Right valve is slightly operculum shaped and left valve (fixed in substrate) is larger than right valve. Muscle impression is purple and adductor muscle is oval central in posterior region. Adductor muscle is posterodorsally located, occupying 1/5 of total size of animal. Color of shell is varying from white, brown, and purple to greenish. Maximum length is 13 cm.

**Biology**

**Food and feeding:** Oyster assemblage provides protection and food for various invertebrates and fishes. It is a filter-feeder, feeding on phytoplankton (green algae, diatoms, dinoflagellates) and substrate particles.

**Reproduction:** Gamete maturation (gonadal tissue maturation) occurs in these oysters at temperatures greater than 22°C.

**Fisheries/Aquaculture:** Potential species for aquaculture in West Africa

**Countries where eaten:** Widely used as food and a source of income for the local population of the estuaries of Northeast Brazil. Mangrove oysters are also eaten in Trinidad and Tobago and Nigeria.

*Crassostrea virginica* (Gmelin, 1791)

**Common name(s):** Eastern oyster, Wellfleet oyster, Atlantic oyster, Virginia oyster, American oyster

**Distribution:** Native to Gulf of Mexico coast of North America

**Habitat:** Estuaries and marine coastal environments

**Description:** Shell is thick, solid, and highly variable in shape. Inequivalve left (lower) valve is convex and right (upper) valve is flat. It grows from round (irregular) to oval and usually bears concentric ridges. Inequilateral, beaks, and umbones are not prominent. Hinge line is without teeth in the adult. Shell margins are smooth. Color is white, dirty white, or brown, sometimes with dark purple markings. Periostracum is thin, dark brown. Interior is bright white with a deep purple or redish-brown muscle scar.

**Biology**

**Reproduction:** It is a very hardy molluscan species and discharges its gametes into the water column in response to a variety of stimuli, including warmer temperatures, pheromones, and suitable phytoplankton species. Fertilized eggs develop into trochophore larvae. After about 24 h, the trochophore develops into a shelled veliger larva. After 2 or 3 weeks, depending on food and temperature conditions, the veliger develops an eyespot and a larval foot (it is now called an eyed larva or pediveliger) with which it looks for potential settlement substrates. Subsequently, it cements its left valve to the substrate and metamorphoses into an oyster spat. It has a lifespan of up to 20 years growing to about 100 mm in length in 2 years. Individuals can reach sexual maturity at 4th month.

**Fisheries/Aquaculture:** Commercially harvested and extensively cultivated in USA, Canada, and Mexico. This species has been reported to accounts for about 85% of total oyster production. This species has been reported to make small, egg-shaped, non-nacreous pearls. These pearls, however, are insignificant in size and of no value (Scarratt et al., 2006).

**Countries where eaten**: Canada, France, Spain, and United States

***Ostrea angasi*** (Sowerby, 1871)

**Common name(s):** Australian flat oyster, common mud oyster, Angasi oyster, Australian mud oyster

**Distribution:** Endemic to southern Australia, ranging from Western Australia to southeast New South Wales and around Tasmania.

**Habitat:** Bays and inlets, on hard surfaces in mud; silty or sand-bottomed estuaries; subtidal; depths between 1 and 30 m

**Description:** This species has large shells which are rounded. Left valve is somewhat concave and right valve is quite flat. Valve surfaces have a flaky, layered (lamellose) appearance and crumble easily (friable). Valve margins are undulating. Like other oyster species, shape of the shell is influenced by the shape of the substrate to which it is attached. Shell is up to 18 cm across.

### Biology

**Food and feeding:** Flat oysters, like all other oyster species, are filter feeders feeding on anything which is small enough to be filtered in their gills. This may include plankton, microalgae, or inorganic material.

**Fisheries/Aquaculture:** Commercially harvested and cultivated

**Countries where eaten:** Australia

***Ostrea chilensis*, Philippi, 1854 (=*Tiostrea chilensis*)**

**Common name(s):** Chilean flat oyster, New Zealand flat oyster, dredge oyster, bluff oyster, kiwa oyster

**Distribution:** Native to Chile and New Zealand; Atlantic and Pacific Oceans

**Habitat:** Nearshore waters and shallow subtidal zone; soft sandy and muddy bottoms, as well as hard substrates; depth range 0–300 m

**Description:** Shells are oval and weakly inflated but extremely variable in shape, depth, color, and sculpture, according to environment. Right valve is almost flat, with only sculpture of wide, brittle lamellae. Left valve is shallowly to moderately cupped, with weak radial costae or no regular sculpture. Chromata are very small and limited to margins near hinge. There is no pallial line. A large, weakly reniform adductor muscle scar is seen slightly in front of middle of valve. It is a small-to-medium-sized species with a maximum length of 10 cm, width of 7 cm, and height of 12 cm.

**Biology:** It is a brooding species where females retain the larvae until they are completely developed.

**Fisheries/Aquaculture:** Commercial dredge fisheries for this species exist around oyster beds in Tasman Bay and in Foveaux Strait in New Zealand. In Chile, it is harvested from the wild. Commercial aquaculture of this species is frustrated by an inability to artificially control its unusual breeding behavior. If the full aquaculture potential of this species is to be realized, further intensive research needs to be directed at identifying the factors involved in controlling female maturation, spawning, and fertilization.

**Countries where eaten:** Pacific coastal waters of Chile

***Ostrea conchaphila*** (Carpenter 1864) (=*Ostrea lurida*)

**Common name(s):** Olympia oyster

**Distribution:** Northern Pacific coast of North America

**Habitat:** Bays and estuaries; 0–71 m

**Description:** Lower (left) valve of this species is shallowly concave and the upper (right) valve fits into the raised margin of the opposite valve. Shell shape is extremely variable. Shells of free-growing oysters are ovate to elongate. Exterior of the shell is without a periostracum and its color varies from white to purplish black and may be striped with yellow or purplish brown. Interior of the valves is white to olive green and scar of the adductor muscle is not much darker than the rest of the shell's interior. It has a maximum length of 8 cm.

## Biology

**Food and feeding:** Olympia oysters are suspension feeders, feeding mainly on phytoplankton.

**Assemblage:** Oyster beds of this species provide shelter for anemones, crabs, and other small marine life.

**Reproduction:** This species spawns between the months of May and August, when the water reaches temperatures above 14°C. During the first spawning cycle, the oyster acts as a male and then switch between sexes during its following spawning cycles. Males release their spermatozoa from their mantle cavity in the form of sperm balls. These balls become free-floating sperm in water. Female's eggs are fertilized in the mantle cavity (brooding chamber) when spermatozoa are filtered into her gill slits from the surrounding water. Fertilized eggs will then move into the branchial chamber (mantle cavity) and will develop into veliger larvae and will stay in the female mantle cavity for 10–12 days for further development. On the first day, the larvae develop into a blastulae (mass of cells with a center cavity); on day two, they develop into a gastrulae (hollow two-layered sac); on the third day, they develop into trochophore (free-swimming, conciliated larvae); and on the fourth day, the valves on the dorsal surface become defined. During the rest of development in the brooding chamber, the valves complete and a straight-hinged veliger larva grows. When the spat (larvae) leave the brooding chamber, they begin to develop an eye spot and a foot. They then migrate to hard surfaces like old oyster shells where they attach by secreting a "glue"-like substance from their byssus gland. Spat swim with their foot superior to the rest of their body. This swimming position causes the larvae to attach to the underside of horizontal surfaces. Brood size is between 2 and 3 lakhs. The amount of larvae produced is dependent on the maternal oyster's size and the amount of reserved nutrients she has at the time of egg fertilization.

**Fisheries/Aquaculture:** It is commercially harvested. Limited aquaculture of this species exists in Canada

**Countries where eaten:** Pacific Canada and USA

*Ostrea edulis* (Linnaeus, 1758)

**Common name(s):** European flat oyster

**Distribution:** Western and southern coasts of Europe from Norway to Morocco; British Isles and Mediterranean coast

**Habitat:** Estuarine and shallow coastal water with hard substrata of mud and rocks; depth range 0–80 m.

**Description:** Shells of this species are oval or pear shaped with white, yellowish or cream coloration. Rough surface shows pale brown or bluish concentric bands on the right valve. Two valves are quite different in shape and size, as the left one is concave and fixed to the substratum, while the right one is almost flat and fits inside the left. Inner surface is smooth, whitish or bluish-gray, and pearly (the opalescent mother of pearl). An elastic ligament holds together these two valves. Meat of these oysters may vary from creamy beige to pale gray. Maximum length is 11 cm.

**Biology**

**Food and feeding:** It is a filter feeder feeding mainly on suspended organic particles and phytoplankton.

**Predators:** It is preyed upon by a variety of species including starfish, the sting winkle/rough tingle (*Ocenebra erinacea*), and American oyster drill (*Urosalpinx cinerea*).

**Reproduction:** This species is a protandric hermaphrodite, changing sex normally twice during a single reproductive season. Oysters function as males early in the spawning season and later change to females before changing to males again. It produces up to 1 million eggs per spawning that are liberated into the pallial cavity where they are fertilized by externally released sperm. Following an incubation period of about 10 days, depending on temperature, larvae (160 μm in size) are released into the environment and spend 8–10 days as a pelagic dispersal stage before settlement. Appropriate larval growth and survival rates are obtained in 20‰ salinity, although they can survive at salinities as low as 15‰. Its fecundity size is 2 million eggs and maximum life span is 15 years.

**Fisheries/Aquaculture:** It is commercially harvested and is extensively cultivated in France, Spain, United Kingdom, Turkey, and Greece

**Countries where eaten**: France, Mediterranean, USA. The meat of this species is considered excellent for eating raw on the half shell.

***Dendostrea folium*** (Linnaeus, 1758)

**Common name(s):** Raccoon oyster, leaf oyster, foliate oyster, zigzag oyster, toothed oyster

**Distribution:** Indo-West Pacific: from East Africa to Melanesia; north to Japan; and south to Queensland

**Habitat:** On rocks or sea whip stems; adheres to the roots of mangrove trees; depth range 0–5 m

**Description:** Shell of this species is slightly inequivalve, elongate ovate, and purple brown in color. Median groove is bordered by spines projections on lower valve. Surface is with many fine concentric growth scales. Many sharp, radial ribs are giving zigzag margins. Maximum length is 10 cm.

**Biology:** Not reported

**Fisheries/Aquaculture:** Small-scale culture of this species exists in Pulau Langkawi, Kedah, Malaysia

**Countries where eaten:** Malaysia; collected by local fishermen for subsistence

***Ostrea puelchana*** (d'Orbigny, 1842)

**Common name(s):** Puelche oyster, Argentina oyster

**Distribution:** Atlantic Ocean and Indo-Pacific

**Habitat:** Depth range 10–100 m

**Description:** It is a large rounded species with a straight margin. Shell is smooth except for a few denticles on each side of the hinge. Muscle scar is centrally located and beaks are almost straight. Exterior of right or free valve is often marked with a light chestnut brown coloration.

**Biology**

**Harm:** Seven borer species have been identified affecting this species in the southwestern Atlantic Ocean. They are the sponges *Clionaopsis platei, Pione angelae,* and *Cliona celata* (Clionaidae); the borer bivalve *Lithophaga patagonica* (Mytilidae); and the polychaetes *Polydora rickettsi* (Spionidae), *Dodecaceria* cf. *choromyticola* and *Caulleriella* cf. *bremecae* (Cirratulidae). *Lithophaga patagonica* was the most harmful borer organism, lowering the condition index of the oyster (Diez et al., 2014).

**Reproduction:** This species has an indirect life cycle. Compared to other larviparous oysters, it has a short period of breeding and produces higher number of embryos of smaller sizes. Their reproductive cycle takes place between the months of November to March. It has a protandrous maturation.

**Fisheries/Aquaculture:** This edible species is extensively cultivated in Chile and New Zealand.

**Countries where eaten:** No report

*Saccostrea cucullata* (Born, 1778)

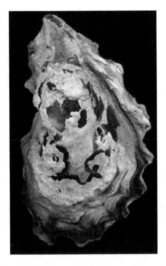

**Common name(s):** Hooded oyster, Natal rock oyster

**Distribution:** Indian Ocean and tropical west Pacific Ocean

**Habitat:** Rocky habitats of intertidal zone; depth range 0–15 m; among seaweed; harbor walls, pilings, and other underwater structures.

**Description:** Shell shape is nearly circular or oblong or roughly oval, often with an irregular outline. Valves are thick and solid. Lower valve is convex and has no sculpturing near the umbo, which is fixed to the substrate. Upper valve is flat and smaller than the lower valve. Margins of the valves are pleated and fit together neatly. Ligament is internal and there are no teeth on the hinge joint. Right valve has some small denticles on its margin which fit into grooves in the left valve margin. A single large adductor muscle holds the valves together, which leaves a large, kidney-shaped scar on the inside of each valve. Color is purplish-brown on the outside of the valves. Inside is white rimmed with black. It grows to a maximum size of 12 cm.

## Biology

**Ecology:** This species is tolerant of a wide range of temperatures and salinities.

**Food and feeding:** It is a filter feeder, pumping water through its gills and removing the phytoplankton.

**Bio-indicator:** In polluted waters, it accumulates heavy metals in its tissues. Hence, it can be used as a bio-indicator for monitoring pollution. Because it selectively removes these metals and is such an efficient biofilter, it has been used in the Persian Gulf to control pollution.

**Association:** *Siphonaria siphon* (false limpet, gastropod) has been found associated with these oysters collected from Sundarbans, India

**Reproduction:** Spawning period is from June to mid-October. Optimum temperature for reproduction is 20–30°C. It can tolerate a wide range of temperatures, between 5 and 40°C. Gametogenesis is concomitant with higher salinity values (34‰) and full maturation is attained when the salinity is maximum (35‰). The larvae are planktonic and preferentially settle out of the water column in locations already occupied by adult oysters.

**Fisheries/Aquaculture:** This species is commercially exploited. Its cultivation is very popular in many parts of the world (India, Australia, West Africa, and Thailand).

**Countries where eaten:** Mediterranean, and Australia

*Saccostrea echinata* (Quoy & Gaimard, 1835)

**Common name(s):** Blacklip oyster, spiny rock oyster, spiny oyster

**Distribution:** Western Central Pacific: Indonesia and Palau

**Habitat:** Rocks or permanent features along tidal areas.

**Description:** Shell of this species is dark to white depending on circumstances and is ragged and sharp. Hinge is very strong. Nacreous surface is present inside the oyster is a smooth white pearl-like surface. It has a distinctive black textured ring or "lip."

**Biology:** Not reported

**Fisheries/Aquaculture:** Cultivated in small scale in Indonesia. It is also commercially exploited.

**Countries where eaten:** Australia

*Saccostrea glomerata* (Gould, 1850) (=Crassostrea glomerata)

**Common name(s):** Sydney rock oyster, New Zealand rock oyster

**Distribution:** Western Pacific: Palau and New Zealand

**Habitat:** Intertidal estuarine habitats such as rocks, mangroves, and man-made structures; subtidally on natural dredge beds.

**Description:** Shell of this species has a very variable in shape. Lower valve is deep and cupped, recessed under the hinge and is moderately fluted. Valve edge is weakly crenulated. Upper valve is flattened and is folded toward lip to fit crenulations of lower valve. Hinge line is moderately wide and ligament is blue. External coloration of valves varies from bluish black to grayish white with indefinite blue-black borders. Internal color is chalky white, with bluish black or sometimes brown border. Muscle scars are usually white but frequently having bluish or creamy markings, particularly on upper valve. Small denticles are seen along edge. It grows to 7–10 cm shell height (measured from hinge to lip). Maximum length is 25 cm.

**Biology**

**Reproduction:** The main spawning and spat settling season for this species in Australia ranges from February (last month of summer) until May (last month of autumn) when water temperatures are generally above 20°C. These oysters are serial spawners and they spawn several times during a season and have a high fecundity, with females producing up to 25 million eggs per spawning. These oysters have the ability to change their sex and it is likely that they spawn for the first time as males or at least develop initially as males like to other cupped oysters. During the active breeding season, there is a consistently higher percentage of females among older oysters when compared with younger oysters (of 1 year or less in age). The percentage of females declines after the breeding season in autumn, with a rise in the number of sexually indeterminate oysters with regressed gonads. Hermaphroditism has rarely been observed in these k oysters. Larvae take around 3 weeks to metamorphose and settle.

**Fisheries/Aquaculture:** Commercially harvested and extensively culti-vated in Australia.

**Countries where eaten:** Australia

***Saccostrea palmula*** (Carpenter, 1857) (=*Ostrea conchaphila*)

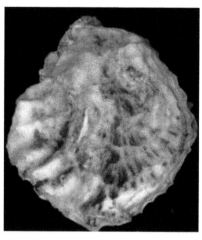

**Common name(s):** Palmate oyster, shell-loving oyster, mangrove oyster

**Distribution:** Western Central Atlantic and Indo-Pacific; Gulf of California to the Galapagos Islands

**Habitat:** Supralitoral and mesolittoral areas higher in exposed rocks and mangrove roots

**Description:** Shell shape is oval. Left valve is very concave and right valve is almost flat. Outer portion of shell is gray and inner portion is green with purple edges. Overall shell length is 37 mm.

**Biology**

**Reproduction:** The mean sex ratio of this species has been reported to be 1.00 male:0.92 females and sexually mature individuals are found throughout the year. Spawning activities was highest between November and January.

**Fisheries/Aquaculture:** This edible species is extensively cultivated in USA.

**Countries where eaten:** USA

*Saccostrea scyphophilla* (Peron & Lesueur, 1807)

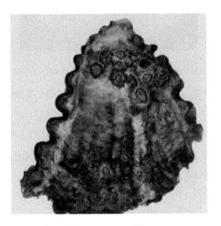

**Common name(s):** Coral rock oyster, milky oyster, northern oyster

**Distribution:** Indo-West Pacific: Philippines, Palau, and Tanzania

**Habitat:** Among clusters of the hooded oysters

**Description:** Shell of this species is solid and oblong with well-marked transverse and fine longitudinal striations. Lower valve is larger than upper valve. Upper valve is flat and opercular. Muscle scaris oblong and denticulations are absent. It grows up to 4 cm.

**Biology:** No report

**Fisheries/Aquaculture:** It is cultivated in Australia

**Countries where eaten:** This species is an important food source in Fiji and Australia.

*Striostrea prismatica* (Gray, 1825)

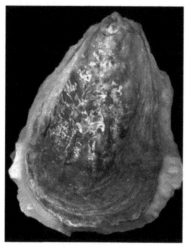

**Common name(s):** Stone oyster, tropical rocky oyster

**Distribution:** Eastern Pacific, Western Central Atlantic, and Southeast Atlantic

**Habitat:** Rocky shores; intertidal; depth range 0–79 m

**Description:** It has a large, heavy, and thick shell which is somewhat elongated. Adult specimens which are about 16 cm are almost straight, wide, and relatively square. Shell is inequivalved. External shell is purple or tan, with concentric lines. Inner surface is of dense texture and its color is whitish, often with an iridescent or metallic luster.

**Biology**

**Reproduction:** These oysters reach highest ripeness in January and February, while spawning occurs in February–March. Gametogenesis is associated to surface seawater temperature. Spawning coincides with warm water temperature fluctuations and a seawater salinity decrease. Sex ratio of this species is 1:1, suggesting that oysters sizing more than 10 cm in shell length present a stable sex proportion in the population.

**Fisheries/Aquaculture:** Overfishing of this species has been reported.

**Countries where eaten**: Mexico

**Remarks:** It is commercially valuable seafood species. This species was also used in the making ornaments from its thick valvas. It has also served as an indicator of pollution (biomonitor) for heavy metals.

### 3.1.2   HAMMER OYSTERS (BIVALVIA: MALLEIDAE)

Hammer oysters are hinged at the top of the "T," where a small byssus emerges at the back. The hinge is held by an oblique ligament rather than teeth, and the shell is partially nacreous. A single large adductor muscle lies at the cross of the T, and the exhalant current is discharged at the hinge. These oyster live mostly in the crevices of coral rocks or on reef flats, in tropical regions.

*Malleus regula* (Forsskål, 1775)

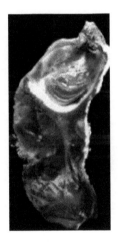

**Common name(s):** Straight hammer oyster

**Distribution:** Indo-West Pacific and the Mediterranean

**Habitat:** Rock crevices or under stones; depth range 0–20 m.

**Description:** Shell is slightly inequivalve and is strongly inequilateral. Outline of shell is variable. Not well-developed wings, and an elongate dorsoventrally curved but irregular shell, are the diagnostic features of this species. Ventral margin is often sinuous and occasionally expanded.

Ligament is in a single triangular pit. Internal shell is nacrous. Color pattern of the shell is variable and it is brown-purple throughout to grayish-yellow with purple brown spots to grayish yellow Maximum length is 12.0 cm.

**Biology:** No report

**Fisheries/Aquaculture:** No report

**Countries where eaten:** Mediterranean, Japan

**Pearl oysters (Winged oysters or feather oysters) (Pteriidae):** Pearl oysters of the family Pteriidae are commercially exploited throughout the world. The two recognized genera of this family are *Pinctada* and *Pieria*. Unlike other bivalves, they have comb-like teeth in the hinge, a feature of this family. They stick to hard surfaces by byssus threads. Some species are collected for food by coastal populations and used as a substitute for true oysters. The pearl oysters that produce commercial precious pearls belong to the genus *Pinctada*.

*Isognomon ephippium* (Linnaeus, 1758)

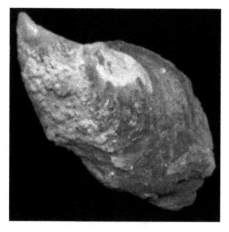

**Common name(s):** Leaf oyster, tree oyster, rounded toothed pearl shell, saddle tree oyster

**Distribution:** Indo-Pacific (excluding Red Sea, Arabian Sea, and NW India)

**Habitat:** Lower intertidal areas of mangroves; roots of mangroves and rocky substrate; sea walls in sheltered areas near freshwater inflows (brackish water)

**Description:** It has a large flat shell (up to 13 cm), which is round but sometimes "hammer" shaped. Hinge is elongated. It has byssal threads similar to mussels. Shell is usually white or grayish and grows up to 13 cm. This species is often found in large groups and are scattered like white coins among mangrove tree roots.

**Biology**

**Ecology:** This species can tolerate salinities down to 15 ppt.

**Feeding:** It is a filter feeder occurring in large clusters or singularly, generally following the distribution of *Saccostrea cucullata*.

**Fisheries/Aquaculture:** No report

**Countries where eaten**: Thailand

***Pinctada maxima*** (Jameson, 1901)

**Common name(s):** Australian South Sea pearl oyster, silver (gold)-lip pearl oyster, silver-lipped pearl oyster, white-lipped pearl oyster

**Distribution:** Ranges from eastern Indian Ocean to the tropical western Pacific; Australia, Fiji, Tahiti, China, Myanmar, Indonesia, and Philippines

**Habitat:** Littoral and sublittoral to a depth of 60 m; sand, mud, gravel, seagrass beds, deepwater reefs, to near sponges, and soft corals or whip corals

**Description:** This species is the biggest pearl-bearing oyster. Shell of this species is rather thick and large, subcircular in outline, with a short, posterior ear which is not drawn out into a wing-like process. Anterior margin is protruding beyond the tip of anterior ear. Outer surface of valves is covered with flattened, imbricating concentric scales bearing large and irregular, flat spines with blunt ends. Hinge is completely devoid of teeth. *Color*: Outside of shell is uniformly fawn, sometimes with radial stripes of darker spots in umbonal region. Internal nacreous area is highly lustrous, silvery with a variably extended golden border. Non-nacreous margin is of plain horny color. There are two different color varieties of this species, namely white-lipped oyster and gold-lipped oyster. It grows to a maximum length of 30 cm.

**Biology**

**Food and feeding:** This species uses its gills to filter tiny food particles of plant and animal matter from the water. Adults can filter as much as 1 t (1 cubic meter) of water per day.

**Growth:** Growth rates are initially fast and they can reach the minimum legal size for collection (120 mm in shell length) in their third year of life.

**Reproduction:** It is a protandrous hermaphrodite (meaning it functions first as a male then transforms into a female). It matures first as a male at 3–4 years of age and at 12 cm length, after which it becomes female. By 17 cm, half are males and half are females. By 19 cm, most are females. This species spawns every year so each pearl oyster can function as a male then a female for several spawning seasons. Primary spawning occurs from mid-October to December. A smaller secondary spawning may occur in February and March. Sperm and eggs are spawned into the water, where fertilization occurs. Fecundity of females is extremely high. A tiny planktonic veliger stage appears from the fertilized egg. After about 30 days, the veliger settles to the bottom. When a suitable substratum is found, it settles on it and metamorphoses into the juvenile stage (spat). Then it starts growing a shell and becomes a sedentary bottom-dweller. This species has a life span of about 40 years.

**Predators:** In the planktonic stage of this species, predators include pelagic fish, such as mackerel and tuna. In the spat stage, the predators include demersal fish, such as snapper. Turtles and boring worms can prey on adults.

**Fisheries/Aquaculture:** This species is cultivated commercially in Southeast Asia and China. White-lip oyster variety of this species produces south

Sea pearls. It can produce pearls ranging from 9 to 37 mm. It takes 3 years for this oyster to produce a pearl and then it takes another 2–3 more years more for it to nurture a pearl. Australia and Tahiti are the largest producers of cultured pearls from this species.

**Countries where eaten:** It is an edible species. Fresh meat from the adductor muscles of this species is a delicacy, particularly in China. It is eaten raw. This species is a specialty in certain Australian and Chinese restaurants. It is quite expensive and can cost about US $500/kg.

*Pinctada margaritifera* (Linnaeus, 1758)

**Common name(s):** Black-lip pearl oyster, Pacific pearl-oyster

**Distribution:** Persian Gulf and southwestern part of Indian Ocean; Australia; Fiji; Tahiti; Myanmar; Baja-California; and Gulf of Mexico

**Habitat:** Intertidal and subtidal zones; reef areas of lagoons; sand, coarse gravel, or dead coral substrates; eel grass beds and other marine plants; depth range 0–40 m

**Description:** Shell of this species is large, heavy almost equivalve and strongly inequilateral. Beaks are seen near anterior margin. Shell is more or less compressed with a triangular projection at anterior end. Outline is nearly circular and as high as broad. Posterior margin is slightly curved and anterior margin is flexous. Rough sculpture is of concentric ridges are usually more projecting at the margins. Hinge is without teeth. Ligament is in a single triangular depression. Externally, shell is grayish green with white or yellowish radial rows of scales. Internally, it is very pearly with a pale blue or violet cast. Margins are brown to black. Maximum length is 25 cm.

**Biology**

**Ecology:** This species requires oligotrophic conditions and low turbidity. The optimal conditions for maximum survival and growth have been reported to be 26–29°C and salinity 28–32‰. It is a continuous spawner with a peak period from December to April (warm season).

**Food and feeding:** This species like other pearl oysters is a suspension feeder feeding mainly on phytoplankton.

**Reproduction:** This species is protandrous hermaphrodite and exhibited different annual cyclical patterns, in which maturity peaked in July and November. It showed two spawning periods a year. The onset of reproduction appears to be regulated by sea surface temperature. Further, females outnumbered males during the spawning season (Hwang, 2007).

**Association:** Sponges, hydroids, polychaetes, lamellibranchs, amphipods, decapods, echinoderms, and fishes have close relationships with pearl oyster beds.

**Fisheries/Aquaculture:** Commercial fisheries exist for this species. It is commonly farmed and harvested for pearls, and the quality of pearls from this species is the highest quality out of all the pearl oysters. Commercial aquaculture of this species is popular in Indo-West Pacific. This species is also used for producing black pearls through culture. The size of produced black pearls may vary from 7 to 22 mm The nacre and pearls of black-lip pearl oysters are generally black or gray with shades of blue, green, silver, and pink. Most black pearls are produced in the sheltered waters of the atolls of French Polynesia and the Cook Islands, although Australia, Indonesia, the Philippines, and the Western Pacific Islands have been growing black pearl industries. Tahitian pearls come from this species.

**Countries where eaten:** This species is an important food source in Fiji. "Mother-of-pearl" (pearl shell) has been used by islanders of the Pacific and other regions as utensils, implements, and ornamentation, while the oyster itself has been a basic food item. Pearl meat is a delicacy in many western cultures.

***Pinctada fucata*** (Gould, 1850) (*=Pinctada imbricata; Pinctada martensii*)

**Common name(s):** Akoya pearl oyster

**Distribution:** Red Sea; Sri Lanka; Persian Gulf; Indian Ocean; Western Pacific Ocean; Australia

**Habitat:** Rocky or dead coral outcrops forming the pearl banks known as "paars;" loose sandy or muddy substratum; attached to the submerged objects in littoral waters.

**Description:** It has two valves connected by a long straight hinge. Length of the shell is slightly greater than its width. Right valve is flatter than the left and there are hinge teeth in both valves. Anterior ear is larger than that in other oysters of the genus and there is a slit-like notch for the byssus threads to pass through at the junction of the ear and the rest of the shell. Posterior ear is large. Outer surface of the valves is scaly and reddish or golden brown with pale radiating streaks. Inner surface of the valve is lined with a thick layer of golden-yellow nacre with a metallic sheen. It attains a length of 72 mm at the end of second year.

**Biology**

**Food and feeding:** Like other bivalve molluscs, this species is a filter feeder. These pearl oysters feed on infusorians, foraminifers, radiolarians, and other small planktonic organisms.

**Reproduction:** Sexes are separate in this species and gametes are released into the sea where fertilization takes place. Spawning peaks from June to

September and again in November and December, during the monsoon periods in India. The developing larvae pass through a veliger stage and after about 24 days settle on the seabed and become juvenile oysters known as spat.

**Fisheries/Aquaculture:** *Commercial importance* — This species extensively cultivated in China for production of pearls. Japan is also the largest producer of cultured pearls from this species. The extensive pearl oyster beds of this species in the Gulf of Mannar and Gulf of Kutch (India) are commercially very important sources for the production of natural "Oriental pearls" or "Lingah" pearls. *Pinctada fucata* occurring in the Persian Gulf and off the coasts of Ceylon also supports excellent fisheries for natural pearls. The Indian pearl fisheries have been famous since ancient times for the most beautiful pearls they yield. The other species of pearl oysters on the Indian coasts are not of any importance as their pearl yield is low or the pearls produced by them are not of high quality.

**Countries where eaten:** This species is an important food source in Fiji

***Pinctada radiata*** (Leach, 1814)

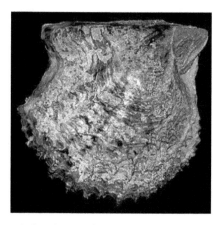

**Common name(s):** Atlantic pearl-oyster, gulf pearl oyster

**Distribution:** Persian Gulf, Red Sea, Mediterranean Sea

**Habitat:** Attached to hard substrata (under stones, in crevices of rocks, algae); depth range 2–25 m.

**Description:** Shell of this species is fragile, compressed, and inequivalve. Outline is nearly quadrate. It is higher than long and its dorsal margin is longer than body of shell. Posterior margin is slightly concave. Beaks are pointing anteriorly. Sculpture of concentic lamellae are often with rows of appressed spines. Hinge line is straight. There are no teeth. Ligament is in a single triangular depression. Margins are spiny. Its coloration varies, though it usually displays a brown or red exterior with a pearly interior and a light brown edge. Rarely, the shell may display a green or bronze exterior. Darker brown or red rays may mark the shell, creating darker areas at the margin. It attains a maximum length of 106 mm.

**Biology:** This species is hermaphroditic and its reproductive maturity is largely influenced by temperature.

**Fisheries/Aquaculture:** Commercially cultivated in Japan, China, and India. This species is harvested for pearls, especially in Qatari waters, where it may constitute up to 95% of the oyster catch. It is also caught for its edible flesh and lustrous shell. This species also serves as a bio-indicator of heavy metals in Persian Gulf waters.

**Countries where eaten**: Japan, China, India

*Pinctada maculata* (Gould, 1850)

**Common name(s):** "Pipi" pearl oyster

**Distribution:** Central to West Pacific and Indian Ocean; Kii Peninsula to Ryukyu Islands; South China Sea to the Philippines, Fiji, and Samoa; New South Wales to Torres Strait, Australia

**Habitat:** Shallow waters; depth range 18–71 m.

**Description**: Shell of this species is rather thin and small, subquadrate in outline, with a short and ill-defined posterior ear which is not drawn out into a wing-like process. Anterior margin is protruding only slightly or not at all beyond the tip of the anterior ear. Outer surface of shell is covered with numerous, flattened, and brittle, imbricating concentric scales bearing slender, radially projecting spines, especially toward the margins. Hinge lines are with two small teeth on each valve. One rounded anterior tubercle is seen just in front of the umbo, and one is slightly slanting posterior ridge behind the ligamental area. *Color*: outside of shell is with a variable coloration, usually white to tan with a number of purple, or brown to black radiating bands. Internal nacreous area is with pale yellow to deep orange-gold tint. Non-nacreous margin is with white porcelaneous patches, generally alternating with irregular, dark purplish brown or black blotches.

**Biology:** Not reported

**Fisheries/Aquaculture:** This species is harvested for its natural pearls.

**Countries where eaten:** Countries of occurrence; meat of this species can be eaten raw or cooked and is very tasty, though a little gritty.

***Pteria penguin*** (Röding, 1798)

**Common name(s):** Penguin's wing oyster, black lip oyster

**Distribution:** Indo-Pacific: Subtropical and tropical Australia; South China Sea, Hong Kong, Hainan; the Philippines; Northern Australia; Arabian Gulf; Red Sea

**Habitat:** Intertidal areas on rocks and corals; sometimes attached to sea whips; depth range 0–35 m

**Description:** Shell is solid, slightly inequivalve and obliquely ovate in outline, with posterior ear drawn out into a narrow, more or less elongated, wing-like expansion. Left valve is slightly inflated and with a weak rounded fold radiating from umbo to posteroventral end of shell. Outline of shell is variable, initially narrowly oblique, later greatly expanding ventrally and almost as high as long, or even higher than long in larger specimens. Interior of shell is with a wide non-nacreous margin ventrally. *Color*: outside of shell is plain dark brown to black. Interior is silvery and brilliantly nacreous, with a broad, posteroventrally expanded, glossy black margin. It grows to a maximum length of 30 cm.

**Biology:** No report

**Fisheries/Aquaculture:** This species is commercially harvested and cultivated. Japan is the largest producers of cultured pearls from this species. Pearl production from this species is often limited by the availability of oysters, particularly in places where it is an exotic species and does not recruit in the wild Hatchery propagation has now become a necessity in areas such as Tonga, where the collection of oysters from the wild can no longer sustain commercial pearl production.

Though this species produces commercial quantities of pearls, they seldom produce precious pearls. The pearls produced by this species are Mabe pearls which have a characteristic semispherical shape and unique rainbow-like luster. These half pearls are produced by inserting a half-sphere between the mantle and the shell. These semispherical "half-pearls may have a size range of 12–20 mm in diameter and possess a uniquely penetrating brilliance, with hues ranging from light pink through deep rose-red to a "rainbow" pink. Sometimes the pearls have gold-pink or other hues of high rarity. This rich variety of lustrous hues combines with nacreous layers of a rarely seen fineness of texture to give the Mabe pearl its perennial appeal and make it our original, popular product. Mabe pearl producing countries are Japan, Hong Kong, Indonesia, Thailand, and Malaysia (Marta Howell, http://www.martahowell.com/new-gallery-1/).

**Countries where eaten:** This species is edible in certain countries of its occurrence but is unpalatable.

## *Pteria colymbus* (Roding, 1798)

**Common name(s):** Atlantic winged oyster

**Distribution:** Atlantic coast of North America: North Carolina to Bermuda and Brazil

**Habitat:** Attached to gorgonians, offshore corals, sea fans, rocks, oil platforms, wharf pilings, and other objects; depth range 3–30 m

**Description:** Shell of this species has a distinctive, asymmetric shape. There is a long straight hinge with one wing drawn out a long way and the other one much smaller. Upper valve is brownish, often mottled with paler markings. Lower valve is flatter and smaller and interior of the shell is pearly gray. Posterior margin is rounded while the anterior margin is elongated and slopes at an acute angle to the hinge. There are a number of indistinct, irregular ribs that flare out from the umbo. These are sparsely covered with blunt spines and there is a fine sculpturing of concentric growth lines. Maximum length is 9 cm.

### Biology

**Food and feeding:** It is a filter feeder feeding on plankton, algae, and other food particles.

**Reproduction:** These oysters are able to reproduce at the age of 3 years. Spawning begins when a male oyster discharges his sperm in the water, other oysters would detect sperm in the water and start their spawning process. From fertilized eggs, the pediveliger stage appears in a period of 2 weeks. This larva attaches itself to a hard surface by secreting a glue at the bottom of the water. The larva then undertakes metamorphosis and turns into a juvenile oyster or spat. This oyster naturally grows up to 2.5 cm each year, depending on environmental conditions. It may change its sex during its lifespan.

**Association:** It often has algae and other organisms growing on the shell which make it well camouflaged.

**Fisheries/Aquaculture:** No report

**Countries where eaten:** Gulf of Mexico

**Gryphaeidae:** Shell is solid, more or less inequivalve and is often irregularly shaped. Both valves are convex and similarly sculptured, with large, irregular radial ribs. Right (upper) valve is flat, with imbricating thin plates of horny material. Ligamental area is with a shallow median groove. Hinge is without teeth. Umbonal cavity is generally very shallow. A single, large, and rounded (posterior) adductor muscle scar is seen closer to the hinge than to the ventral margin. Pallial line is without a sinus. Internal margins are with long, branched, sinuous chomata on either side of the ligamental area.

*Hyotissa hyotis* (Linnaeus, 1758)

**Common name(s):** Giant honeycomb oyster, foam oyster

**Distribution:** Indo-Pacific: from East Africa to eastern Polynesia; north to Japan and south to Western Australia and Queensland.

**Habitat:** Intertidal and shallow subtidal; rocky shores, coral reefs and shipwrecks in warm, high-salinity environments; depth range 0–50 m

**Description:** It is the largest of all the edible oysters and has two unequal valves. Shell structure is foam-like and shape is irregular, according to the substrate and conditions of growth. Animal is cemented to the substrate by the left valve. Both valves are solid and thick. Shell is very deeply ridged, with ridges extending to the shell margin, making it strongly saw-toothed. Arches of the valves interlock. Area of attachment of the ligament has a shallow median groove. Hinge is without teeth. External coloration of the shell is purplish black, while the interior is bluish-white in the center, and bluish-black toward the margin. There is a single purplish adductor muscle scar, located toward the hinge. Living animals are frequently heavily over-grown with fouling organisms. This species has been reported to reach 30 cm in size.

**Biology:** No report

**Fisheries/Aquaculture:** This species is commercially exploited and is a potential species for aquaculture in Philippines and Vietnam

**Countries where eaten**: Caribbean and Indo-Pacific; South Africa to the Red Sea, out to New Guinea, Indonesia, and the Philippines; Malaysia.

*Neopycnodonte cochlear* (Polish, 1795)

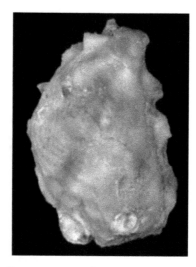

**Common name(s):** Spoon oyster, deep sea oyster

**Distribution:** Indo-Pacific: Eastern Africa to Japan, Hawaii, and Easter Island

**Habitat:** Deep-water; sessile; continental margins and on the slope; depth range 125–200 m

**Description**: Shell of this species is inequivalve, suboval to subcircular, and solid but brittle. Chalky layer of shell is honeycombed. Upper valve is flat sitting inside deeply concave lower cemented valve. Upper valve is rather regular but lower valve is with irregular wavy margins. Upper valve is with concentric ridges and irregular laminar layers. Lower valve is similar but surface drawn out into irregular extensions especially around the attachment point. Inner margin is with vermiculate (irregular ridges) and chromata are seen on the anterior dorsal edge. Simple chromata are present on the posterior dorsal edge. Adductor scar is subcircular. Color of shell is buff to greenish yellow to dirty white. It can reach a maximum size of about 12 cm.

**Biology:** No report

**Fisheries/Aquaculture:** Not known

**Countries where eaten:** Europe and Red sea

**Thorny oysters (Spondylidae):** The thorny oysters are considered to be the daughters of the sea. Shell is stout, highly variable in shape but generally subequilateral, rounded, and higher than long. It is usually inequivalve and cemented to substrate by the right (lower) valve, which is higher and more convex than the left (upper) valve. Hinge line is straight, with a small, more or less equal ear on either side of the median umbo. Outer surface is with more or less spinose to scaly irregular radial ribs, often brightly colored (at least at the left valve). Umbones are seen on top of a trigonal cardinal area. Ligament is mainly internal and lodged in a deep median pit of the hinge plate. Hinge is stout, with two strong curved teeth and two deep sockets in each valve. Interior of shell is porcelaneous. A single, rounded (posterior) adductor muscle scar is present. Pallial line is without a sinus. Spondylus shells are much sought after by collectors, and there is a lively commercial market for them. The meat of these bivalves is edible (WILD Fact Sheets, nd; Anon, ndu).

***Spondylus americanus*** (Hermann, 1781)

**Common name(s):** Atlantic thorny oyster

**Distribution:** Atlantic coast of North America, ranging from North Carolina to Brazil

**Habitat:** Fouler on sea walls, man-made structures and wrecks; depth range 0–140 m

**Description:** Valves of the shell are roughly circular and upper one is decorated with many spiny protuberances up to 5 cm long. Color varies but is usually white or cream with orange or purplish areas making it well camouflaged. Lower valve is flat and is attached to the substrate. Flat tree oyster and Lister's tree oyster are its common epibionts. Its size ranges from 76 to 230 mm.

**Biology**

**Reproduction:** The sexual organs of this species which has separate sexes are located beneath the foot within the visceral mass. Reproductive morphology of this species is very similar to that of members of the family Pectinidae. The reproductive organs of this species swell when sexual activity is imminent. Ducts from the organs connect to the kidneys, which then work to disperse eggs or sperm into the water—this may be accompanied by a flapping of the shell to further spread out the sperm and eggs. During this process, this species appears docile and sedentary, reacting little to the surroundings. Spawning in Spondylidae will occur in warm water during the summer, sometimes as late as August.

**Fisheries/Aquaculture:** No report

**Countries where eaten**: Spondylidae are generally edible and may be consumed where they are commonly found. The shell is something of a collector's item because of the large adorning spines.

***Spondylus crassisquama*** (Lamarck, 1819) **(=*Spondylus princeps*)**

**Common name(s):** Pacific thorny oyster

**Distribution**: Peru, Mexico

**Habitat:** Deep-dweller; below 60 m of ocean surface

**Description**: It is a small shell with a crimson inner rim. Size is 115 mm, including spines

**Biology:** No report

**Fisheries/Aquaculture:** This edible species has been an item of ritual importance in Peru for well over thousand years. Further, its fisheries were traded by Spaniards for their coral-colored sea shells. This species has been reported to produce non-nacreous natural pearls (Ho and Zhou, nd).

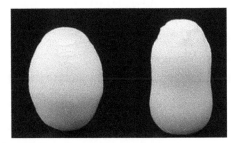

Non-nacreous, natural pearls from *Spondylus crassisquama*

**Countries where eaten:** This species is fished commercially for food in Peru

*Spondylus gaederopus* (Linnaeus, 1758)

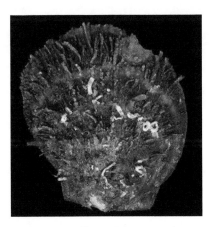

**Common name(s):** Spinous scallop, European thorny oyster

**Distribution:** Western Indian Ocean and the Mediterranean Sea: Turkey and Greece

**Habitat:** Benthic; depth range 4–200 m

**Description:** Shell of this species is elongated-ovate, almost equivalve. On both valves, there are 8/9 obscure ribs, ornamented with irregularly shaped plugs which vary from flat to pointed to spatulate. In adult specimens, this sculpture tends to be reduced. Exterior color of the upper valve is dark burgundy, pink-purple, crimson, orange, yellow-brown, white. Exterior color bottom valve is clearer than the upper valve and can be completely white or light orange, with yellow-orange or violet colored ends. Interior color is white porcellanaceous, quite polished, often mottled with brown, greenish, purplish. This species has a maximum length of 12 cm.

**Biology**: No report

**Fisheries/Aquaculture:** No report

**Countries where eaten:** Montenegro coast. It is also a Croatian cuisine.

*Spondylus groschi* (Lamprell & Kilburn, 1995)

**Common name(s):** Not designated

**Distribution:** Origin is Indian Ocean; East coast of Africa, Red Sea, and Persian Gulf

**Habitat:** Attached to corals or rocks to 150 m

**Description:** Shell of this species is inequivalve, solid, and ovate to elongate-ovate. Area of attachment is rather large. Lower valve is moderately convex. Sculpture is of 6–10 well-defined principal radial ribs with strong, depressed, overlapping, spatulate spines. In adults, umbonal area is usually worn. Wide interspaces, with one secondary rib, which bears sharp or slightly flattened overlapping spines. Cardinal area of the upper valve (right valve) is triangular and shallow under hingle plate. Inner margin is finely crenulate. Shell is purple-brown, orange red, or deep brick-red. Ribs and spines are the same color as the shell but rarely yellow or lilac colored. Umbonal area is often with irregular brown spots, blotches, or undulating marks. Internal shell is white. Its common size is 7.5 cm.

**Biology:** Unknown

**Fisheries/Aquaculture:** No report

**Countries where eaten**: It is an edible species in areas of its occurrence.

*Spondylus leucacanthus* (Broderip, 1833)

**Common name(s):** White daisy spine sea clam

**Distribution:** Sea of Cortez (Gulf of California)

**Habitat**: Deep water species; depth range 25–70 m

**Description:** Shell is more ovate with an orange line or band within the inner margin. Length and width are 13 and 10 cm, respectively.

**Biology:** The population of this species has an annual pattern of reproduction. The start of gametogenic phase and maturity in the population coincide with the abundance of phytoplankton which are the primary food item of this species.

**Fisheries/Aquaculture:** It is fished commercially and artisanal overfishing has been reported.

**Countries where eaten:** Meat is quite tasty as it is a close relative of the scallops and is served in restaurants in Mexico and Baja California.

*Spondylus limbatus* (Sowerby, 1847)

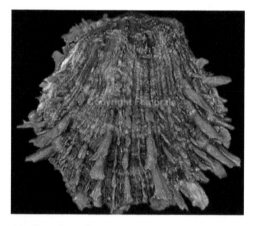

**Common name(s):** Donkey thorny oyster

**Distribution**: Indo-West Pacific

**Habitat**: Living on dead corals; depth range 10–30 m

**Description:** Distinguishing features of this species are a pair of nearly equal-size teeth in each valve and a centrally placed resilifer. Cardinal area appears as a smooth or horizontally grooved triangular area between beak and hinge line. Shells are nearly equilateral and inequivalve. Shell is solid, oval in shape, smooth, and flattened. Scaly projections are present in the periostraccum. Cardinal area is prominent and triangular in shape. Internally it is yellow white with a broad purple margin. A single adductor muscle scar is present near the posterior margin of the shell. Maximum size is 45 × 36 mm.

**Biology:** Not known

**Fisheries/Aquaculture:** No report

**Countries where eaten:** It is an edible species in areas of its occurrence.

*Spondylus multisetosus* (Reeve, 1856)

**Common name(s):** Not designated

**Distribution**: Indo-Pacific, Philippines; Mediterranean

**Habitat:** Attached to rocks, corals, and other debris

**Description**: Shell of this species is solid, inequivalve, subequilateral, and ovate. It is cemented to substrate by the right (lower) valve. Attachment area is usually small. Right valve is more convex than the left one. Sculpture of numerous radial ribs is with upstanding spines. Ribs are dense with spins that are regularly spatulate and hollow underneath. Interstices are arrayed with smaller sharp spins. Umbones are seen on top of a triangular cardinal area which is higher in the right valve. Ligament is internal on the hinge plate. Inner margin is crenulate. Shell is mauve or brown, ivory-yellow, and yellow on ribs and spines. Umbonal area is yellow-white. Internal is blue-white. Crenulated margin is with tones of dark brown and yellow. Common length is 14 cm.

**Biology:** Unknown

**Fisheries/Aquaculture:** No report

**Countries where eaten:** It is an edible species in its areas of occurrence.

*Spondylus sinensis* (Schreiber, 1793)

**Common name(s):** Chinese thorny oyster

**Distribution:** Indo-West Pacific

**Habitat:** Intertidal areas on coral and shell debris; attached to living corals; depth range 0–30 m

**Description:** This species has a thick shell which is also large. Lower valve is with a wide umbonal triangle. There are 8–10 high axial ribs which are with short spines and often wide spatulas. Lower axial ribs are scaly and uneven. Color of shell is red brown. Maximum size is 47 × 45 mm.

**Biology:** Not known

**Fisheries/Aquaculture:** No report

**Countries where eaten**: It is an important food source in Fiji.

*Spondylus spinosus* (Schreibers, 1793)

**Common name(s):** Spiny oyster

**Distribution:** Indo-West Pacific and the Mediterranean Sea

**Habitat:** Intertidal on shell debris; attached to subtidal rocks and corals; depth range 0–20 m

**Description**: Shell of this species is equivalve and is generally higher than long and irregularly oval. Area of attachment is variable. There are 6–16 strong, raised, principal radial ribs with dense strong, moderately appressed spines of various lengths. In adults, umbonal area is usually worn. Wide, smooth interspaces are, with one or two secondary ribs, which bear smaller spines. Inner margin is finely crenulate. Interstices are dark brown with white ribs and spines. Internally it is blue-white with purple red margin. Hinge is brown. It grows up to 70 mm in height.

**Biology:** This species shows valve flapping as a reaction to attacks by muricid gastropods. This antipredatory behavior is comparable to the swimming escape mechanism shown by living pectinids.

**Fisheries/Aquaculture:** No report

**Countries where eaten**: The species is collected by diving and is served in restaurants in Jbail, Lebanon.

*Spondylus squamosus* (Schreibers, 1793)

**Common name(s):** Ducal thorny oyster, scaly thorny oyster

**Distribution:** Indo-West Pacific, from India to Melanesia; north to Japan and south to Queensland and New Caledonia

**Habitat**: Attached to rocks or dead corals; littoral and sublittoral to a depth of 30 m.

**Description:** Shell is highly variable in shape but roughly rounded-ovate to elongate-ovate in outline, and inequivalve. Right (lower) valve is somewhat more convex and higher than left (upper) valve, with a well-developed cardinal area. Attachment area is from moderately big to small, with a discrepant, mainly concentric sculpture. Outer surface of valves is with low radial undulations forming a series of appressed, rather broad white ribs bearing generally a few short, flattened, and often arched imbricate spines. Interstices of main ribs are broad, radially ridged, and sometimes with small and sharp fine spines. Main spinose ribs are often slightly more numerous on right valve (8–12 ribs on right valve, instead of 5–8 on left valve). Hinge teeth of right valve are more or less bifid on top. Internal margins are provided with crenulations corresponding with the outer radial sculpture. Outside of shell is white to cream, with a few purple to blackish brown blotches on umbonal area. Interior is whitish and is often tinged pale brown on ears and hinge. Internal margins are marked with brown and purple on the crenulations. Maximum shell height is 10 cm.

**Biology:** No report

**Fisheries/Aquaculture:** Unknown

**Countries where eaten**: Collected for food in Fiji Islands and the Philippines. Shell of this species is used for shell craft.

*Spondylus versicolor* (Schreibers, 1793)

**Common name(s):** Golden thorny oyster

**Distribution:** Indo-West Pacific: from East Africa, including the Red Sea to eastern Indonesia; north to Japan and south to northern Queensland

**Habitat:** Littoral and sublittoral to a depth of 25 m; attached to rocks or dead corals

**Description:** This species is heavy and fairly large. There are very unequal radial ribs, with fairly long and wide spines, but never spatulas. Interior is with a yellow-orange border. Maximum length is 14 cm.

**Biology:** No information

**Fisheries/Aquaculture:** It is commercially exploited for food.

**Countries where eaten:** Locally collected for food and shell trade in Indonesia and Philippines

**Remarks:** The presence of a pennate diatom species, namely, *Pseudonitzschia multiseries* was found to be responsible for the presence of domoic acid and associated "amnesic shellfish poisoning" in this species.

**Saddle oysters (jingle shells, mermaid's toenails) (Anomiidae):** Shells of anomiids are inequivalve, thin and brittle, approximately circular in outline and often irregular. They are attached on a rock or a large shell of another creature by a byssus passing through a hole in right valve, with the left valve uppermost. Shells are inequilateral and umbones rounded and indistinct. Left valve is convex and right valve is flat or concave. There are no hinge teeth. Ligament is tinternal, in a deep pit below umbone of right valve. Inner surface of shell is nacreous, glossy, and there is no pallial line.

*Anomia peruviana* (d'Orbigny, 1846)

**Common name(s):** Peruvian Jingle, Peruvian jingle shell

**Distribution:** Native to eastern Pacific Ocean ranges from southern California to Peru and the Galapagos Islands

**Habitat:** Intertidal zone to a depth of 110 m

**Description:** Shell of this species is very thin and pearly; white or coppery brown on the upper valve and bluish green internally and on the central part of the lower valve. It is sessile on other shells or smooth objects adhering by a prominent byssus which passes through a large hole in the lower valve. In this species, the left valve is with four muscle scars (one adductor, three byssal muscles).

**Biology:** Unknown

**Fisheries/Aquaculture:** Unknown

**Countries where eaten**: These oysters are not, as far as known, used for food, though edible.

**Pen shells (fan clams) (Pinnidae):** Members of the Pinnidae family occur in shallow water in most parts of the world. They are distinguishable by their fan like shells. Animal buries itself, pointed end down. These clams are usually found well spaced apart from one another. Like other bivalves, fan clams are filter feeders. All seaweeds and encrusting animals often settle on the portions of the fan clam that sticks out above the sand. These provide food and shelter for small animals. The tiny Pea crab (*Pinnotheres* sp.) is sometimes found living inside these clams. This crab not only gains shelter but also eats some of the food gathered by the fan shell host. Size of animals is 10–75-cm long. Fan clams are edible. But being large, the muscular meat is rather chewy, and although it can be prepared in the style of a scallop, is more suited for making soup, although in Southeast Asia, its chopped and fried with coriander.

*Atrina fragilis* (Pennant, 1777) (=*Atrina pectinata*)

**Common name(s):** Comb pen shell, fan shell, brittle pen shell

**Distribution:** Mediterranean, Black Sea, and Indo-West Pacific: from southeast Africa to Melanesia and New Zealand; north to Japan and south to New South Wales

**Habitat:** Muddy-sand bottom; depth range 0–200 m

**Description:** Shell is reaching a large size, usually rather thin, fragile, moderately inflated and triangularly wedge-shaped in outline. It has a highly variable sculpture. Dorsal margin is nearly straight or slightly concave and posterior margin is generally truncate. Ventral margin is widely convex posteriorly, straightish to shallowly depressed anteriorly. Outer surface of valves has 15–30 radial ribs which may be smooth to densely set with short, open spines. Inner surface of shell has shallow grooves corresponding to the external radial ribs. Internal nacreous layer is rather thin, undivided, occupying the anterior 3/4 of valves. Posterior adductor scar is completely enclosed within the nacreous area. Coloration: Exterior is slightly shiny, translucent, and olivaceous tan. It is often tinged with darker purplish brown or gray toward the umbones. Interior is similarly colored and iridescent on nacreous area. Maximum length is 37 cm.

**Biology:** In this species, differences exist not only in shell shape but also in comparative size of back adductor muscle among four types of pen shell, namely, green pen shell, yellow pen shell, thorny pen shell, and scabrous pen shell. Among of them, scabrous pen shell is especially distinct from green pen shell and yellow pen shell. It is suggested that these four types of pen shell should be categorized to two or more species taxonomically.

**Fisheries/Aquaculture:** It is commercially exploited and it is a potential species for aquaculture in Japan.

**Countries where eaten:** Adriatic, Mediterranean, United Kingdom, Morocco, Japan and China. In Japan, it is served as sashimi, a Japanese cuisine. Meat is frozen and sold in markets.

*Atrina maura* (Sowerby, 1835)

**Common name(s):** Fan shell

**Distribution:** Pacific, from Baja California to the Peru.

**Habitat:** Estuaries, sandbars, or mud in bays and coastal lagoons

**Description:** Its shell is triangular, fan-shaped, slightly flattened and thin. Coloration is cream, white or brown, and translucent. Trailing edge is wide. There are 18 rows of thin spines. On an average, valve length of adult specimens is 18 cm and dia. of 5 cm.

**Biology:** No report

**Fisheries/Aquaculture:** This species is commercially exploited in Mexico and it is a potential species for aquaculture. It is threatened by overexploitation and habitat degradation. Cultivation is an alternative for preserving this resource and increasing its production. It was cultivated in a suspension system which was followed by a bottom phase. Results indicated that this species is suitable for aquaculture and the cultivation system adopted here should be tested in pilot-scale ventures. This system may work with other species of pen shells.

**Countries where eaten:** This species is economically important and is consumed as food in areas of its occurrence.

*Atrina oldroydii* (Dall, 1901)

**Common name(s):** Oldroyd penshell

**Distribution:** *Eastern Pacific* — Baja California, Gulf of California, and southern Sinaloa and Nayarit coasts

**Habitat:** Muddy substrate at 30-m depth

**Description:** It has a large, long, more trigonal, relatively thick and grayish-brown shell with finer sculpture. *Size*: Maximum length 32 cm; height 14 cm; and dia. 7 cm.

**Biology:** No report

**Fisheries/Aquaculture:** It is commercially exploited in Mexico. It is also a potential species for aquaculture.

**Countries where eaten**: Mexico

***Atrina rigida*** (Lightfoot, 1786)

**Common name(s):** Rigid pen shell, stiff pen shell, wing shell

**Distribution:** Western Central Atlantic; from the North Carolina coast through the West Indies and to Brazil

**Habitat:** shallower waters; soft bottoms; depth range 0–27 m

**Description:** Shell of this species is large, fan-shaped, and triangular. Surface sculpture is of about 15 narrow radial ribs separated any larger interspaces. Ribs are bearing regularly spaced, fluted spines. Large muscle scar seen inside shell touches border of nacreous area. Hinge area is straight, representing larger side of triangular shell outline. Coloration of shell is dark olive brown. Mantle color is bright golden orange. Maximum length is 30 cm.

**Biology**

**Behavior:** Majority of the shell of this animal is buried. Animal points down leaving only a few inches exposed. It is held in position ace by an extensive net of byssal threads.

**Food and feeding:** It is a suspension feeder, taking in water, and nourishment through the gap between the valves at the posterior (upper) edge.

**Association:** Commensal crustaceans may live in this opening, apparently feeding upon excess food. Barnacles and other creatures may also attach themselves to the exposed exterior of the shells.

**Predators:** This species is preyed upon by starfish, and other carnivorous gastropods, including the horse conch, *Pleuroploca gigantea*. This species is noted for its ability to heal breaks and holes made in its shell.

**Fisheries/Aquaculture:** This species is commercially harvested in and around Campeche, Mexico. It has also been reported to produce black pearls (stiff pen shell pearls) of 4.5 mm size.

**Countries where eaten:** In Mexico, it is consumed in soups, marinated, or grilled.

*Atrina seminuda* (Lamarck, 1819)

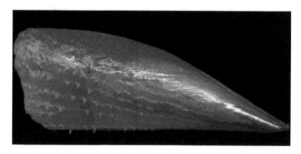

**Common name(s):** Half-naked pen shell, saw-toothed pen shell

**Distribution:** Western Atlantic; Atlantic coast of North America, ranging from North Carolina to Texas and south to Argentina

**Habitat:** Rocks and gravels; shallow-water seagrass beds; burrows in fine sand leaving only the broad posterior region exposed to the outside environment; depth range 0–2 m

**Description:** Shell of this species is large, fan-shaped, and triangular. Surface sculpture is of about 15 narrow radial ribs separated by larger interspaces. Ribs are bearing regularly spaced, fluted spines. Muscle scar is completely surrounded by nacreous layer. Hinge area is straight, representing larger side of triangular shell outline. Byssus at pointed extremity anchors pen shell into seagrass bottom. Coloration of shell is dark olive brown and mantle color is pale yellow. Maximum length is 24 cm.

**Biology**

**Association:** Associated with the outer faces of the leaflets of this species usually inhabit a range of fouling organisms including gastropod molluscs, bivalves, and chitons.

**Fisheries/Aquaculture:** It is commercially exploited.

**Countries where eaten:** Edible in countries of occurrence.

*Atrina tuberculosa* (Sowerby, 1835)

**Common name(s):** Pen shell

**Distribution:** Gulf of California to Panama

**Habitat:** Muddy banks

**Description:** Shell of this species is very thick, more strongly convex and ham shaped. Dorsal border is initially concave at the anterior end, then almost straight. Ventral margin is also concave at the anterior end, convexly and convex toward the posterior end. Rear edge is well rounded. Outline changes in the course of growth. Shell is dark brown to almost black. Adult specimens are usually without ribs and thorns, and only margin-parallel growth lines are present. *Size*: length is 29 cm, 20 cm; width 12.5 cm; and dia. 4 cm.

**Biology**

**Association:** An association between the monogamous and gonochoric shrimp *Pontonia margarita* found in the mantle cavity of the pen shell

*Atrina tuberculosa* has been reported. The positive relationship between shrimp size and pen shell size, a prevalence of male–female pairs of shrimp (sex ratio was 50% males and 50% females) and morphological measurements, suggest that a long-term symbiotic association exists between these species, and that the mating system of the shrimp involves social monogamy (Góngora-Gómez et al., 2015).

**Fisheries/Aquaculture:** It is commercially exploited in the region of Bahia de Kino, Sonora (Mexico).

**Countries where eaten:** Mexico

*Atrina vexillum* (Born, 1778)

**Common name(s):** Flag pen shell, sawtooth pen shell

**Distribution:** Indo-Pacific: from East Africa, including Madagascar, the Red Sea and the Persian Gulf, to eastern Polynesia; north to Japan and Hawaii, and south to Queensland and New Caledonia.

**Habitat:** Intertidal areas in sand and mud; depth range 0–50 m

**Description:** Shell of this species is reaching a very large size, thick and solid, inflated, and variable in shape from triangular to hatchet-shaped or subglobular. Dorsal margin is usually nearly straight and posterior margin is broadly oval to somewhat truncate in outline. Ventral margin is broadly convex posteriorly and concave near the umbones. Outside of valves is with 10–17 main radial ribs, often bearing scale-like spines, and with weaker interstitial riblets. Internal nacreous layer is moderately strong, undivided and is occupying the anterior half or 2/3 of valves. Hind margin of posterior adductor scar is slightly protruding beyond the nacreous area. *Color of shell*: outside of shell is dark reddish brown to nearly black, usually dull. Interior is dark brown to black and is iridescent on nacreous area. Maximum length is 48 cm.

**Biology:** No report

**Fisheries/Aquaculture:** This species is commercially exploited. By virtue of its large size, it is one of the most economically important members of the family in the Indo-West Pacific. Beautiful but fragile black pearls are sometimes produced by this species.

**Countries where eaten**: All the species of the genus have been reported to be edible in the areas of their occurrence.

***Pinna bicolor*** (Gmelin, 1791)

**Common name(s):** Two-colored pen shell, bicolor pen shell, bicolored razor clam.

**Distribution:** Indo-West Pacific: East Africa to Melanesia, including southern Japan, the Philippines, Australia and Hawaii; Malaysia and Singapore

**Habitat:** Littoral and adjacent subtidal shallow waters; buried in hard substratum and associated with seagrass beds; depth range 0–50 m

**Description:** Shells are triangular in shape. Posterior margin of the shell is usually rounded and attenuated to anterior margin. It is usually translucent, yellowish, and light horn to dark brownish purple in color. Posterior margin is rounded and is showing signs of repeated breakage. Dorsal margin is straight posteriorly and concave anteriorly but ventral margin is often convex and straight posteriorly. Maximum length is 50 cm.

**Biology**

**Association:** Epifauna on the shells of this species include encrusting bryozoans, serpulid polychaetes; uncommon tunicates and sponges

**Fisheries/Aquaculture:** This edible bivalve is of minor economic importance and is collected for subsistence fisheries.

**Countries where eaten**: It is consumed in the areas of its occurrence.

***Pinna carnea*** Gmelin, 1791

**Common name(s):** Amber pen shell

**Distribution:** Western Atlantic and Western Central Pacific; from southern Florida across the Caribbean and the West Indies to Brazil

**Habitat:** Medium to coarse sand or mixed substrata (sand, gravel, and rocks); fine calcareous sandy mud of eelgrass (e.g., *Zostera* spp.); sandy substrata of turtle grass (e.g., *Thalassia* spp.) and other mixed-species seagrass meadows; coastal waters at depths between 0 and 5 m.

**Description:** Shell of this species has a pair of long, thin translucent valves held together by ligaments that run along the entire dorsal side of the bivalve. Bivalve is triangular with 8–12 low ribs radiating from the pointed anterior end (or umbo) to the large posterior edge. Exterior of the shell is usually a dull orange amber-like color and may have fragile, scale-like spines. Anterior end is usually buried and attached by byssal threads, whereas its wider posterior gaping end extends above sea bottom surface to facilitate its filter-feeding. Maximum length is 29 cm.

**Biology**

**Feeding:** As a filter feeder, this species draws water into the shell from above and passes it over the ctenidium before expelling it into the open water at its exposed portion of the shell. During this process, oxygen is absorbed and food particles are captured and transferred to the mouth in balls of mucous.

**Association:** This species may host a number of symbionts in its mantle cavity. Shrimp *Pontonia mexicana*, cardinalfish *Astrapogon stellatus*, pea crabs (Pinnotheridae), and sea anemones (Actiniaria) have been found sheltered inside its shell.

**Reproduction:** This species is a hermaphrodite and the gonads are producing both sperm and ova. The larvae are planktonic and drift with the currents. When they settle on the seabed, they undergo metamorphosis into juveniles.

**Fisheries/Aquaculture:** No report

**Countries where eaten**: It is consumed in areas of its occurrence.

***Pinna deltodes*** (Menke 1843)

Image not available

**Common name(s):** Deltoid razor clam

**Distribution**: Indo-Pacific: extends off eastern Africa to eastern Australia and as far north as Pakistan and probably the Red Sea.

**Habitat:** Intertidal zone to 140 m; hard substrate, either within cobble flats or beneath boulders, coral blocks, and bombies; associated with seagrasses

**Description:** Mean total length and width of shells of this species are 23 and 12 cm, respectively. Shell is extremely flared posteriorly so that width nearly equals the length. Color of shell is translucent, light horn to dark brownish purple. Posterior shell margin rises in high, symmetrical arch, with highest point and greatest shell length midway between dorsal and ventral margins. Dorsal margin is as usually longer than ventral margin, posteriorly straight, and concave anteriorly. Ventral margin is posteriorly convex and concave anteriorly. Nacreous layer is iridescent and well separated by longitudinal sulcus. Dorsal lobe of nacreous is higher than ventral lobe and posterior margin of dorsal and ventral lobes is truncated. Posterior adductor muscle is medium in size and located near or touching onto posterior edge of dorsal lobe. An anterior adductor muscle is small to moderate in size, subapical, and located just anterior to end of longitudinal sulcus.

**Biology:** No report

**Fisheries/Aquaculture:** No report

**Countries where eaten:** Edible in areas of its occurrence

*Pinna nobilis* (Linnaeus, 1758)

**Common name(s):** Noble pen shell, fan mussel, sea-wings

**Distribution:** Mediterranean and Black Sea

**Habitat:** Buried beneath soft-sediment areas (fine sand, and mud); associated with seagrass of the species, *Posidonia oceanica* and *Cymodocea nodosa*; depth range 125–200 m; prefers more or less closed, protected bays with low water depths up to 40 m.

**Description:** Shell shape of this specie differs depending on the region it inhabits. Inside of shell is lined with brilliant mother-of-pearl. Maximum length is 120 cm. It is an endangered species.

**Biology**

**Ecology:** The main factors that characterize the habitat of this species are good lighting, clean water, low seasonal variations in salinity (3.4–4%), moderate temperatures (7–28°C) and a uniform, slow flow with enough nutrients.

**Behavior:** Like all pen shells, it is relatively fragile to pollution and shell damage. It attaches itself to rocks using a strong byssus composed of many silk-like threads which used to be made into cloth. These keratin fibers that the animal secretes by its byssus gland are even 6 cm.

**Food and feeding:** The animal lives on plankton, which it filters from about 6 l of water every hour.

**Reproduction:** As a hermaphrodite, it produces male and female germ cells, which it alternately releases in the water from June to August. After fertilization, free-swimming larvae are developed, which, after a few days and once they have formed a thin calcareous shell, drop to the ocean floor. Now, the shell produces the first byssus threads and grows into place. In the first year, it reaches a size of 10–15 cm. When it is fully grown, the shell reaches a size of 120 cm at an age of 20 years.

**Association:** There is a symbiotic relation between this species and a little crayfish which is 3–7 cm long and totally transparent.

**Remarks:** A new protein, tropomyosin (paramyosin) has been isolated from the smooth adductor muscles of this species (Saxena, 2006).

**Fisheries/Aquaculture:** Not known

**Countries where eaten:** Mediterranean (endemic) Montenegro Coast. It is a Croatian cuisine.

## *Pinna rudis* (Linnaeus, 1758)

**Common name(s):** Rough pen shell, spiny fan-mussel

**Distribution:** Eastern Central Atlantic and the Mediterranean Sea: Cape Verde Island to Congo

**Habitat:** Subtidal; in small patches of sand in rocky bottoms and in rock crevices, at depths 0–60 m

**Description:** Rough pen shell of this species has a pair of very fragile, long, triangular, wedge-shaped valves, covered with large, protruding scales arranged in quite regular rows. These scales are more prominent close to the opening of the shell. Six low ribs radiate from the pointed end and run the length of the valves. Valves are nearly symmetrical, toothless, and transparent on the ends. Their color is usually reddish brown. Maximum length is 57 cm.

### Biology

**Behavior:** This species lives with the pointed anterior end of its shell vertically anchored to rock or firm sediment by its byssus threads. The rear edge of the shell is rounded and free.

**Food and feeding:** It is a filter feeder. Water is drawn into the shell from above and passed over the ctenidium before it is expelled into the open water at the exposed part of the shell.

**Reproduction:** It is a hermaphrodite and the gonads are producing both sperm and ova. The larvae are planktonic and drift with the currents.

**Fisheries/Aquaculture:** No report

**Countries where eaten**: Black Sea, Adriatic, Mediterranean

*Pinna rugosa* (Sowerby, 1835)

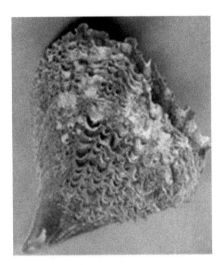

**Common name(s):** Wrinkled pen shell

**Distribution:** Southern Baja California, Pacific and southern Gulf of California, to Guaymas, Mexico, and south to Panama

**Habitat:** Mud-sand bars in protected bays.

**Description:** This species has a flattened, fan-shaped shell with eight rows of tubular spines that help anchor the shell in the sand. Shell is dark blackish or dark brown color. Shell length is up to 59 cm.

**Biology:** Unknown

**Fisheries/Aquaculture:** It is commercially exploited in Mexico.

**Countries where eaten:** The adductor muscles and mantle of this species are consumed in Mexico. The large posterior adductor muscle of this species is widely marketed domestically.


## 3.2 MUSSELS (MYTILIDAE)

Mytilids include the well-known edible sea mussels. Shell is elongated, equivalve, and inequilateral. Beaks at anterior end are terminal or subterminal and they are rarely. Periostracum is conspicuous and is darker than shell, with or without spines. Hinge line is without teeth and rarely with small crenulations which are continuous with shell margin. Ligament is internal or external in a narrow groove and inconspicuous. Adductor muscle scars are unequal and anterior muscle scar is reduced. There is no pallial sinus. It is attached by a diffuse, fibrous byssus.

*Arcuatula arcuatula* (Hanley, 1843)

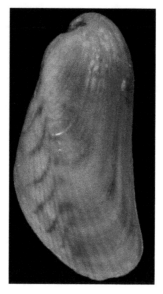

**Common name(s):** Arcuate mussel

**Distribution:** Indo-West Pacific: from northwest Indian Ocean, including Red Sea to Indonesia; north to Vietnam and south to New Caledonia

**Habitat:** Intertidal and shallow subtidal; soft bottoms; occurring gregariously

**Description:** Shell of this species is rather small, thin, and fragile. It is somewhat translucent, roughly elongate trapeziform in outline, with very long and rather narrow shape. Anterior margin of valves is quite short and sharply rounded, extending beyond the umbones. Ventral margin is very long and broadly concave medially. Outer surface of valves is smoothish with only fine concentric growth lines and a few small radial grooves in front of the umbones. Periostracum is smooth and shiny. Hinge line has very small crenulations, mainly appearing behind ligamental margin. Anterior adductor scar is present. Internal margins are very thin and smooth.

Outer coloration is variable and is light tan to olive-green or brown. Interior is slightly pearly, pale bluish gray. Maximum length is 5 cm.

**Biology:** No report

**Fisheries/Aquaculture:** Unknown

**Countries where eaten**: Thailand

*Aulacomya atra* (Molina, 1782) (=*Aulacomya ater*)

**Common name(s):** Ribbed mussel, magellan mussel

**Distribution**: Southeast Pacific, Southwest Atlantic, and Antarctic Indian Ocean

**Habitat:** Intertidal beds; sessile; attached to rocky substrates; associated with holdfasts of juvenile brown algae; depth range 0–30 m

**Description:** Shell has a concave central edge. Dorsal edge is noticeably more prominent toward the back half of leaflet. Outwardly, it presents concentric grooves and radial ribs. Periostracum is bright to dark brown black-bluish color. Umbones are curved and pointed. Hinge has a single tooth in the left valve. Interior valve is pearlescent. Maximum length is 16 cm.

**Biology**

**Food and feeding:** During most of the year, benthic, and pelagic micro-algae contributed equally to the diet of the ribbed mussel. However, during dinoflagellate blooms, the dinoflagellate cells dominated the ribbed mussel stomach contents.

**Fisheries/Aquaculture:** This species is commercially cultivated in Peru and Chile.

**Countries where eaten:** Peru and Chile

**Other uses:** Mussel species such as *Mytilus californianus*, *Choromytilus chorus*, and *Aulacomya ater* have been reported to produce a glue (an adhesive protein), a polyphenolic protein rich in amino acids lysine, 3,4-dihydroxyphenylalanine, serine, threonine, and hydroxyproline. This glue may have applications as a bio-adhesive (Waite, 1986; Burzio et al., 2000).

Recent findings have shown that this species has distinctive anatomy and physiology to respond to environmental stress. Hence, it may serve as a complementary biomonitor to the blue mussel to assess the impact of pollutants and climate change.

*Aulacomya maoriana* (Iredale, 1915)

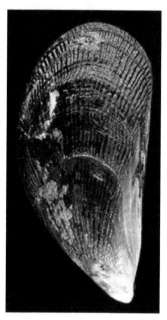

**Common name(s):** Ribbed mussel

**Distribution:** Southwest Pacific: New Zealand; from Magellan Straits northward to Chile and Peru

**Habitat:** From mid-to-low intertidal to 30 m; attached to rocks; found mixed with blue or green lipped mussels

**Description:** Shell of this species is of moderate size, straight, ovate-elongate, ventricose, anteriorly attenuated, and subpyriform. Surface is feebly concentrically sculptured. Anterior half of the shell is with more or less distinct radiating grooves and ridges. Beak is narrowly pointed and umbo is anteriorly placed. Pallial line is prominent, but there is a single large adductor scar. Shell is black to bluish purple, which is often eroded to whitish purple. Interior is with a bluish nacre which is somewhat distributed in zones. Maximum length is 12 cm.

**Biology:** No report

**Fisheries/Aquaculture:** Unknown

**Countries where eaten:** It is an edible species in areas of its occurrence

***Choromytilus chorus*** (Molina, 1782)

**Common name(s):** Choro mussel, giant mussel, rainbow mussel

**Distribution:** Southeast Pacific and Atlantic Ocean: introduced in the United Kingdom

**Habitat:** Attached to hard substrates like rocks or stones; depth range 0–100 m

**Description:** It is the largest of all mussels. Shell of this species is with slightly curved umbones. Outwardly, only it shows concentric grooves. It has a black periostracum. Its hinge is provided with a tooth in the right valve and two on the left. Dorsal edge of shell is angled in central portion, while ventral end is slightly concave. Internally, in upper and central sector, it has a pearly white color. Posterior adductor muscle footprint is large; maximum length is 16 cm.

**Biology:** Unknown

**Fisheries/Aquaculture:** It is commercially exploited. Commercial aquaculture of this species exists in Chile.

**Countries where eaten**: Chile; it is regarded as the best of the edible shellfish

*Crenomytilus grayanus* (Dunker, 1853)

**Common name(s):** Black shell, giant mussel

**Distribution:** Sea of Japan, in the southern part of the Sea of Okhotsk, and the Japanese islands

**Habitat:** Different soils at depths of 1–60 m; occupies huge masses of the coastal zone, forming mussel beds

**Description:** It has a long life expectancy (126 years). Otherwise, nothing is known.

**Biology:** The growth rate of this species was found to vary with the habitat conditions and was markedly lower than in the warmer waters.

**Fisheries/Aquaculture:** Unknown

**Countries where eaten**: Edible in areas of its occurrence

**Other uses:** This species has medicinal values. A 150 amino acid residue lectin (multivalent marine lectin) with anticancer activity has been isolated from this species (Liao et al., 2016).

*Lithophaga lithophaga* (Linnaeus, 1758)

**Common name(s):** Date mussel, date shell, European date mussel

**Distribution:** Mediterranean Sea

**Habitat:** Benthic; wave-beaten areas, within limestone rocks into which it digs tunnels; depth range 125–200 m.

**Description:** Shells of this species are yellowish or brownish, almost cylindrical, and rounded at both ends. Interior is whitish iridescent purple with a pink tinge. These shells are relatively thin. Surface is nearly smooth and is covered with growth lines, which may be quite rough. Maximum length is 8.5 cm.

**Biology:** Growth of this species has been reported to be very slow, and it reached the 5 cm length in 15–35 years. They feed on plankton, algae, and debris by filtering them from the water. They reach the sexual maturity after about 2 years. The maximum number of eggs that are released in a season is about 4.5 million. The fertilization takes place in the open water.

**Fisheries/Aquaculture:** It is commercially harvested. Fishing and merchandising of Lithophapagas is prohibited in Croatia, because of the ecosystem devastation.

**Countries where eaten:** It is consumed in Italy, Spain, and Montenegro Coast. In the past, it used to fetch the price of caviar on the markets of Provence. It is a Croatian cuisine. Meat of this species is a delicacy; cooked and served in a broth of white wine, garlic, and parsley.

*Gibbomodiola adriatica* (Lamarck, 1819)

**Common name(s):** Adriatic horse mussel, tulip mussel

**Distribution**: Northeast Atlantic and the Mediterranean

**Habitat:** Partly buried in the bottoms of the relatively coarse sand; depth range 25–200 m

**Description:** This species has relatively a thin shell which is slightly more elongated and is without rough hair or thorn-like projections on periostracum. The latter is very finely hairy. Further sitting umbo is slightly further from the front edge of the shell. Coloration is yellowish with a number of clear reddish streaks running from the umbo toward shell periphery. Animal is bright reddish with brown gills. Maximum length is 5 cm.

**Biology:** No report

**Fisheries/Aquaculture:** Unknown

**Countries where eaten:** Black Sea, Mediterranean

*Modiolus auriculatus* (Krauss, 1848)

**Common name(s):** Ear mussel, eared horse mussel

**Distribution:** Indo-Pacific and Mediterranean Sea

**Habitat:** Intertidal reefs in rock pools and crevices; sandy mudflats; depth range 0–25 m

**Description:** Shell of this species is equivalve and inequilateral with beaks very close to the anterior end. Outline is modioliform. Ligament and dorsal margins are distinctly disjunct. Dorsal and ventral margins are parallel. Dorsal margin is concave. Sculpture is smooth with growth lines. Periostracum has hairs. Shades of orange-brown to olive-brown are seen under periostracum. Interior shell is with shades of purple. Maximum length is 7.5 cm.

**Biology:** This species is able to adapt over a large range of environmental parameters. This high tolerance to changes might be an acquired physiological ability inherent to this species or to the *Modiolus* genus. It is a suspension feeder.

**Fisheries/Aquaculture:** It is commercially exploited.

**Countries where eaten:** It is an important food source in Fiji.

**Other uses:** This species can be used as a biological indicator to monitor chemical contaminants in tropical marine ecosystems.

*Modiolus barbatus* (Linnaeus, 1758)

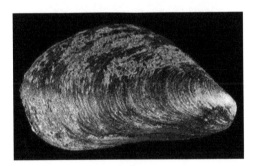

**Common name(s):** Bearded horse mussel

**Distribution:** British Isles to Mauritania, West Africa, Mediterranean, and Black Sea

**Habitat:** Lower eulittoral–sublittoral fringe; found among rocks and stones; depth range 125–200 m

**Description:** Shell of this species is thin and brittle and elongate-oval. Umbones are subterminal, above a curved anterior margin. Hinge line is at a steep angle to umbones. Ventral margin is slightly concave. Umbonal ridges are very prominent, rounded, and are almost parallel with ventral margin. Periostracum is produced into dense fringes of flattened, serrated spines. Sculpture is of fine concentric lines. Shell is yellowish white to light yellow or reddish brown. Periostracum is glossy, yellow, reddish brown, or light mahogany and lighter along umbonal ridges. Inner surface of shell is pale bluish white, with tinges of red or light. Maximum length is 6 cm.

**Biology:** Not known

**Fisheries/Aquaculture:** Not known

**Countries where eaten:** Not known. This species however has a tasty meat, which is usually eaten raw.

*Modiolus capax* (Conrad, 1837)

**Common name(s):** Fat horse mussel, capax horse mussel

**Distribution:** Pacific Ocean: from California to Peru

**Habitat:** Undersides of concrete shelters

**Description:** This species has a bright orange-brown shell under a thick periostracum. Thin shell is covered with a heavy rich-brown periostracum bearing coarse hairs. Umbones are seen at anterior end. It attains a maximum length of 15 cm.

**Biology**

**Reproduction:** Advanced gametogenesis and spawning match the maximum increases in water temperature and phytoplankton abundance. It is a stable gonochoric species, with sporadic and protandry hermaphroditism. Greater reproductive potential has been reported in sizes between 60 and 100 mm in length, when they have more than 2 years. The monthly average size of females is higher than that of males. The sex ratio by weight ranges from 1:0.97. They also show that the growth of the organic part of the body (soft tissue) stops during the winter. It was assumed that the higher average temperature has favored the reproduction, recruitment and growth of the species (Baez and Isabel, 1987).

**Fisheries/Aquaculture:** Not known

**Countries where eaten:** It is an edible species in areas of its occurrence.

*Modiolus modiolus* (Linnaeus, 1758)

**Common name(s):** Horse mussel, northern horsemussel

**Distribution:** Pacific: North Sea: from northern parts of Norway to the Bay of Biscay

**Habitat:** Buried or partly buried in gravel; aggregated in large groups; depth range 10–150 m

**Description:** Shell of this species is thick, tumid, and irregularly oval or rhomboidal in outline. Umbones are anterior but subterminal, with a projecting, rounded, anterior margin. Dorsal margin is markedly convex and ventral margin is slightly concave. Umbonal ridges are well-developed. Young specimens have numerous long, smooth, slender periostracal spines. Sculpture is of fine concentric lines and ridges. Shell is bluish white to slate-blue, darkening in larger specimens. Periostracum is very glossy. Inner surface of shell is white. It has a common length of 5–10 cm. Its maximum life span is more than 50 years.

**Biology**

**Maturation:** Sexual maturity occurs in this species at about 40 mm in a period of 3–8 years depending on the environmental conditions.

**Predators:** Predators, largely crabs and starfish, play an important role in the population structure of horse mussel beds and determine the survival of juveniles to adulthood. Predation may also limit the ability of this species to colonize other habitats, such as hard substrata.

**Fisheries/Aquaculture:** It is commercially harvested.

**Countries where eaten**: Norway. A delicious chowder (a rich soup typically containing fish, clams, or corn with potatoes and onions) is prepared from the meat of this species.

*Modiolus modulaides* (Röding, 1798) (=*Modiolus metcalfei*)

**Common name(s):** Yellow-banded horse mussel, brown mussel

**Distribution:** Indo-West Pacific: from East Africa to the Philippines; north to Japan and south to Indonesia

**Habitat:** Attached to pebbles or to mangrove prop roots; on muddy bottoms of sheltered bays, especially in areas under the influence of freshwater supply; littoral and sublittoral to a depth of 25 m.

**Description:** Shell of this species is rather inflated, roughly triangular and elongate-ovate in outline. Anterior margin is short, slightly protruding anteriorly beyond the umbones. Posterodorsal margin is straightish, forming a rather sharp angle with the produced, roundly wedge-shaped posterior margin. Ventral margin is long and nearly straight in the posterior part. Outer surface is smoothish, with only fine concentric growth marks. Periostracum hairs are not branched. Hinge line is smooth and is without teeth or crenulations. Anterior adductor scar is present. Internal margins are smooth. *Color*: outside of shell is dull olive-brown, often with a median yellowish radial band. Interior is pearly, pale grayish blue to purple. Maximum length is 8.0 cm.

**Biology**

**Fisheries/Aquaculture:** It is commercially exploited. It is also a potential species for aquaculture in Taiwan and Philippines.

**Countries where eaten**: It is consumed in areas of its occurrence.

## *Modiolus philippinarum* (Hanley, 1843)

**Common name(s):** Philippine horse mussel

**Distribution**: Indo-Pacific: from eastern Africa to eastern Indonesia; north to Japan and south to Queensland and Western Australia

**Habitat:** On muddy and gravely mudflats; littoral and sublittoral to a depth of 40 m

**Description:** Shell of this species is relatively thin but solid, swollen, elongate-ovate, and roughly trapeziform in outline. Anterior margin is short, protruding anteriorly well beyond the inflated umbones. Posterior margin is broadly rounded. Ventral margin is long and slightly sinuous, with a shallow concavity at about midlength of shell. Outer surface is sculptured with numerous concentric growth marks. Periostracum is smooth. Hinge line is smooth, without teeth or crenulations. Anterior adductor scar is present. Internal margins are smooth. *Color*: outside of shell is yellowish-brown. Interior is pearly and off-white to purplish red. Maximum length is 13.0 cm.

**Biology:** Not known

**Fisheries/Aquaculture:** It is commercially exploited and is a potential species for aquaculture in Malaysia and Philippines.

**Countries where eaten**: In areas of its occurrence

## *Modiolus rectus* (Conrad, 1837)

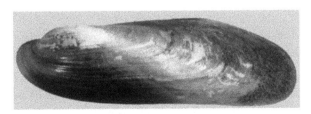

**Common name(s):** Straight horsemussel

**Distribution:** Northern and Central Pacific and Northeast Atlantic: United Kingdom

**Habitat:** Solitary; found partially buried in mud or gravel; Subtidal; depth range 0–15 m.

**Description:** This species may easily be distinguished from the other mussels by its umbo which is not at the extreme end and that the color is not black but brownish. Elongated shells are also either bearded or markedly ribbed. The horsemussel's byssus is composed of many thousands of long fine filaments, estimated to number in excess of 25,000 and possibly more than 50,000. Maximum length is 23 cm.

**Biology:** Unknown

**Fisheries/Aquaculture:** No report

**Countries where eaten:** On the Atlantic coast, this species is eaten though it is considered inferior to *Mytilus edulis*.

*Mytilus edulis* (Linnaeus, 1758)

**Common name(s):** Common mussel, blue mussel

**Distribution:** Native to the North Atlantic and North Pacific, ranging north to Newfoundland, Iceland, and Norway, and south to Virginia and Spain

**Habitat:** Marine and brackish waters; from the high intertidal to shallow subtidal; rocky shores of open coasts; attached to the rock surface and in crevices, and on rocks and piers in sheltered harbors and estuaries; occurring as dense masses

**Description:** Shell of this species is solid and equivalve. It is inequilateral and beaks are present at anterior end. Shape is triangular in outline. Hinge line is without teeth but with 3–12 small crenulations under the umbones. Margin is smooth. Pallial line is wide. Anterior adductor scar is very small and posterior one is large. External ligament is much concealed, extending more than half-way from the beaks to the highest point of the shell. Sculpture is of fine concentric lines. Color of shell is purple, blue, and sometimes brown, and it is occasionally with prominent dark brown to purple radial markings. Periostracum is almost black, dark brown, or olive. Interior is pearl-white with a wide border of purple or dark blue.

## Biology

**Ecology:** Blue mussels can withstand wide fluctuations in salinity, desiccation, temperature, and oxygen tension, and its salinity range is from estuarine areas to fully oceanic seawaters. Highly tolerant of a wide range of environmental conditions, it is euryhaline and occurs in marine as well as in brackish waters (Baltic) down to 4‰. These mussels are also eurythermal, even standing freezing conditions for several months. Its climatic regime varies from mild, subtropical locations to frequently frozen habitats.

**Food and predators:** Blue mussel is a filter-feeder and it filters bacteria, plankton, and detritus from the water. Predators of this species include dog-whelks (*Nucella lapillus*), crabs, sea urchins, star-fish, and birds such as the oystercatcher (*Haematopus ostralegus*). Organisms that attach to mussels, such as seaweeds and barnacles, may increase the risk of the mussel becoming detached by wave action.

**Reproduction:** Like clams, mussels have growth rings, which show their age, and full maturation takes from 1 to 5 years. The sexes are separate and fertilization occurs externally and spawning peaks in spring and summer. The larval stage is free-swimming and planktonic for about 4 weeks before settling first on filamentous organisms such as seaweeds. After growing for some time, they detach and drift in the water on a long byssal thread. After about 4 weeks, the young mussel will have settled again on a mussel bed.

**Fisheries/Aquaculture:** This species is commercially harvested aquaculture and is widely cultivated in the Netherland, Spain, France, United Kingdom, and China.

**Countries where eaten:** Canada, China, France, Italy, Korea, Spain, United Kingdom, and USA

**Other uses:** The carotenoids alloxanthin and mytiloxanthin have been isolated from this species. These compounds may have applications in food industries and as food and feed additives (Sri Kantha, 1989). The byssal threads of some mussels (Mytilidae) are being tested as possible glue in surgery (Anon., ndn).

## *Mytilus galloprovincialis* (Lamarck, 1819)

**Common name(s):** Mediterranean mussel

**Distribution:** Native to the Mediterranean coastline; Europe

**Habitat:** Hard substrates; attached to rocks and piers, within sheltered harbors and estuaries and on rocky shores of the open coast; intertidal to 40 m deep

**Description:** This species has a smooth shell roughly shaped like an elongate triangle, with the beak forming the apex. Anterior margin is straight, while posterior margin is broadly rounded. Surface is marked by concentric growth lines. Exterior color is black to bluish-black, or brownish, while interior is white, with a violet margin, and a distinct muscle scar. Shell can reach a length of 15 cm. Dimensions of the species is greatly influenced by its biotope: intertidal shells often remain small, rarely exceeding 6 cm, while deep-water shells easily measure 9 cm.

**Biology**

**Diet and mode of feeding:** It is a filter feeder and eats a wide range of planktotrophic organisms and detritus. It feeds by pumping water through enlarged sieve-like gills. Food particles are accumulated on the gill lamellae and are then transported by cilia toward the mouth. Fine particles are ingested and travel into the esophagus and stomach.

**Growth:** This species grows rapidly and can attain 70 mm within its first year at favorable conditions. Results from a study show that greatest recruitment and growth rates are achieved at exposed sites in comparison to sites sheltered from direct wave action or at sites exposed to extreme wave action.

**Reproduction:** These mussels are gonochoristic broadcast spawners. The gonads extend throughout the body and are cream colored in males and pink in females. Millions of gametes are released during spawning events with fertilized eggs developing into free swimming planktotrophic larvae capable of dispersing large distances. Upon reaching sexual maturity (1–2 years), reproduction takes place more than once each year. The life cycle includes a pediveliger (larval) stage—where the animal is able to crawl using its foot; and a plantigrade (spat, seed, juvenile) stage—the postlarval stage following metamorphosis.

**Fisheries/Aquaculture:** It is commercially harvested and the countries with the largest catches are Italy and Greece. It is also extensively cultivated in Italy, Netherlands, France, Spain, Canada, Japan, China, and Greece.

**Countries where eaten:** Mediterranean, USA, and Canada. This species is marketed fresh, frozen, and canned.

*Mytilus platensis* (d'Orbigny, 1842)

**Common name(s):** Chilean mussel, River Plata mussel, Chilean blue mussel

**Distribution:** Southern Pacific, Southwest Atlantic, and the Antarctic

**Habitat:** Brackish water; intertidal and subtidal rocky zone; epifaunal species; forming extensive mussel beds; soft sandy and muddy bottoms; depth range 0–100 m

**Description:** Maximum length is 18 cm. No other information is available

**Biology**

**Ecology:** This species is a semisessile epibenthic bivalve that can tolerate wide variations in salinity, desiccation, and temperature and oxygen concentration. Owing to this, it has the ability to occupy a large variety of microhabitats.

**Association:** An association between *Mytilus chilensis* and the isopod *Edotea magellanica* has been reported. *Edotea magellanica* is a nonobligatory commensal, nourished on food that is associated with the host's activity (formation of pseudofeces) (Jaramillo et al., 1981).

**Fisheries/Aquaculture:** It is commercially exploited in Chile and intensively cultivated in Chile, Netherlands, France, Spain, and Canada. The main target market for this species is Spain followed by other European markets and the USA.

**Countries where eaten:** Australia

*Mytilus trossulus* (Gould, 1850)

**Common name(s):** Baltic mussel, Pacific blue mussel

**Distribution:** Native to the North Atlantic and North Pacific; ranging north to Alaska, Siberia, and Labrador; Baltic Sea

**Habitat:** Hard substrates of intertidal and nearshore habitats (wharves, docks, pilings); quiet bays; depth range 0–40 m

**Description:** It has an elongated shell that reaches about 10 cm in length. Outside of shell is almost smooth and ranges in color from bluish black to brown. Shell is somewhat pointed at the anterior end and round on the posterior end. It has three small teeth adjacent to the hinge.

**Biology**

**Food and feeding:** The animal is a continuous filter feeder when immersed in water. Mucus and cilia on filibranch gills capture particles as small as 4–5 µm. As food enters the digestive tract, it is pulled into the stomach by a crystalline style.

**Life history:** It reaches sexual maturity when the shell reaches about 20 mm in length. Gonad development normally occurs in the cooler months. Depending on location, the gonad development period varies. Gametes are released through the gonoducts into the mantle cavity near the excretory structures, called nephridia, and finally into the surrounding water. Within

2 days of fertilization, eggs (60–65 μm) form veliger larvae which exist as plankton for 4–7 weeks. During this period, the larvae develop two valves. At the end of the larval period, the larvae seek a place to settle, such as hydroid colonies, filamentous algae, other mussels' byssal threads, barnacle shells, wharf pilings, or bare rock. As the larvae settle, they shed the velum and metamorphose into juveniles.

**Fisheries/Aquaculture:** It is commercially harvested and cultivated in Canada, Alaska, Russia

**Countries where eaten**: USA and Canada

***Mytilus unguiculatus*** (Valenciennes, 1858) (=*Mytilus coruscus*)

**Common name(s):** Far eastern mussel, Korean mussel, hard-shelled mussel, silk mussel

**Distribution:** Pacific ocean: Yellow Sea and Sea of Japan

**Habitat**: Upper part of the sublittoral zone; depth range 0–100 m

**Description:** Shells of this species are ovate. Posterior side is rather convex and apices are distant and slightly curved. Hinge is channeled with a tooth in one valve. Shell is dirty white color and is coated with a blackish-brown epidermis and inside is pearly. Maximum length is 10 cm.

**Biology:** Unknown

**Fisheries/Aquaculture:** It is commercially harvested. It is also heavily exploited as a food item via mariculture in Korea and China

**Countries where eaten**: Korea and China

**Other uses:** This species has the carotenods, pectenol, and pectenol B possessing applications in food industries (Sri Kantha, 1989).

## *Mytilus zonarius* (Lamarck, 1819) (=*Mytilus californianus*)

**Common name(s):** California mussel, blue mussel

**Distribution:** Native to the west coast of North America; from northern Mexico to Aleutian Islands of Alaska and Baja California

**Habitat:** Intertidal; found clustered together, in very large aggregations, on rocks in upper intertidal zone on the open coast; depth range 0–24 m.

**Description:** Shell of this species is thick and has a maximum length of 25 cm. Shell is blue on outside with a heavy brown periostracum. Beaks of the shell are often eroded. Shell has coarse radial ribbing and irregular growth lines on the outer surface. Inner surface of the shell is blue and faintly pearly.

**Biology**

**Ecology:** This species forms extensive beds, which may be multilayered. These beds form habitat for many species of invertebrates including polychaete worms, snails, crabs, and a blackish sea cucumber *Cucumaria pseudocurata* and algae.

**Food:** California mussels open their shells slightly to eat fine organic detritus (plankton) suspended in the seawater. Mussels of average size filter about 3 l of water per hour during feeding. Mussels have been observed to feed at temperatures at 7–28°C and survive at salinities of 50–125% that of seawater.

**Reproduction:** The breeding season of these mussels extends throughout the year although peaks have been reported in California during July and December. These mussels broadcast sperm or eggs, depending on their sex, into the sea where fertilization takes place. Sexes are separate. A California mussel grows to full size in about 3 years.

**Predators:** Ochre sea stars are California mussels' main predator. They affect the distribution of the mussels because the sea star generally occurs lower on the shore than the mussels and moves up to feed mussels during high tide. Mussels are also eaten by shorebirds, crabs, and snails (they drill holes in shells).

**Symbiotic association:** This species has been reported to harbor the symbiotic pea crab, *Fabia subquadrata* which lives within the mantle cavity of the host. Public health codes usually prohibit the marketing or serving of parasitized animals but since the pea crab is very tasty, organisms with this crab are sometimes sold.

**Fisheries/Aquaculture:** It is commercially harvested. This species has the potential for cultivation in Mexico.

**Countries where eaten:** Alaska to central Mexico. California mussels are an important food source for the Native Americans. The flesh of these mussel which is orange in color can be baked, boiled, or fried. While these mussels are usually edible, care needs to be taken during times of red tide. These mussels may contain harmful levels of the toxins which can cause paralytic shellfish poisoning.

*Perna canaliculus* (Gmelin, 1791)

**Common name(s):** New Zealand mussel, New Zealand green-lipped mussel

**Distribution:** Endemic to New Zealand

**Habitat:** Intertidal zone

**Description:** Shells are more elongated and curved. Color of juveniles is bright green and wild adults are dark purple to black. Farmed adults are, however, green to yellow-brown, often with brown stripes radiating from

the umbo or brown spots. Inside of shell is milky white and slightly iridescent. These mussels possess a distinctive green lip along inside margin of
shell. Individuals are either male or female throughout their life. Imma-
ture and mature male gonads are creamy white. Mature female gonads are
vivid orange to pink. Foot is golden-brown. Shell is up to 26 cm long, 11
cm wide, and 9 cm deep.

**Biology:** Unknown

**Fisheries/Aquaculture:** This edible species has economic importance as
a cultivated species in New Zealand.

**Countries where eaten:** New Zealand

***Perna indica*** (Kuriakose & Nair, 1976)

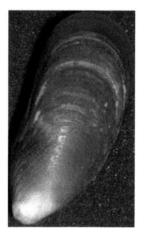

**Common name(s):** Indian brown mussel

**Distribution:** India

**Habitat:** Rock coasts; from intertidal regions to depths of 20 m.

**Description:** Shell of this species is thick, equivalve, inequilateral, elongate, and triangularly ovate in outline. Umbos are terminal and umbonal
peaks are poorly developed. Hinge plate is narrow and thin with a well-
developed tooth on left valve. Posterior margin is rounded and ventral
margin is straight. Periastracum is thick and shining. Sculpture consists of
concentric ridges and growth lines. External color is dark-brown. Foot is
finger shaped. Byssus threads emanating from byssus stem are elongate
and strong. It reaches 12 cm length and 5 cm height.

**Biology:** Unknown

**Fisheries/Aquaculture:** It is commercially harvested and has the potential for aquaculture in India.

**Countries where eaten:** India

***Perna perna*** (Linnaeus, 1758)

**Common name(s):** South American rock mussel

**Distribution:** Native to the tropical and subtropical regions of the Atlantic Ocean; off the west coast of Africa and the coast of South America up to the Caribbean

**Habitat:** Marine, freshwater, and epifaunal; depth range 1.5–2 m

**Description:** This species is easily recognized by its brown color but its identifying characteristic is the "divided posterior retractor mussel scar." Its pitted resillal ridge also differentiates it from other bivalves. Maximum length is 12 cm.

**Biology**

**Food and feeding**: It is a filter feeder and feeds on phytoplankton, zooplankton, and suspended organic materials.

**Life history:** This species has external fertilization and the spawning season is between May and October/December. Both male and female release eggs and sperm to the water during spawning to produce the veliger larvae. Fifteen hours after fertilization, the larvae have well-developed hinge teeth. Ten days after fertilization, the larvae undergo metamorphosis where byssal threads are secreted. The larvae then settle on rocky surfaces.

**Predators:** It is often preyed upon by the whelk *Nucella cingulata*, lobsters, octopuses, gulls, and the African black oystercatcher. On the South American coastline, it is a food source for *Callinectes danae*, *Cymatium parthenopeun*, *Chicoreus brevifrons*, *Thais haemastoma*, and *Menippe* sp.

**Fisheries/Aquaculture:** It is harvested as a food source in Africa and South America. It is also considered for cultivation as it can grow quickly to the commercial size of 60–80 mm in just 6 or 7 months. It is also well-suited to tropical and subtropical regions for aquaculture.

**Countries where eaten**: Africa and South America

**Remarks**: This mussel can harbor saxitoxin from consumed dinoflagellates. Its consumption has caused outbreaks of paralytic shellfish poisoning (PSP) in Venezuela.

**Commercial importance:** It is harvested as a food source but is also known to harbor toxins and cause damage to marine structures. Further, the brown mussel is known to aggregate in such large amounts that it is able to sink navigational buoys. It also coexists with the Asian green mussel in fouling water pipes and marine equipment. It is less resistant to chlorination and thus easier to control.

*Perna viridis* (Linnaeus, 1758)

**Common name(s):** Asian green mussel, green-lipped mussel, Philippine green mussel

**Distribution:** Native to Asia-Pacific and Indo-Pacific regions; introduced to coastal Australia, Japan, the Caribbean, and North and South America

**Habitat:** Attaches to hard surfaces; depths of less than 10 m

**Description:** It is a large (>8 cm) bivalve, with a smooth, elongate shell. It has visible concentric growth rings and a ventral margin which is concave on one side. It has its characteristic green coloration in periostrocum. It is uniformly bright green in juveniles, but dulls to brown with green margins in adults. Inner surfaces of the valves are smooth and iridescent blue to bluish-green in color. A prominent, kidney-shaped retractor muscle scar is present, but it lacks anterior adductor muscles. A pair of hinge teeth is present on left valve that interlock with a single hinge tooth on right valve. Maximum size is 16 cm.

## Biology

**Reproduction:** Peak spawning activities normally occur once a year. Eggs and sperm are released into the water. Fertilized eggs develop into free-swimming larvae within a day. After 2–3 weeks, these larvae settle onto a hard substrate and attach to the surface using byssus. At 2–3 months of age or approximately 20 mm in length, they are sexually mature. Life span is about 3 years.

**Fisheries/Aquaculture:** It is commercially exploited and cultivated in India, Malaysia, China, Taiwan, Philippines, Thailand, and Singapore.

**Countries where eaten:** Philippines, Australia, and China

**Threat to humans:** They can be carriers of diseases and parasites harmful to native species. *Perna viridis* has also been recorded with high levels of accumulated toxins and heavy metals and is linked to shellfish poisoning in humans.

*Xenostrobus pulex* (Lamarck, 1819)

**Common name(s):** Little black horse mussel, flea mussel

**Distribution:** Indo-West Pacific: Australia and New Zealand

**Habitat:** Coastal rocks; intertidal rocks; living in large clusters where sand and rock meet on medium to high energy coasts; jetty piles above mid-tide level.

**Description:** It is almost triangular in shape, with one edge almost straight while the other edge has a distinct hump. Shell is long, inequilateral, and fattened. Shell is mostly smooth, with a sculpture of fine concentric lines. The little black horse mussel's shell is blue-black, covered by a shiny black covering (periostracum). Inside the shell, it is slightly iridescent blue. Typical shell length is 2.5 cm.

**Biology:** Unknown

**Fisheries/Aquaculture:** No report

**Countries where eaten:** New South Wales, southern Australia to Yanchep, just north of Perth, Western Australia, and New Zealand

## 3.3   SCALLOPS AND THEIR ALLIES (PECTINOIDA)

### 3.3.1   SCALLOPS (PECTINIDAE)

This group of bivalves has a single, fused adductor muscle, and a hinge bearing a socket-like arrangement. Foot is greatly reduced and no siphons are seen along the mantle edge. All scallops possess well developed but tiny eyes which are set along the edge of the fleshy mantle. Some species of scallops have the ability to swim. Normal swimming is in the direction of the valve opening, but the scallop can sharply change direction with its velum. Scallops are found in all seas, from shallow water to great depths. All scallops are edible (Kim, 2014).

*Aequipecten opercularis* (Linnaeus, 1758) (=*Aequipecten audouini*; *Chlamys opercularis*)

**Common name(s):** Queen scallop

**Distribution:** South of Norway to the Mediterranean and the Canary Isles

**Habitat**: Found between tidemarks, to depths of 100 m and on sand or gravel; occurs amongst beds of horse mussels *Modiolus modiolus.*

**Description:** In this species, both valves of the shell are convex. Outline of shell is rounded with conspicuous ribs and projections, ears, on each valve. Shell is variable in color but often light-pink to brown, orange, or yellow and often with bands, zigzags, rays, and spots of darker or lighter shades. Right valve is often paler and flatter than the left. Ears of the shell have fine ribs and concentric corrugations. Sculpture of shell consists of about 20 bold, radiating ridges, and the margins of the shell are strongly crenulate. When viewing the inside of right valve (more convex shell), left (or anterior) ear is larger than right. Left forms a notch below the ear and possesses small teeth. It is from this notch young scallop release byssal threads to attach to hard substrata. Spines are absent. Shell is often overgrown with encrusting sponge. It can grow up to 9 cm in diameter.

**Biology:** The queen scallop feeds on a diet of plankton.

**Fisheries/Aquaculture:** It is commercially harvested and is extensively cultivated in United Kingdom and France.

**Countries where eaten**: Mediterranean, Norway, and Morocco

*Aequipecten tehuelchus* (d'Orbigny, 1842)

**Common name(s):** Vieira

**Distribution:** Brazil: Espirito Santo, Rio de Janeiro; Argentina: Buenos Aires, Rio Negro, Chubut

**Habitat:** Depth range 10–120 m

**Description:** It is a medium sized pectinid with 14–17 ribs, each covered with 3–5 radial rows of small scales. Shell is almost circular. Coloration of exterior shell is brown to purplish brown or yellow brown. Interior is smooth and glossy often with wide red-stripe violet. Maximum reported size is 6.4 cm.

**Biology:** No report

**Fisheries/Aquaculture:** It is commercially harvested.

**Countries where eaten**: Argentina

*Argopecten gibbus* (Linnaeus, 1758)

**Common name(s):** Atlantic calico scallop

**Distribution:** Western Atlantic

**Habitat:** Lives in beds in shallow-to-moderately deep water; usually on (or buried in) sandy bottoms; seagrass and algae beds; depth range 0–366 m

**Description:** Shell outline of this species is almost circular and valves are very inflated. Wing-like projections are relatively poorly developed. Surface sculpture is of about 20 ribs. Hinge is straight. *Color*: upper valve is bright and variable, ranging from brown to red to lavender rose to whitish with purplish or reddish mottling. Color of lower valve is much lighter and whitish with lighter markings. It reaches 4–6 cm shell height and approximately 8 cm in shell length.

**Biology**

**Reproduction:** This species reaches sexual maturity at approximately 4 months of age, or when they reach a size of 19 mm shell height. Spawning and recruitment in these scallops occur throughout the year with healthy individuals spawning as much as 3–4 times. Maximum reproductive effort occurs from late fall through spring. Calico scallops are sequential hermaphrodites, first releasing sperm into the water column followed by eggs, in response to changes in water temperature. Eggs develop into free-swimming trochophore larvae within 48 h of fertilization. The larval stage persists for about 15 days before settlement, typically on hard substrata. Juvenile scallops attach to substrata via byssal threads and will remain attached until they reach 2.5 cm in shell length in approximately 3 months. Individuals that settle in spring generally reach a size of 35 mm shell height by the following fall and are fully able to reproduce.

**Predators:** Predators of *Argopecten gibbus* juveniles and adults include sea stars, gastropod molluscs, squid, octopods, crabs, sharks, rays, and several species of bony fishes

**Fisheries/Aquaculture:** Highly commercial fisheries exist for this species. It is also a potential species for aquaculture in Taiwan.

**Countries where eaten:** USA, Atlantic and northern Gulf coasts

*Argopecten irradians* (Say, 1822) (=*Argopecten circularis*; *Pecten circularis*)

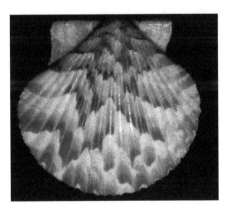

**Common name(s):** Speckled scallop

**Distribution:** Atlantic coastal waters from New Jersey to Florida and in the Gulf of Mexico

**Habitat:** Seagrass beds

**Description:** Both valves of this species are convex and inequivalve (left valve is less inflated) with 17–22 raised ribs. Interstices between ribs are with fine imbricate sculpture. Hinge line length is about 3/4 of disc diameter (except in juveniles less than 15 mm disc diameter). Right valve is usually lighter colored than left. Color of shell is tan or pinkish and chocolate bands or maculations over white, waxy, yellowish, or orange base. These scallops range in size from 5.5 to 9.0 cm.

**Biology**

**Ecology:** Adult *Argopecten irradians* exhibits a small increase in its oxygen intake when seawater temperatures increase. They exhibit an increase in its ammonia content in response to decreasing salinities.

**Food and feeding**: The bay scallop is a filter feeder feeding primarily on phytoplankton.

**Locomotion:** Adult bay scallops swim by "clapping" pulsed expulsions of water from the mantle cavity. They exhibit a "zigzag" swimming pattern by alternating water expulsions between the anterior and posterior gapes in the shell.

**Reproduction:** The bay scallop is a functional hermaphrodite, having both male and female sex organs. It is protandrous, releasing male gametes before the female gametes, to avoid self-fertilization. Spawning will not occur until seawater temperatures reach 20°C. Individuals of this species have one reproductive cycle during their lifetime. Oocyte development begins in August and spawning usually occurs by October. Bay scallops have planktotrophic veliger larvae that appear within 2 days after fertilization. The pediveliger, with a hinged shell and a fully developed foot stage, emerges at about 10 days. Settlement and metamorphosis occurs in less than 2 weeks. The growth of juvenile bay scallops is rapid during the spring months and slows down as water temperatures decrease in the fall.

**Associated species:** *Argopecten irradians* is the most common host for *Pinnotheres maculatus* (squatter pea crab). This symbiosis causes a decrease in the growth of this species.

**Age and life span:** The average life span of this species is 12–26 months. The larval mortality rate is somewhat high and it can range from 10% to 50% within the first 8 days of settlement. High water temperatures in the summer months may also lead to a shortened lifespan in adults.

**Fisheries/Aquaculture:** It is commercially harvested. In areas where fishing is allowed, the bay scallop season is usually from autumn to spring, outside of the reproductive periods. The *Argopecten irradians* fishery along the east coast of the United States has collapsed since the 1950s. This species is very sensitive to natural (i.e., hurricanes, red tide events) processes. The decline in seagrass beds on the west coast of Florida may be a chief contributing factor for the loss of bay scallop populations. In 1995, the Florida Marine Fisheries Commission stopped all commercial fishing of the bay scallop within state waters and put a cap on recreational fishing in areas near the mouth of the Suwannee River.

**This species is cultivated in China.**

**Countries where eaten:** USA, North shore of Cape Cod and Massachusetts to New Jersey

*Argopecten purpuratus* (Lamarck, 1819)

**Common name(s):** Peruvian scallop, Chilean scallop

**Distribution:** Southeast Pacific: Chile and Peru

**Habitat:** Shallow bays; sedimentary grounds in sheltered areas; depth range 0–500 m

**Description:** Shell of this species is orbicular, moderately convex, subequivalve, and rather thin. It is with about 26 flat-topped ribs which are laterally fringed and separated by channeled interspaces. Color of shell is white and rose with and different shades of purple distributed in an irregular manner. Interior zone is with blackish purple. Maximum length is 12 cm.

**Biology:** It is a continuous spawner with spawning peaks that are only partially reflected in the recruitment pattern, indicating that spawning intensity and success of subsequent recruitment are not closely related.

**Fisheries/Aquaculture:** It is commercially exploited and cultivated in China and USA.

**Countries where eaten:** It is eaten in China and USA. The large adductor muscle of this species is a delicious when delicately cooked.

*Argopecten ventricosus* (Sowerby, 1835)

**Common name(s):** Pacific calico scallop, catarina scallop

**Distribution**: East Pacific: California to Mexico, also in the Gulf of California.

**Habitat:** Brackish water; shallow bays, sloughs and calm offshore areas; associated with eelgrass beds depth range 12–50 m

**Description**: It is a medium sized (30–70 mm), solid, and very colorful pectinid. Both valves are very convex. Maximum length is 10 cm.

**Biology:** Smaller scallops are generally attached to the substrate by byssal threads. Animal moves by expelling water between valves propelling to a distance of 1 m upward or several meters sideways, that is, flying motion.

**Fisheries/Aquaculture:** Commercial scallop fisheries and aquaculture exist in Peru, Chile, Baja California Sur, Mexico, and Gulf of California areas since the 1980s with peak catches at Magdalena Bay in 1990 accounting for 53% of total Mexican scallop production.

**Countries where eaten**: Peru, Chile, Baja California Sur, Mexico, and Gulf of California

*Azumapecten farreri* (Jones & Preston, 1904) (*=Chlamys farreri*)

**Common name(s):** Farrer's scallop, Chinese scallop, Zhikong scallop

**Distribution:** Pacific Asian subtropical species; Italy, Greece, Algeria, Chile, Japan and China

**Habitat:** Subtidal areas down to 30 m in depth; gravel and pebbled grounds

**Description:** Shell height is up to 112 mm, but individuals measuring 70–80 mm predominate. The maximum age of this species is 9 years.

**Biology:** It is a dioecious animal. The age of sexual maturity is 2 years old where shell's height is 40–45 mm. The approximate sex ratio (male:female) is 1 to 1.4. The spawning period comprises 65 days. In the reproductive season, the male and female gametes are extruded out of the spawner's gonads directly into the water, where egg fertilization and development take place. The fecundity of a mature scallop is about 6 million eggs. At 16–19°C, sperms retain their fertilizing capacity for 6 h in seawater.

**Fisheries/Aquaculture:** It is harvested in East Asia for food. Commercial aquaculture exists for this species in Eastern Pacific. In China, it is the most important among the species cultured. This species is farmed at an industrial level, but production was devastated by a series of epidemics in the 1990s. It is now thought that this die-off was the result of infections with a herpes virus.

**Countries where eaten:** It is one of the most important edible scallops in Asia.

*Crassadoma gigantea* (Gray, 1825) (*=Hinnites giganteus*)

**Common name(s):** Giant mule scallop, purple-hinge rock scallop, giant rock scallop

**Distribution:** Northeast Pacific: Alaska, USA, and Canada

**Habitat:** Intertidal and subtidal zones; attached on rocks and coral branches; depth range 0–80 m

**Description:** This species has a fan-like shell with two valves. Adults of this species have one valve (right valve) fixed to a base, often a vertical rock face. This valve often becomes deformed to fit the contours of the rock. Left valve is roughly circular but irregular in outline, thick and heavy, with deep radial ribs. Every third or fourth rib is clad with blunt spines that overhang each other. These often get abraded and worn smooth as the scallop ages. There are flaps of shell known as auricles on either side of the straight hinge. The general color is brown and interior of the shell is glossy white with a purple patch at the hinge and a large, central adductor muscle scar. It grows to a dia. of 15 cm in intertidal zone and 25 cm in subtidal zone. Between the valves, margin of the orange mantle can be seen, with a row of tiny blue eyes and a sparse fringe of short tentacles.

**Biology**

**Ecology**: Unlike other scallops, adults of this species are firmly attached to the substratum. After passing a free-living juvenile life, attachment is achieved by temporary byssal threads. Permanent attachment occurs once the young scallop reaches a size of about 2.5 cm through deposition of shell material by the right valve.

**Food and feeding**: Like other scallops, it is a suspension feeder, filtering phytoplankton from water as it passes over the gills. The particles are moved by cilia along grooves to the mouth where edible matter is separated from sediment particles. The waste is incorporated into mucous balls which are removed from the mantle cavity periodically by a clapping of the valves.

**Reproduction:** The sexes are separate in this species and spawning takes place in the summer. The veliger larvae that develop from the eggs form part of the zooplankton for about 40 days before settling, undergoing metamorphosis and beginning to form a shell. Juveniles are free living, are able to swim, and can attach themselves temporarily to the substrate by byssus threads. By the time they have grown to 4.5 cm in diameter they have usually cemented themselves to a hard surface and become sessile.

**Life span:** Although it is very slow growing, this scallop can live for 50 years and reach a diameter of 25 cm.

**Association and predators:** It is difficult to detect this species because of the sponges, sea anemones, hydroids, barnacles, bryozoans, worms, and algae which tend to grow on the shell. The parasitic boring sponge *Cliona celata* damages the shell by making hollows in it. Other sponges that grow on the shell may be considered mutualistic, granting some protection against predators such as the sunflower starfish (*Pycnopodia helianthoides*) and purple sea star (*Pisaster ochraceus*). Further, it can be easily depleted by human harvesters.

**Fisheries/Aquaculture:** It is a potential species for aquaculture in Iceland, Norway, Canada, and in the west coast of USA.

**Countries where eaten:** Alaska (Prince William Sound) to Baja and USA. This species is highly prized as food. The single large adductor muscle is eaten either raw or fried.

***Equichlamys bifron*** (Lamarck, 1819)

**Common name(s):** Bifrons scallop, queen scallop, queen fan scallop

**Distribution**: Endemic to southeastern Australia; Tasmania, Victoria and New South Wales

**Habitat:** Strongly associated with macroalgae and seagrass cover; subtidally on sand and mud; depth range 0–36 m.

**Description**: Shell of this species is ovate. Sculpture of left valve is with about 9 strong, rounded ribs surmounted by a single ridge. Right valve is with 8–9 strong, smooth, flattened ribs which are radially ridged marginally. Interstices are as broad as ribs with shagreen sculpture. Color of left valve is usually purple, white toward the umbones. Right valve ribs are white, pale purple marginally and in the interstices. Interior is purple or orange. Maximum shell length is 12 cm.

**Biology**: Unknown

**Fisheries/Aquaculture:** It is commercially cultivated.

**Countries where eaten**: Australia

*Flexopecten flexuosus* (Polish, 1795)

**Common name(s):** Not designated

**Distribution**: Mediterranean Sea: Turkey and Greece.

**Habitat:** Moderate depths

**Description:** This species is known for its tremendous polymorphism. There are dozens of hues of color and about 15 different methods of pattern formation. Shell ribs of this species are generally strong and smooth and are separated by deep furrows that are radially striated. Crenulations can appear on margins. Anterior auricle of this species looks like that of *Mimachlamys varia*, with well-defined trench and byssal notch.

**Biology:** Unknown

**Fisheries/Aquaculture:** No report

**Countries where eaten:** Adriatic, Mediterranean

***Flexopecten glaber*** (Linnaeus, 1758) (=*Chlamys glabra*)

**Common name(s):** Smooth scallop, proteus scallop

**Distribution:** Eastern Atlantic, Mediterranean and Black Sea: Portugal to Morocco

**Habitat:** Sandy bottoms and detritus; lagoons; depth range 0–215 m

**Description:** It has a remarkable morphological and chromatic variability. Shape of shell is round, with a dia. of about 7 cm. Five to 12 ribs are present. These ribs are accompanied by thin radial cords that invest the entire surface of the shell. Left valve is more lively, while the right is often white. Maximum length is 8.5 cm.

**Biology:** Unknown

**Fisheries/Aquaculture:** It is commercially harvested in Italy and Turkey. It is also cultivated.

**Countries where eaten**: Italy, Mediterranean, Portugal, and Morocco. It is an edible species of commercial interest especially in Italian markets. It is marketed lively and chilled shucked.

*Leptopecten latiauratus* (Conrad, 1837)

**Common name(s):** Kelp-weed scallop, wide-eared scallop, broad-eared pectin

**Distribution:** From Point Reyes, California south to Cabo San Lucas, Baja California Sur, and Gulf of California, Mexico

**Habitat:** From the low intertidal zone to a depth of 250 m; attached to kelp, hard objects (rocks, oil well rigs) or invertebrates such as hydroids.

**Description:** Shell of this species is mostly circular with two flat auricles or ears that extend off of the hinge. It usually has ridges that radiate out from hinge and toward mantle. Ribs when present are perpendicular to the ridges. Shell is extremely thin (compared to other scallops) with one side extending out more so than the other. Some of the common colors seen in these scallops are orange, brown, red, and black. Color may be in a zigzag pattern. Shell size is between 3 and 5 cm.

**Biology:** In this species, around the edges of the mantle, there are about 20 eyes on top and another 20 eyes on the bottom. The animal also has roughly 50, 1–5 mm, somewhat transparent tentacles on each side of the rim of the mantle. These tentacles are believed to contain chemical sensors that are used to detect predation. These scallops are eaten by perches and probably by other fish species.

**Fisheries/Aquaculture:** No information

**Countries where eaten**: Point Reyes, California, and Gulf of Mexico

*Manupecten pesfelis* (Linnaeus, 1758)

**Common name(s):** Cat's paw scallop

**Distribution**: Worldwide distribution

**Habitat:** Deeper waters; depth range 125–200 m

**Description:** Shells are brown-reddish or yellowish-brown with a maximum length of 8 cm

**Biology:** No report

**Fisheries/Aquaculture:** No information

**Countries where eaten:** Mediterranean and Spain

*Minnivola pyxidata* (Born, 1778)

**Common name(s):** Box scallop

**Distribution:** Indo-West Pacific: from Madagascar and Sri Lanka to Indonesia; north to China and Taiwan Province of China, and south to Queensland.

**Habitat:** On sandy bottoms; gregarious; sublittoral and shelf zones; depth range 5–100 m

**Description:** In this species, left (upper) valve is flat to concave and right valve is very convex. Maximum length is 5 cm.

**Biology:** Not reported

**Fisheries/Aquaculture:** It is incidentally caught from shrimp trawlers.

**Countries where eaten:** Indo-Pacific. It has medicinal values.

*Mizuhopecten yessoensis* (Jay, 1857) (=*Patinopecten yessonensis*)

**Common name(s):** Yesso scallop, giant ezo scallop, ezo giant scallop, Japanese scallop

**Distribution:** Eastern Asian coast: China, Korea, Japan and Aleutian Islands

**Habitat:** Sheltered, shallow bays, and inlets adjacent to rocky shores; open sea areas; depth range 30–40 m

**Description:** Shell is almost circular and umbones are in center between two almost equal ears. Exterior of right valve is purplish brown with 20 broad, flattened ribs. It has a reticulated structure between the ribs. Interior valve is whitish, furrowed, with a single adductor muscle scar. Maximum length is 22 cm.

**Biology**

**Ecology:** These scallops have been reported to be most abundant within the bathymetric range 4–10 m and they occur mainly at salinities of

32–34‰. Optimal growth temperature is 4–8°C and the tolerance range is from −2 to 26°C.

**Reproduction:** Unlike many species of scallop, sexes are separate with hermaphrodites rarely observed. They are protandrous hermaphrodites maturing initially as males and changing sex to female as they grow. Spawning occurs in spring as water temperature rises and reaches 7–12°C. Males dominate in younger year classes and females in older year classes. Females of about 15 cm shell height produce a maximum of 18 million eggs. Spawning begins in March and peaks in April at 7–12°C (in Japan). Larvae, which are planktonic feed on phytoplankton and develop over a period of 30–40 days when they are fully developed and ready to settle and metamorphose. Following growth over a period of about 4 months, when the juveniles (spat) are >10 mm in shell height, they detach and disperse on suitable bottom substrate. Life span of these scallops is about 10 years, when scallops will have grown to about 20 cm and weigh 1 kg.

**Harm:** Its tissues have been reported to accumulate algal yessotoxins.

**Fisheries/Aquaculture:** It is commercially cultivated in China, South Korea, Japan, Canada, USA, Australia, and Russia

**Countries where eaten**: Northern Pacific and Australia

**Other uses:** The carotenoid 3R,3′R-astaxanthin isolated from this species could be used in food and feed additives (Sri Kantha, 1989).

*Nodipecten nodosus* (Linnaeus, 1758)

**Common name(s):** Lion's paw scallop

**Distribution:** Eastern Pacific and Western Atlantic; Atlantic coast of North America: from Cape Hatteras to the West Indies, including Brazil and Bermuda

**Habitat:** At the limit between rocks and sandy or gravel bottoms; attached to rocks; inside small rocky caves and crevices or on sandy or calcareous patches adjacent to rocks; depth range 0–150 m

**Description:** This species has very colorful shell with big nodules on the ribs. Color pattern of shell consists of a red, brown, or orange background with some spots. Maximum length is 15 cm.

**Biology:** This species is a simultaneous functional hermaphrodite with an asychronous pattern of reproduction.

**Fisheries/Aquaculture:** It is commercially cultivated.

**Countries where eaten:** It is an edible species in areas of its occurrence.

*Nodipecten subnodosus* (Sowerby, 1835)

**Common name(s):** Giant lion's paw, lion's paw scallop

**Distribution:** Eastern Pacific Coast from Baja California to the northern coast of Peru

**Habitat:** Offshore neritic form

**Description:** It is one of the largest pectinids with a maximum length of 22 cm and height of 17 cm. Valves of this species are equally convex and their height is equal to length. Auricles are unequal, and radially costate. Right valves are with 10–11 ribs. Left valves are with 10 ribs. Macrosculpture consists of fine lirae and striae. Adult interspaces contain three to five radials each. Color of shells vary from red to purple or may be mottled. Largest shell measures 10 cm high and 10.2 cm long.

## Biology

**Reproduction:** This species is a functioning hermaphrodite and gonad gametogenesis and maturity occurs from May to September. It maintains a high rate of growth and somatic production beyond a single growth year. An average overall growth rate of 0.22 mm/day for the 18-month culture duration and a final adductor muscle weight averaging 55 g have been recorded.

**Associated species:** Epibenthic pycnogonid species *Eurycyde bamberi, Eurycyde hispida, Eurycyde spinosa, Nymphopsis duodorsospinosa, Callipallene californiensis, Nymphon lituus,* and *Pycnogonum rickettsi* have been collected from this species (de León-Espinosa and de León-González, 2015).

**Fisheries/Aquaculture:** The desire to develop sustainable small-scale aquaculture on the Baja California peninsula has focused attention toward this native scallop species which is characterized by rapid growth, large size, and high domestic market price. The cool upwelling waters around Bahía Magdalena seem to provide suitable conditions for the culture of this species. The only commercial fishery of this species in Mexico occurs in Laguna Ojo de Liebre in northern Baja California.

**Countries where eaten:** Baja California

**Remarks:** A decrease in hemocytes has been recorded in this species after 24 h of injection of paralyzing shellfish poisons (PSPs) (gonyautoxin 2/3 epimers, GTX2/3) of toxic dinoflagellate *Gymnodinium catenatum* in the adductor muscle in the lions-paw scallop Nodipecten subnodosus (Estrada et al,, 2014).

*Patinopecten caurinus* (Gould, 1850)

**Common name(s):** Giant sea scallop, Pacific Weathervane scallop, giant Pacific sea scallop

**Distribution**: Northwest Atlantic and Northeast Pacific: Canada, USA, and Alaska

**Habitat:** Small depressions on sandy-gravely substrates; depth range 10–200 m

**Description:** The largest species of scallop in the world, reaching about 30 cm, 12 inches in dia. Shell color is usually brown to purple with white. Max length is 28.0 cm; big pectinid, up to 20 cm in height.

**Biology:** Unknown

**Fisheries/Aquaculture:** Commercial fisheries exist in Alaskan waters from Yakutat to the eastern Aleutians. It is a potential aquaculture species in France and United Kingdom.

**Countries where eaten**: Pacific Canada (Alaska to Oregon)

*Pecten albicans* (Schröter, 1802)

**Common name(s):** Japanese baking scallop

**Distribution:** Western Pacific; Japanese and South China Seas

**Habitat:** Shallow inshore reef areas; sand and mud; depth range 10–150 m.

**Description:** It has a shell reaching a size of 9.5 cm, with about 12 radiating ribs. Color of the surface usually ranges from light brown to dark brown, but it may be also orange or purple. Lower valve of this species is less convex.

**Biology:** Unknown

**Fisheries/Aquaculture:** This species is of commercial value for fishing in Japan.

**Countries where eaten:** Japan

*Pecten fumatus* (Reeve, 1852)

**Common name(s):** Commercial scallop

**Distribution:** Indo-Pacific: Australia, Lord Howe and Kermadec islands, and New Zealand

**Habitat:** Soft sediments ranging from mud to coarse sand; buried with only the flat valve visible; aggregated into beds; depth range 0–80 m.

**Description:** Shells of this species are characterized by a flat left valve, and a strongly convex right valve, each with 12–16 strong radial ribs. External shell color is extremely variable among specimens, ranging from light brown to pink and orange. Shell size is up to 14 cm across.

**Biology**

**Locomotion:** As with other scallop species, these scallops are capable of swimming by rapidly closing the valves of the shell, which expels a jet of water and propels the animal along.

**Reproduction:** The commercial scallop is hermaphrodite and the gonads are divided into both male and female portions. Individuals usually mature at 12–18 months of age. Once maturity has been reached, spawning occurs from winter to spring (June to November). The larvae are planktonic and go through a number of larval changes before final settlement and meta-morphosis to the adult form. These scallops grow to marketable size in about 18 months.

**Life span:** It has been reported to live up to 16 years.

**Predators:** The starfish (*Cosctinasterias calamaria*) and the whelk (*Pleuroploca australasia*) have both been observed to predate on these scallops.

**Fisheries/Aquaculture:** This species commercially harvested throughout their natural range and a potential species for aquaculture.

**Countries where eaten**: Australia

***Pecten jacobaeus*** (Linnaeus, 1758)

**Common name(s):** Great Mediterranean scallop, Saint James scallop

**Distribution:** Eastern Central Atlantic and Mediterranean Sea

**Habitat:** Offshore; depth range 0–500 m

**Description:** Valves of this species have different shapes. Lower valve, with which the animal rests on the bottom, is very convex and light-colored, while upper valve is flat and brown. They show 14 to 16 ribs (radial wrinkles) with a more or less rectangular cross section. Interior of the valves is porcelain-like smooth. Max length is 25 cm.

**Biology:** At the edge of the mantle, it has many short tentacles, between which there are a total of 60 blue lens eyes. By quickly closing of its valves, it can swim away several meters in case of danger.

**Fisheries/Aquaculture:** It is commercially harvested for human consumption.

**Countries where eaten**: Mediterranean, Montenegro Coast, and Canary Islands

*Pecten maximus* (Linnaeus, 1758)

**Common name(s):** Great Atlantic scallop, St. James' shell, great scallop, king scallop

**Distribution**: Eastern Atlantic and the Mediterranean Sea

**Habitat**: Sand, mud, and gravel bottoms; depth range 0–250 m

**Description:** Shell of this species is solid, equilateral with equal ears. It is inequivalve with a convex right (lower) valve which is slightly overlapping the flat left (upper) valve. Valves are almost circular in outline. Ears are prominent occupying at least half the width of the shell. Sculpture is of 12–17 broad radiating ribs and numerous concentric lines. Ears are with a few thin ribs radiating from the beaks. Margin is crenulated. Left valve is red-brown and right valve is white, cream, or shades of light brown with pink, and red or pale yellow tints. Both valves may carry zigzag patterns and may be adorned with bands and spots of red, pink, or bright yellow. Maximum length is 17 cm.

**Biology**

**Behavior:** The young scallops live attached with their byssus to a hard substrate, but when they become adult they are free-swimming. They spend most of the time resting on the lower (right or convex) valve. Sand, mud, gravel, or living organism cover the upper valve so that only the margin of the shell (with all tentacles and eyes) is visible. It is most active during the day. When disturbed, the animal retracts with a quick movement into its valves and becomes virtually undetectable.

**Food and feeding**: The great scallop feeds by filter feeding; pumping water through a filter in the gill chamber to remove particulate organic matter and phytoplankton.

**Reproduction**: Size at reproductive maturity is about 6 cm and full maturity is attained at an age of 3–5 years. Fecundity size is 15–20 million eggs when it is 3 years old. Larval phase is for about 1 month. Adults that are benthic have limited swimming mobility.

**Lifespan:** It is about 20 years.

**Predators:** These include a range of starfish, commonly *Asterias rubens*, large crabs, and cephalopods.

**Fisheries/Aquaculture:** It is commercially cultivated. It is also intensively cultivated in China, Japan, and Russia.

**Countries where eaten:** Europe and Atlantic coasts. The edible parts are the white muscle and the egg sac. Giant scallops can be eaten raw lightly steamed, lightly fried, or gratinated in any number of different ways. They are often served in their shells, which are very decorative. Carotenoids of industrial importance such as diatoxanthin, alloxanthin, pectenolone, and astaxanthin have been isolated from this species (Matsuno et al., 1999).

***Pecten novaezealandiae*** (Reeve, 1852)

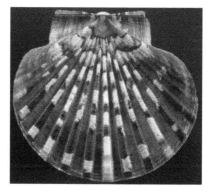

**Common name(s):** New Zealand scallop

**Distribution:** Southwest Pacific: New Zealand

**Habitat:** Sand, silt, and mud; soft sediments; depth range 0–500 m

**Description:** Two valves of this species are asymmetric. Left valve is convex while the right is flat. Concave valve has approximately 16 ribs. Color is variable; however, the valves are usually a whitish pink, but sometimes it may be a dark reddish brown. Maximum length is 14.0 cm.

## Biology

**Feeding and locomotion:** This species is a filter feeder and is completely free-living, mobile, and somewhat migratory.

**Reproduction:** Sexually mature individuals of this species are hermaphrodites. They are broadcast spawners. Spawning is at peak from November to March especially in Tasmania. Fertilization occurs and a planktonic larva forms. This life stage is conserved for approximately 3 weeks. Metamorphosis occurs and the larvae changes from planktonic to a benthic existence by attaching itself to a suitable substratum on the seafloor. The attachment lasts until the individual is at least 5 mm long. The individual then detaches from the substrate and begins to grow into an adult. Maturity is usually attained in 18 months.

**Predators:** Juveniles of this species are normally preyed upon by sea stars, gastropods, octopus, snapper, and large rays.

**Fisheries/Aquaculture:** It is commercially harvested and is a potential species for aquaculture in France, United Kingdom, and Ireland.

**Countries where eaten:** This species has a large industry and is a valuable export product of New Zealand. The large white adductor muscle is eaten. Sometimes, the orange and white gonad is also eaten. *P. novaezelandiae* is considered as a fine food and it may be expensive to purchase.

*Placopecten magellanicus* (Gmelin, 1791)

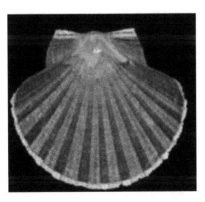

**Common name(s):** Atlantic deep-sea scallop, American sea scallop, deep sea scallop

**Distribution:** Western Central Pacific and Northwest Atlantic: Palau, Canada to USA

**Habitat:** Continental shelf; estuaries and embayments; depth range 0–300 m

**Description:** Shell of this species is large, subcircular, and compressed. Two valves are subequal, hinged dorsally, and meet along the ventral margin. Right valve rests on the bottom. Radial ribs are present along with concentric lamellae or growth lines. Sculpturing is more prominent on the left valve. Wings on the left valve are nearly equal. Wings on the right valve differ in that the anterior wing contains a byssal notch. Inner surface is lustrous and smooth with distinct adductor muscle and pallial scars. Adductor muscle scar is located slightly posterior and dorsal to the center of the valves and is slightly larger on the left valve. Left valve is usually reddish-brown in color but may be lavender or yellow and the right valve is pale cream or white. Maximum length is 20 cm.

**Biology:** This species is semimobile and normally inhabits waters with salinities of oceanic waters. It is an opportunistic filter feeder, utilizing both pelagic and benthic organisms as food. The optimum temperature for growth of adult is about up 10°C. Feeding on plankton and organic detritus. It spawns at temperatures between 14 and 16°C (in USA).

**Fisheries/Aquaculture:** It is commercially harvested along the Atlantic coast of North America from about Cape Hatteras northward. It is a potential species for aquaculture in New Zealand.

**Countries where eaten:** It is the most often eaten scallop in the USA. Other countries where this species is eaten include Argentina, Canada, Chile, Iceland, Japan, and Russia.

*Chlamys islandica* (Muller, 1776)

**Common name(s):** Icelandic scallop

**Distribution:** Predominantly subarctic (low-arctic and high-boreal) species distributed in the subarctic transition zone; Atlantic coast of North America, ranging from Greenland to Massachusetts

**Habitat:** Hard bottom with variable substrate composition consisting largely of sand, gravel, shell fragments, and stones; outer regions of fjords; areas characterized by strong tidal currents; depth range 20–60 m.

**Description:** Shell height of this species is varying between 6 and 11 cm. Further information is not available.

**Biology:** This species is known for its minor morphometric variation between areas of its occurrence. The sex ratio is generally 1:1 and they become sexually mature from 30 to 55 mm shell height at an age of 4–9 years. Seasonal and areal variations in both the muscle and gonad weights have also been observed.

**Fisheries/Aquaculture:** Fisheries for this species occur in Iceland, Norway, West Greenland, Canada, and to a lesser extent the USA. It is also a potential species for aquaculture in China and Japan.

**Countries where eaten**: Northern Atlantic, Newfoundland, and Labrador

***Chlamys rubida*** (Hinds, 1845)

**Common name(s):** Smooth pink scallop, reddish scallop, Pacific pink scallop, swimming scallop

**Distribution:** Alaska Peninsula, Alaska to San Diego, California

**Habitat:** On rocky or soft bottoms; most common on gravel/mud bottoms; low intertidal to 300 m; mainly subtidal

**Description:** The pink scallop has two convex valves joined together by a hinge joint. Each valve has an umbo or knoblike protuberance from which 20–30 shallow ridges radiate to the margin at the other side of the shell. Left valve is usually uppermost as it lies on the seabed and is colored with some shade of red intermixed with white streaks. Annual growth rings can be seen and there is concentric sculpturing parallel to the margin, often in a different shade of pink. Lower valve is either a paler shade of pink or dull white. There is a large auricle or flap on one side of the umbo. It grows to a diameter of about 6 cm.

**Biology**

**Feeding:** When the animal is feeding, it holds the valves apart and the mantle becomes visible, fringed with short tentacles and with a ring of tiny eyes near the margin of each valve.

**Association:** The small free-swimming scallops, *Chlamys hastata* and *Chlamys rubida*, are often encrusted by the sponges *Mycale adhaerens* and *Myxilla incrustans*. It is unclear why this association exists.

**Fisheries/Aquaculture:** It is commercially harvested.

**Countries where eaten**: Japan, Aleutians, Alaska to San Diego, and California

***Mimachlamys crassicostata*** (Sowerby, 1842) (=*Mimachlamys nobilis*)

**Common name(s):** Noble scallop, purple glory scallop

**Distribution:** Japan

**Habitat:** Down to a depth of several hundred meters

**Description:** This species is an extremely colorful big pectinid. Its fragile shells are circular in shape with square ribs composed of scales and threads that act as a brace. Its secondary sculpture is less well developed. It is very bright in color and usually is a shade of purple, orange, or yellow. Size is 9 cm wide, 9 cm high, and 3 cm thick.

**Biology:** Unknown

**Fisheries/Aquaculture:** It is commercially harvested and is a potential species for aquaculture in Mexico.

**Countries where eaten**: Japan and Australia

*Mimachlamys sanguinea* (Linnaeus, 1758)

**Common name(s):** Senatorial scallop

**Distribution:** Indo-West Pacific: from East Africa, including Madagascar to Melanesia; north to the Philippines and south to Queensland, New Caledonia and New Zealand

**Habitat:** On sandy or muddy-sand bottoms with gravel, coral rubble, shell debris, or rocks; attached by its byssus to hard elements, even in adult stages; adheres to corals or rocks among coral rubble on soft sediments; depth range 0–610 m.

**Description**: Shell of this species is solid and medium sized and is higher than long and rounded-ovate in outline. Both valves are convex and subequal. Right (lower) valve is a little flatter than the left (upper) valve. Ears are markedly unequal in size and the anterior ones are more than twice the length of the posterior ones. Ventral side of right anterior ear is with a deep byssal notch and a ctenolium. Main sculpture of each valve is of 20–26 rounded often with fine secondary radial threads. Ears are strongly ribbed and the dorsal most rib of right anterior ear somewhat protruding and adorned with erect spines. Interior is shiny, with low, rounded radial ribs corresponding to the outer sculpture. Outside of shell is dull purple, brown, or orange, frequently variegated with paler blotches. Interior is similarly colored, suffused with white on hinge and umbonal cavity. Maximum length is 8 cm.

**Biology**: No report

**Fisheries/Aquaculture:** It is commercially harvested

**Countries where eaten**: Japan

*Mimachlamys scabricostata* (Sowerby, 1915)

**Common name(s):** Not designated

**Distribution**: Antarctic species: South Pacific Ocean, Australia

**Habitat:** Depth range, 2–5 m

**Description**: In this species, the valves are both convex, though the left valve is more convex than the right. Ear-like projections (auricles) on either side of the hinge, are inequal in size, with the anterior always being much larger than the posterior. Byssal notch is deep, and the valves are generally similar in sculpture.

**Biology:** It is a stationary epifaunal suspension feeder.

**Fisheries/Aquaculture:** It is commercially harvested and cultivated.

**Countries where eaten:** Australia

*Mimachlamys varia* (Linnaeus, 1758) (*=Chlamys varia*)

**Common name(s):** Variegated scallop

**Distribution:** Northeast Atlantic and the Mediterranean and Black Sea: Norway to Senegal and Turkey.

**Habitat:** Shallow-living, intertidal species; rocks or stones and algae of great stature; depth range 125–200 m

**Description:** Shell of this species is thin and more or less oval. It is inequivalve and right valve is with a distinct, rounded byssal notch in anterior ear and with fine teeth on lower border. Shell is inequilateral and both valves have anterior ear more pronounced, twice as long as posterior. Sculpture is of 25–35 bold ribs. Ears are similarly ribbed. Growth stages are usually clear. Sculpture is visible on inner surface and margin is crenulate. Color of the shell is very variable, from off-white, through yellow and orange to brick-red, purple, or brown, often with bands or patches of darker or lighter color. Inner surface of shell is glossy and is often colored same as exterior. Maximum length is 9 cm.

**Biology:** This is a sessile-burrower, suspension feeder.

**Fisheries/Aquaculture:** It is commercially harvested and is a potential species for aquaculture in China.

**Countries where eaten:** Adriatic, Italy, Mediterranean, Norway and Senegal

***Talochlamys multistriata*** (Poli, 1795) (=*Chlamys multistriata*)

**Common name(s):** Dwarf fan shell, lined scallop

**Distribution:** Mediterranean Sea: Turkey and Greece

**Habitat:** Sandy or rocky bottoms, under stones and among algae; depth range 125–200 m

**Description:** Shells of this species are more irregular, spherical, and rounded. It is a small species with a single ear-like protrusion at the hinge. Its shell is finely ridged longitudinally and variably colored from yellow to purple. This animal grows up to 4 cm in length.

**Biology:** These scallops live free (not attached to the substrate) and are able to "swim" through beats valves to move or flee a predator. Their activity is mainly nocturnal. If disturbed, it can swim by clapping its valves together in less than 20 m of water.

**Fisheries/Aquaculture:** No information

**Countries where eaten:** Adriatic, Mediterranean, Straits of Gibraltar Atlantic, Canary Islands, and Mediterranean

*Zygochlamys patagonica* (King, 1832)

**Common name(s):** Patagonian scallop

**Distribution:** Cosmopolitan. Subtropical to polar; Antarctic

**Habitat:** Epifaunal species; forms large banks in deeper waters; depth range 1–270 m

**Description:** Shell of these shellfish is very characteristic, because of its large size, varied colors and presence of regular radial folds. Side of the hinge, which has no teeth bears planar wings. Maximum length is 7.8 cm.

**Biology:** Adults live on the bottom supported by their right valve while juveniles appear as epibionts of algae and other invertebrates. Larvae are free-living.

**Association:** The demosponge, *Lophon proximum*, was the most frequent and abundant epibiont of this species. *Polydora* sp. was registered as a parasite of these shells.

**Fisheries/Aquaculture:** Commercial fisheries exist for this species. It is a potential species for aquaculture in USA, Canada, Chile, and Argentina.

**Countries where eaten:** Chile

*Euvola laurenti* (Gmelin, 1791)

**Common name(s):** Laurent's moon scallop, Asian moon scallop

**Distribution:** Northern Caribbean, from Honduras to Greater Antilles

**Habitat:** On sandy mud bottoms around 20–30 m

**Description:** Shell of this species is thin but strong, circular, moderately inflated, and inequivalve. Wing-like projections are small. Surface is smooth and glossy, but internally, it is with 30–40 paired radial ribs. Lower (right) valve is more convex than upper (left) valve. Hinge is straight. Lower valve is cream colored with light brown rays and upper valve is reddish-brown mottled with white. Maximum length is 8 cm.

**Biology:** No information

**Fisheries/Aquaculture:** It is commercially harvested.

**Countries where eaten**: Venezuela

*Amusium pleuronectus* (Linnaeus, 1758)

**Common name(s):** Asian moon scallop

**Distribution**: Indo-West Pacific

**Habitat:** Occurs in large schools in shallow waters with sand and muddy bottoms; clean sand with silt and shell rubble where the lower right valve is buried in soft sediment; depth range 10–160 m

**Description:** Shell of this species is thin, medium sized, laterally compressed, almost circular in outline and is gaping anteriorly and posteriorly. Both valves are somewhat convex. Right (lower) valve only is a little more inflated and large than the left (upper) valve. Ears are moderately small and subequal in size and shape, with the right anterior ear slightly sinuated anteroventrally. Outside of shell is polished and nearly smooth, with only many faint concentric and radial lines. Interior of both valves is with distinct radial ribs which are usually in pairs, much narrower than the flat interstices and becoming obsolete on umbonal area. Right valve is with 22–34 internal radial ribs. *Color*: outside of left valve is with light to deep pinky brown of varying shades along concentric growth marks, and with darker radial lines and tiny white dots on umbonal area. Interior of left valve is whitish, often with a pinkish hue on margins and central area, and with a brown blotch under the hinge. Right valve is white externally and internally. Maximum length is 10 cm.

**Biology:** Spawning occurs throughout the year.

**Fisheries/Aquaculture:** Commercial fisheries exist for this species. It is a potential species for aquaculture in Caledonia.

**Countries where eaten**: Philippines, China, Korea, Taiwan, Thailand and Australia

***Ylistrum balloti*** (Bernardi, 1861)

**Common name(s):** Ballot saucer scallop, southern saucer scallop, tropical or northern scallop, Queensland saucer scallop, Australian scallop

**Distribution:** New Caledonia; western, northern, and eastern Australia

**Habitat:** Continental shelf; sediment bottoms; depth range 15–50 m

**Description:** Shell shape of this species is disc-like, thin, externally smooth, and suborbicular to orbicular. Valves are very slightly convex, compressed near umbo, and are gaping at anterior and posterior sides. Auricles are small and equal in size. Exterior color of left valve is reddish, covered with a concentric pattern from the umbo to ventral margin of many thin brown lines and violet-brown spots; and light brown or pale reddish stripes transverse valve height. Auricles are often similar in color to base color of left valve. Exterior of right valve is white or pale brown, with concentric, irregularly sized violet-brown spots; valve interior is white, becoming a pink, brown, or yellow tint along the margins of one or both valves; and interior ribbing is seen on both valves 30–38 on left valve, 36–49 on right valve. Size of this species is moderate, up to a maximum of 11 cm in height.

**Biology**

**Behavior:** This species swims actively but rests on sandy bottoms. These scallops have been reported to show a negligible to weak response when touched with the sea star *Pentaceraster regulus* and the red portunid crab *Portunus rubromarginatus*. It showed a consistent and vigorous swimming response to contact with the slipper lobster, *Thenus orientalis*, the blue swimmer crab *Portunus pelagicus* and the coral crab *Charybdis cruciata*. This is the first report of a scallop that has a strong swimming escape response to contact with decapod crustacean predators (Himmelman et al., 2009).

**Reproduction:** This is a gonophoristic species and it spawns in winter and spring. It has a 12–25-day larval phase with settlement of post larvae. Successful settlement appears to create aggregations or beds of scallops. This species is characterized by exceptionally rapid growth and a growth rate of 2.2 mm/week has been reported.

**Fisheries/Aquaculture:** This species is important in commercial scallop fisheries especially in Queensland, Australia. Unlike other scallop species, it is an active swimmer; so the scallops are trawled rather than dredged. Sea ranching of this species has recently been developed to compensate for the fluctuations in catch that naturally occur between seasons.

**Countries where eaten**: Australia, New Caladonia. The flesh is white to cream when raw and white when cooked. The roe varies from white to orange or pinkish purple depending on species and condition.

*Ylistrum japonicum* (Gmelin, 1791) (=*Amusium japonicum*)

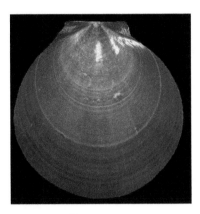

**Common name(s):** Saucer scallop

**Distribution:** Indo-West Pacific

**Habitat:** Benthic; on sandy to muddy bottoms of lagoons; associated with brown seaweeds; depth range 0–80 m

**Description:** Shell of this species is thin, medium to large sized, laterally compressed, almost circular in outline and is gaping anteriorly and posteriorly. Both valves are somewhat convex. Right (lower) valve only has a little more inflated and larger than the left (upper) valve. Ears are small and subequal in size and shape, with the right anterior ear faintly sinuated at anteroventral margin. Surface of left valve is with two broad and very shallow depressed areas radiating from the umbo to anteroventral and radial lines. Interior of both valves has distinct radial ribs which are usually in pairs. Right valve is with 42–54 internal radial ribs. *Color*: outside of left valve is reddish brown, with variable shades along the concentric growth marks. Interior of left valve is glossy to pale yellow externally and internally. Maximum length is 14 cm.

**Biology:** These scallops can actively swim by clapping the valves when disturbed, with a speed of about 2 knots and for a distance of some 10 m. Sexes are separate. Spawning occurs in the cool season, from June to November. Growth is relatively rapid (about 7–8 cm in the first year).

**Fisheries/Aquaculture:** This species is commercially exploited. It is a potential species for aquaculture in Japan and Australia.

**Countries where eaten**: China, Japan, and South Korea

## 3.3.2 WINDOWPANE SHELLS (PLACUNIDAE)

Shell is thin and brittle, nearly equilateral, rounded to saddle-shaped and is very compressed laterally. It is slightly inequivalve; right (lower) valve is nearly flat or a little concave and left (upper) valve is weakly inflated. Umbones are low and submedian. Outer surface is smoothish, with slightly lamellate lines of growth and sometimes fine radiating threads. Periostracum is inconspicuous. Ligament is mostly internal, forming an inverted V-shaped structure under the umbones. Hinge line is straightish, without teeth. Interior of shell is subnacreous. A single, centrally situated, rounded (posterior) adductor muscle scar is seen. Pallial line is obscure, without a sinus. Internal margins are smooth. Gills are of filibranchiate type. Foot is long, narrow, and cylindrical. Mantle is widely open ventrally, with marginal tentacles.

**Placuna ephippium** (Philipsson, 1788)

**Common name(s):** Saddle oyster, saddle window shell

**Distribution:** Indo-West Pacific: from Thailand to Indonesia; north to Taiwan Province of China and the Philippines, and south to Australia

**Habitat:** Benthic; lying free on fine soft bottoms with its right valve underneath, in shallow waters; depth range 0–10 m

**Description:** Shell of this species is, thin, fragile, and flat. It is moderately large, and its outer surface is slightly rough, and concentric lines are widely developed. Shell is almost transparent. Inner surface of the shell is smooth and glossy. Umbo is small, and there are two well-developed hinge teeth. Adductor muscle scar is present at center of valve. Shell is light brown in color with black patches are present on both side. It has shiny and silky appearance. Maximum length is 21 cm.

**Biology:** The sexes are separate and the larval stage is free-swimming.

**Fisheries/Aquaculture:** It is commercially exploited.

**Countries where eaten:** This species is collected for food in Indo-China. In India, its flesh is eaten by the coastal people.

**Other uses:** The shells of this species were originally used as a glass substitute in glazing, but nowadays they are mainly used in the manufacture of trays, lampshades, and numerous decorative items.

***Placuna placenta*** (Linnaeus, 1758) (=*Placenta placenta*)

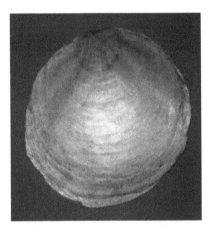

**Common name(s):** Windowpane oyster, common window shell

**Distribution:** Gulf of Aden, India, Malaysia to the southern South China Sea, and Philippines

**Habitat:** Lying on the right valve on the surface of soft muddy to sandy-mud bottoms, from low tide levels to a depth of about 100 m.

**Description:** Shell of this species is thin, more or less translucent and is almost circular in outline. Dorsal margin is somewhat flattened to widely curved. Valves are greatly compressed laterally. Lower (right) valve is flat and upper (left) valve is with slight convexity. Outer surface is nearly smooth, excepting numerous, minute radiating threads forming tenuous wrinkles on the finely lamellate concentric lines of growth. Hinge line is straight and rather short, not raised. External ligament is forming a ventrally undulated narrow band on both sides of the umbo. Interior of shell is smooth and glossy, with only very weak radiating lines. Inner side of anterodorsal and posterodorsal margins is often slightly rugose. *Color*: outside of shell is silvery white with a dull finish, occasionally with pale brown or light purplish rays umbonally. Interior is nacreous white. Maximum shell length is 18 cm.

**Biology**

**Food and feeding:** They consume plankton filtered from the water passing through their slightly opened shell. The shell closes if the bivalve is above water during low tide.

**Reproduction:** Males and females are distinguished by the color of the gonads. Fertilization is external and larvae are free-swimming like plankton for 14 days or attached to surfaces by byssal thread during meta-morphosis. These larvae are eventually settling on the bottom.

**Fisheries/Aquaculture:** This species commercially cultivated in Japan, Taiwan, and Philippines.

**Countries where eaten:** Japan, Taiwan, and Philippines

**Other uses**: Apart from its food value, this species is valued for its shells and pearls. The shells have been used for thousands of years as a substitute for glass owing to their durability and translucence. They have been used in the manufacture of decorative items like chande-liers and lampshades. Shells are also used as raw materials for glue, chalk, and varnish. The pearls produced by this species are spherical and ivory white in color. They are used in pharmaceutical preparations of medicines and ornaments (Achuthankutty et al., 1979; Rahman et al., 2015).

## 3.4 SOFT-SHELL CLAMS AND GEODUCKS (MYOIDA)

### 3.4.1 SOFT-SHELL CLAMS (MYIDAE)

The shells are inequivalve and possessing a projecting chondrophore in the left valve. The hinge is desmodont. Only young specimens have a byssus. The pallial sinus is large and the siphons are bound together in a tube which is made from the periostracum. The *Mya* live deep in the mud, and when disturbed, they may eject water before burrowing at high speed down in their tunnel of more than one meter deep.

## *Mya arenaria* (Linnaeus, 1758)

**Common name(s):** Soft-shell clam, eastern soft shell clam, mud clam, soft clam, long-necked clam, long clam steamer clam, nanny nose, sand gaper

**Distribution:** Coast of New England in the Western Atlantic Ocean; Canada and south to the Southern states; Eastern Atlantic Ocean; North Sea Wadden Sea; and Mediterranean Sea

**Habitat:** Muddy, sandy, and gravelly bottoms; intertidal from the shore and estuaries in shallow water down to 75 m

**Description:** Shell of this species is large, strong but chalky, and inequivalve with smaller lower valve. Outline of shell is ovate-elongate. Anterior end is rounded. Posterior end is slightly pointed. Gapes are at both ends. Rough sculpture is marked by concentric lines (growth lines). Distinctive hinge is with an erect spoon-like tooth (chondrophore) located under the beak in left valve. Pallial sinus is deep, reaching to the middle of the shell. *Color*: chalky white shell with thin grayish/brownish periostracum. Interior of shell is brown. *Common size*: 15 cm in length and 6 cm in height.

### Biology

**Ecology:** This species has great adaptability and tolerance to low salinities and pollution. It lives buried as much as 20 cm deep in mud and sand, extending only its long siphon to the water. It rapidly retracts its siphon below the surface, if disturbed, and ejects a stream of water.

**Food and feeding:** It feeds on microscopic plankton (flagellates, diatoms, and bacteria) and organic detritus. It extends its paired siphons up to the surface, which draw in seawater, filter it for food, and expel it.

**Reproduction:** Mya arenaria spawns once or twice a year in the spring or summer, at temperatures of 10–15°C. A female may release about a million of planktonic eggs in a single spawning. After about 12 h, the eggs hatch out larvae that typically spend about 3 weeks, drifting in the plankton before settling to the bottom as tiny clams. The newly settled clams then spend 2–5 weeks crawling and sometimes attaching to stones or other objects, before burrowing into the sediment. This species reaches sexual maturity at 1–4 years at a length of 2–5 cm. It normally lives up to 28 years.

**Association:** Five species of small commensal crabs native to the Pacific Coast have been found living inside *Mya arenaria*.

**Predators:** *Mya arenaria*'s predators include snails, crabs, rays, sharks, flounder, sculpin, ducks, cormorants, gulls, shorebirds, sea otter, and raccoons; while jellyfish, comb jellies and fish feed on the larvae.

**Fisheries/Aquaculture:** It is commercially harvested and cultivated. It is a potential species for aquaculture in France. It has been introduced for mariculture in the Baltic Sea and the Black Sea.

**Countries where eaten:** Atlantic USA (main commercial sources are Maine, Cape Cod, and Maryland) Mediterranean and Greece. It is a common ingredient of soups and chowders.

*Mya truncata* (Linnaeus, 1758)

**Common name(s):** Truncate soft shell clam, blunt gaper, truncate soft shell

**Distribution:** Arctic, Northern Atlantic, and Eastern Pacific

**Habitat:** Intertidal; continental shelf (to 200 m); depth range 2–100 m

**Description:** Shell of this species is thick, strong, and quadrate. Umbones are present on midline. Anterior end is regularly rounded and posterior is abruptly truncate. Sculpture is of coarse concentric grooves and growth stages are clear. Left valve is with a prominent spatulate chondrophore, projecting at a right angle to hinge, bounded anteriorly and posteriorly by a distinct ridge. Right valve has a concave, spatulate chondrophore. Pallial line and adductor scars are distinct. Anterior scar is elongate and posterior scar is rounded. Shell is dull white to light brown and periostracum is light olive to brown. Inner surfaces are white. Maximum length is 8 cm.

**Biology:** It is a suspension feeder. It burrows at considerable depths in mixed sand, sandy mud, or gravel substrata from the lower shore to about 70 m.

**Fisheries/Aquaculture:** No information

**Countries where eaten:** It is an edible species in areas of its occurrence.

### 3.4.2 GEODUCKS (HIATELLIDAE)

Shell of this family is quivalve and inequilateral. Ligament is external. Right valve is with one cardinal toothand left valve is with two or none. There are no lateral teeth. Pallial line comprises a series of disjunct scars, with a small sinus.

***Cyrtodaria siliqua*** (Spengler, 1799)

**Common name(s):** Northern propeller clam, banks clam

**Distribution:** Northern Atlantic: USA, Canada, and Norway

**Habitat:** Benthic; deep waters; completely buries itself in fine sand; depth range 0–600 m

**Description:** Shell of this species has a slightly twisted valves gape (i.e., valves do not completely meet) at both ends. Shell is heavy, chalky, and bluish-white with dark, flaky periostracum. Shell surface is filled with concentric grooves. Maximum length is 11 cm.

**Biology**

**Food:** It is a suspension feeder and it mainly depends on phytoplankton and detritus material for nutrition.

**Reproduction**: Sexual maturity was estimated as being reached at 28.6 mm in length and 4.7 years in age.

**Fisheries/Aquaculture:** A minor commercial fishery exists for this species. This species is a common bycatch in the Arctic surfclam, *Mactromeris polynyma* fishery on Banquereau Bank in Eastern Canada.

**Countries where eaten:** Atlantic Canada and USA. It is a favorite item in clam chowders.

***Panopea abrupta*** (Conrad, 1849) (=*Panopea generosa*)

**Common name(s):** Pacific geoduck clam, geoduck clam, king clam

**Distribution:** Pacific Ocean

**Habitat:** Buries in a variety of substrates, mud, sand and gravel, along the intertidal zone; depth range 0–110 m.

**Description**: This gigantic clam has fused siphons which are too large to fit into the mantle cavity. Siphons can project up to a meter from the shell and cannot be retracted into the shell. The shell is dirty white to cream with a small amount of yellow periostracum. It has well-developed concentric growth lines but there are no radial ribs. Each valve has one hinge tooth. Shell gapes are widely on all sides except the hinge, and very widely on the posterior (siphon) end. Its pallial line is continuous and wide. Long siphons are light brown with darker brown on the end. There are no thick pads on the end of the siphons. Shell is quadrate (approximately rectangular in outline) and is more rounded on the anterior than the posterior end. Size is up to 23 cm long. It may weigh as much as 3.6 kg.

**Biology**

**Locomotion:** This species is the largest burrowing clam. Adults of this species are poor diggers and do not seem to be able to dig themselves back into the mud if removed.

**Food and feeding:** These clams consume both suspended detritus and plankton residing within the water column. Thus, geoduck clams are an important element of a healthy trophic pyramid. It is extremely long siphons (more than 1 m in length) helps in drawing and expelling marine water.

**Reproduction:** Adults reach full sexual maturity at 3–8 years of age, with males typically beginning to spawn at 3 years and females at 4 years. They are broadcast spawners, with individual females producing 1–2 million eggs during a spawning cycle. Larger geoduck clams produce more gametes, leading to a greater chance of spawning success. Spawning occurs annually from March to July.

**Life span:** Geoducks are one of the longest lived animals and they are living up to 168 years.

**Predators:** As adults, geoducks have very few predators other than humans. These clams are a food source for many organisms, including sea otters, dungeness crab, halibut, sea stars, and species of infaunal nematarian worms. Juveniles are largely preyed upon by fish, starfish, crabs, and snails.

**Association:** The copepod *Herrmanella panopeae* (Cyclopoida: Lichomologidae) has been reported to be the symbiont inside the shell of this species.

**Fisheries/Aquaculture:** The Southeast Alaska geoduck clam fishery has expanded greatly. These clams are wild-captured by divers and are also a popular recreational species. They are also farm-raised in Alaska.

**Countries where eaten:** Alaska to Baja and California. The meat of the siphon of this species is prized for its sweet, sea-like flavor. Geoducks are popular in Asia and many are exported.

*Panopea bitruncata* (Conrad, 1842)

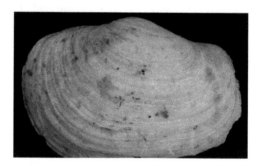

**Common name(s):** Atlantic goeduck

**Distribution:** North America; From the coast of North Carolina to the Gulf of Mexico

**Habitat:** Marine, coastal; estuary, bay; shorefae; depth range 0–48 m

**Description:** Shells are short, rhoimboidal, ventricose, contracted, and obliquely truncated anteriorly. Posterior margin is oblique and slightly emarginated. Cardinal tooth is in right valve. Pallial sinus is widely and obtusely rounded. Shell is white with a maximum length of 19 cm.

**Biology**: It is a stationary deep infaunal suspension feeder and is deeply burrowing (up to 1 m).

**Fisheries/Aquaculture:** No report

**Countries where eaten**: North Carolina to the Gulf of Mexico

**Remarks:** It is considered to be a fossil clam and the collected individuals need a thorough investigation about its taxonomy.

*Panopea globosa* (Dall, 1898)

**Common name(s):** Cortes geoduck, Mexican geoduck

**Distribution:** Eastern Central Atlantic: Gulf of California

**Habitat:** Subtidal zone at 10–25 m depth

**Description:** Maximum length of this species is 20 cm.

**Biology:** No report

**Fisheries/Aquaculture:** Commercial fisheries exist for this species. It is farm-raised in Mexico.

**Countries where eaten:** It is eaten in areas of its occurrence.

### 3.4.3 PIDDOCKS, WINGCLAMS, ANGELWINGS (PHOLADIDAE)

Piddocks have uniquely evolved shells. They burrow a cavity into wood, rock, and other shells for their protection. To accomplish this, the main shell halves each have formed into separable, movable, grinding plates. These rounded plates have stubby external spikes on the anterior sides. Each half-shell also has a unique spoon shaped apophysis on the inside surface. This serves as a muscle attachment to allow dorsal/ventral movement in addition to anterior/posterior movement. Several additional points of attachment mark other muscle bundles that facilitate rotational movement. Ciliary currents of water flush out the debris while the shell assembly operates against a rock surface to grind hard materials. All the pholades have been reported to be edible.

*Barnea candida* (Linnaeus, 1758)

**Common name(s):** White paddock, white angel wing, white barnea

**Distribution:** Western Pacific, Eastern Atlantic, and the Mediterranean

**Habitat:** Brackish water; intertidal areas in soft mud and sand

**Description:** Shell of this species is thin and brittle, elongate-oval and shaped rather like a date. Anterior and posterior margins are rounded with

posterior gaping. Sculpture is of corrugated concentric ridges and radi-
ating lines which are developed as sharp tubercles. Umbonal reflections
are prominent anteriorly. An elongate, oval protoplax is present dorsally.
Apophysis is slender, extending one-third of distance to ventral margin.
Adductor scars and pallial line are indistinct. Pallial sinus is deep and
U-shaped. Shell is white and periostracum is yellowish or light brown.
Inner surfaces are white. Maximum length is 6.5 cm and width is 2 cm.

**Biology:** White piddocks dig themselves into wood, blocks of peat, and
even soft stone. They do not eat the wood, but filter out the plankton from
the seawater.

**Fisheries/Aquaculture:** No information

**Countries where eaten**: Black Sea, Mediterranean, Spain, and Montenegro Coast

*Cyrtopleura costata* (Linnaeus, 1758)

**Common name(s):** Angel wing clam

**Distribution:** Northwest Atlantic

**Habitat:** Intertidal zone; just below low water mark; seabed; burrows to
a depth of 1 m

**Description**: It has a pair of brittle, translucent, asymmetric white valves.
Anterior end is elongated and has a rounded point which is used for digging
through the substrate. Posterior end is truncated and rounded and near the
beak has an apophysis, a wing-like flange, which helps provide an attachment for the foot muscles. On the anterior side of the beak, margin is smooth
and bent slightly upward. Whole valve has finely sculptured radial ribs which
intercept with a series of concentric growth rings parallel with the margin.
In the living animal, valves are covered by periostracum, a thin gray protective layer. Surface of shell is chalky, white, and has radial ribs bearing fluted

scales. Ribs are higher and more widely spaced toward the front. Hinge is toothless and the pallial sinus is broad and deep. Maximum length is 18 cm.

**Biology**

**Behavior:** It is able to bore through sand, mud, wood, clay, and even soft rock using a twisting motion of its pointed, anterior end assisted by jets of water ejected from the mantle cavity.

**Food and feeding**: It is a filter feeder. The siphons extend to the surface of the substrate and water is drawn in through one and expelled through the other with microalgae and zooplankton which are filtered out as the water passes through the gills. Respiration takes place at the same time. Cilia then waft the food particles to the mouth.

**Reproduction**: Spawning usually takes place in summer. Gametes are passed out of the exhalent siphon and fertilization takes place externally. After hatching, eggs develop into veliger larvae which are planktonic. After about 20 days, these larvae undergo metamorphosis into a pediveliger stage and settle out onto a soft substrate such as sand where they become juveniles and start burrowing.

**Fisheries/Aquaculture:** It is commercially harvested and is a potential species for aquaculture in Italy.

**Countries where eaten:** It is an edible species in Great Britain and Ireland.

*Penitella penita* (Conrad, 1837) (=*Pholadidea penita*)

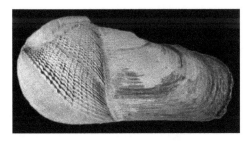

**Common name(s):** Common piddock, flat-top piddock, flat-tipped piddock, Piddock clam, rock clam

**Distribution:** Northeast Pacific

**Habitat:** Open coast; mid-intertidal to 22 m; buried in clay and soft rock

**Description:** In this species, much of the anterior portion of the shell is roughened, so that the animal can rasp a hole in the rock or clay much like an augur bit. Anterior portion of the shell, which is higher and more globose than the posterior portion, is not nearly globular. Anterior portion occupies less than half the length of the valve and is separated from the posterior by a groove which runs from the dorsal to the ventral side. It has a myophore (apophysis) in both valves. Posterior end of the shell, though narrower than the anterior, does not taper to a point like a bird's beak. Siphons are white, smooth, and fused together. Anterior end has a thick, shield-like plate (protoplax) which is dorsal to the anterior rasping portion. Shell is white with brown periostracum. Shape and length of the shell varies with the hardness of the rock it is boring into. Maximum shell length is 7.6 cm.

**Biology**

**Behavior:** It can bore into shale, clay, or firm mud. These clams bore 4–5 mm/year, depending on the hardness of the rock and may burrow to depths of 15 cm.

**Reproduction:** Sexual maturity is reached when the animal stops drilling and a callum covers the anterior gape in the shell. In soft shale, the animal may mature in 3 years, while in harder rock it may not mature until 20 years or later. Spawning occurs in July.

**Association:** The empty holes of piddock clams may contain small porcelain crabs, the flatworm *Notoplana inquieta*, as well as other crabs, worms, and sipunculids.

**Predators:** Predators of this species include the leafy hornmouth, *Ceratostoma foliatum*.

**Fisheries/Aquaculture:** No information

**Countries where eaten**: It is an edible species

***Pholas dactylus*** (Linnaeus, 1758)

**Common name(s):** Common piddock

**Distribution:** Northeast Atlantic and the Mediterranean: Greece and Germany

**Habitat:** Intertidal; depth range 0–200 m

**Description:** It is a bioluminescent clam-like, boring bivalve species. Shell is thin and brittle, and elongate oval. Its sculpture is of concentric ridges and radiating lines. Siphons are joined and at least one to two times the length of the shell, white to light ivory in color. Adductor scars and pallial line are clear and pallial sinus is deep and U-shaped. Shells are dull white or gray and periostracum is yellowish, often discolored. Inner surfaces are white. It has phosphorescent properties, the outlines of the animal glowing with a green-blue light in the dark. Maximum length is 15 cm.

**Biology:** This species bores into soft rock, clay peat and submerged wood. It is a suspension feeder. Light in this animal could be obtained by the addition of $Fe^{2+}$ to purified luciferin from this species in the absence of luciferase. Light emission could also be obtained by the addition of $H_2O_2$ in the presence of luciferin.

**Fisheries/Aquaculture:** No information

**Countries where eaten**: It is an esteemed food in Europe and is eaten in Black Sea, Mediterranean, Spain, British Isles, and Morocco

***Pholas orientalis*** (Gmelin, 1791)

**Common name(s):** Oriental angel wing

**Distribution:** Indo-West Pacific: from Pakistan to the Philippines and Indonesia; north to South China Sea and south to Queensland

**Habitat:** Found in intertidal areas in clay; burrows into soft mud; depth range 0–20 m

**Description:** Shell of this species is elongate-ovate and inaequilateral. It is longer than high and widens toward the umbo. Anterior end is round and posterior end is narrow. Both anterior and posterior ends are gaping. Anterior and ventral halves of valves are densely ridged and strongly spinose. Posterior and dorsal half of valves are smooth but densely covered with tiny granulations. It has a reduced ligament and hinge margin lacks teeth. Outside of the shell is dirty white and interior is milky white and porcelaneous. Maximum length is 12 cm.

**Biology:** It is a burrowing and infaunal species. It has been reported as an endangered species.

**Fisheries/Aquaculture:** It is a potential species for aquaculture in Spain, France, USA, Canada Mexico, and Philippines.

**Countries where eaten**: Thailand

*Zirfaea crispata* (Linnaeus, 1758)

**Common name(s):** Atlantic great paddock

**Distribution:** Northern Atlantic: USA, Germany, and North America

**Habitat:** Intertidal; depth range 0–73 m

**Description:** Shell of this species is thin and brittle and obliquely oval. Both valves are strongly convex. Umbones are anterior to midline. Anterior margins are obliquely truncate above a broad pedal gape; posterior margin is gaping widely. Sculpture is of fine concentric lines, raised as coarse ridges anteriorly. A median groove extending from umbones to ventral margin separates the two areas. Pallial line and adductor scars are clear and pallial sinus is deep, extending to beyond midline of shell. External median groove is visible as a distinct ridge. Color of shell is dull white with a periostracum which is light yellowish brown. Inner surfaces are white. Maximum length is 9 cm.

**Biology:** This species has been reported as an indicator of low water level in Minas Basin, Nova Scotia.

**Fisheries/Aquaculture:** No information

**Countries where eaten:** Great Britain and Ireland

*Zirfaea pilsbryi* (Lowe, 1931)

**Common name(s):** Pilsbry piddock, rough piddock, Pacific rough piddock

**Distribution:** Eastern Pacific: Canada and USA

**Habitat:** Low intertidal to 126 m

**Description**: The anterior portion of the shell is not nearly globular. Anterior rasping portion comprises about half the shell and is separated from the posterior nonrasping portion by an oblique groove. Posterior shell tapers a little. Siphons cannot be withdrawn into the shell. Valves have coarse concentric ridges. On the anterior end these ridges which have file-like projecting spines or teeth. Anterior end of the valves is also often has projecting spines or teeth on the valve margin and also has a gape through which the foot protrudes. Periostracum is brown and may be seen extending onto the base of the siphons. Siphons have small chitinous spots on the surface. Interior of the valves is chalky with well-defined muscle scars. Maximum length is up to 15 cm.

**Biology:** Unlike most piddock clams, this species can live outside its burrow for long periods of time and can burrow to 50 cm depth. It can extend its siphons up through as much as 48 cm of clay plus 30 cm of sand. Life span is estimated as 8 years.

**Fisheries/Aquaculture:** No information

**Countries where eaten:** It is an edible species: It is an edible species.

## 3.5 ARK CLAMS AND THEIR ALLIES

### 3.5.1 ARK CLAMS (ARCIDAE)

Shells of this family are thick and strong, tumid, oval, and quadrate or irregular in outline. They are equivalve, and inequilateral. Umbones are anterior to midline and are widely separated. Ligament is external, extending between umbones across a broad, grooved, diamond-shaped cardinal area. Hinge line is straight, with a continuous series of identical teeth, alternating with sockets. Adductor muscle scars are almost equal. There is no pallial sinus. These clams are an important source of protein in many tropical, subtropical, and warm temperature areas. These ark clams are relished in many local favorites. They are also farmed in some places for sale as seafood. However, ark clams may be affected by red tide and other harmful algal blooms. They are also linked to cholera, hepatitis A, and dysenteric shellfish poisoning.

*Anadara antiquata* (Linneaus, 1758)

**Common name(s):** Antique ark, mud ark clam

**Distribution:** Indo-Pacific: Eastern Africa to Japan, Australia, eastern Polynesia, and Hawaii.

**Habitat:** Inertidal, estuarine species; sand and muddy bottoms; depth range 0–25 m

**Description:** Shell of this species is inequivalve, solid, inequilateral, obliquely ovate, and elongate in outline, with an extended posteroventral part. Umbones are much inflated and are situated rather forward. Cardinal area is narrow and elongate. About 40 radial ribs are seen at each valve.

Ribs are usually with a narrow median groove on top, most visible toward the anterior ventral margin of valves in mature specimens. Periostracum is coarse and velvety, often eroded on umbones. Internal margins are with strong crenulations corresponding with the external radial ribs. There is no byssal gape. Color of outside of shell is grayish white and is often stained darker gray on umbonal and posterior areas. Periostracum is dark brown. Inner side is white, sometimes light yellow in the umbonal cavity. Maximum length is 10.5 cm.

**Biology**

**Food:** This species feeds on a variety of algae belonging to Bacillariophyceae, Chlorophyceae, Pyrophyceae, and Prasnophyceae. *Scenedesmus, Chlamydomonas, Palmeria, Exuviella,* and *Coscinodiscus* have been reported to be the major food sources of this species.

**Sex:** In this species, hermaphrodites have been reported to occur extremely rarely, that is, less than 1%. In addition, both male and female gametes are present within the same individual follicles (Afiati, 2007).

**Fisheries/Aquaculture:** Commercial fisheries exist for this species in Philippines. It is also a potential species for aquaculture in Indonesia and Sri Lanka

**Countries where eaten:** Indonesia, Philippines, and Sri Lanka. These cockles are eaten after boiling with rice or raw on the half-shell with condiments or lime juice.

*Anadara broughtonii* (Schrenck, 1867) (=*Scapharca broughtonii*)

**Common name(s):** Blood clam, inflated ark, arkshell, and small clam.

**Distribution**: Northwest Pacific

**Habitat**: Intertidal; depth range 0–60 m

**Description:** Shell of this species is obliquely ovate, thin, inflated and is nearly equivalve. Sides are slightly angulated at upper part and rounded beneath. It is white covered with a brown, horny epidermis and is scaly in interstices between ribs. It is radiately ribbed and there are about 40 ribs. Umbones are swollen. Size is 7–9 cm long and 6–8 cm high.

**Biology:** Its shell contains a large amount of calcium carbonate and a small amount of calcium phosphate. Besides, it also contains aluminum silicate and inorganic elements, such as chlorine, chromium, copper, iron, potassium, manganese, sodium, nickel, phosphorus, sulfur, silicon, strontium, and zinc. It also has a specific enrichment capacity of nuclide manganese. Dimeric hemoglobins have been isolated from this species.

**Fisheries/Aquaculture:** It is harvested on an intensive commercial basis in South Korea. It is also cultured in China, Japan, and Vietnam.

**Countries where eaten:** It is eaten in China. These cockles are eaten after boiling with rice or raw on the half-shell with condiments or lime juice.

*Anadara corbuloides* (Monterosato, 1878)

**Common name(s):** Tennis shoe ark, basket ark

**Distribution:** Mediterranean Sea: Greece; along the West African coast and south to northern Angola

**Habitat:** Demersal; buried in soft bottoms; depth range 20–100 m

**Description:** Shell is inequivalve, and left valve is larger than right valve. It is inflated and is normally a little greater than half the length. Shell is also inequilateral with beaks located in the anterior half. Outline of shell is subelliptical and longer than high. Ventral margin is curved lacking angles with anterior and posterior margins. Anterior margin is broadly rounded to obliquely curved. Dorsal margin is relatively long and straight. Hinge plate is rather narrow and teeth are in two series. Teeth are small and pointed. Sculpture is of 30–35 radial ribs. Right valve is with a few anterior ribs with very weak nodules or ridges. Periostracum is a thin coating and its coloration is rust to olivaceous brown except for interspaces. Shell is white. Adductors scars are subequal. The posterior is a little larger. Inner margin is deeply crenulate. Maximum size is 7 cm.

**Biology:** It is a filter feeder feeding on suspended matter and food particles from water.

**Fisheries/Aquaculture:** It is a bycatch species and is caught rarely with bottom trawls and beach seines.

**Countries where eaten**: It is eaten in Black Sea and Mediterranean

*Anadara cornea* (Reeve, 1844)

**Common Name**: Not designated

**Distribution:** Fiji, Philippines, India

**Habitat:** Coastal waters; sand bottoms in the littoral and sublittoral zone

**Description:** Shell of this species is thick and solid, inflated, inequilateral, somewhat transversally elongate in shape, and with height greater than inflation and is roughly quadrate to trapezoidal in outline. It is slightly inequivalve, and its left valve is slightly overlapping right valve along posteroventral margin. Anterior margin is rounded and ventral margin is widely convex and meeting the oblique posterior margin at a blunt angle. Umbones are moderately prominent. Ligament is external and is often with V-shaped grooves. Cardinal area is rather narrow and elongated. There are 30 radial ribs at each valve. Periostracum is well developed. Internal margins are with strong crenulations corresponding with the external radial ribbing. Outside of shell is white, frequently tinged deep bluish-green posteriorly. Periostracum is dark grayish-brown. Umbonal area is smooth and velvety. Interior is whitish. Common size is 3 cm.

**Biology:** No information

**Fisheries/Aquaculture:** This species is harvested in Fiji on subsistence basis

**Countries where eaten**: Widely eaten in Fiji

*Anadara diluvii* (Lamarck, 1805)

**Common name(s):** Diluvial ark

**Distribution:** Mediterranean, Black Sea and Eastern Central Atlantic: Mediterranean to Madeira

**Habitat:** Shallow subtidal, coastal, marine, reef, buildup or bioherm, offshore, lagoonal, deep subtidal, and deep-water; depth range 5–500 m

**Description:** Shell of this species is heavy, medium sized, moderately elongate, strongly inflated, inequilateral, and equivalved. Valves are closed along ventral margins. Umbos are high and full. Sculpture is consisting of strong, narrow, flattened radial ribs separated by squarely channeled interspaces. Hinge is uninterrupted but is consisting of two series of teeth which are unequal in length. Posterior muscle scar is quadrangu. Margin of valves is deeply fluted. Common length is 7 cm.

**Biology:** It is facultatively mobile semi-infaunal suspension feeder.

**Fisheries/Aquaculture:** Not reported

**Countries where eaten:** Spain and Black Sea

*Anadara formosa* (Sowerby, 1833)

**Common name(s):** Not designated

**Distribution**: From Baja California to Peru

**Habitat:** Depth range 11–82 m.

**Description:** Shell is elongated and its posterior dorsal area is rather broad. There are 35–38 flat-topped ribs. Posterior ones are wider and anterior ones are finely nodulous and grooved down in middle. Large specimens measure 121 mm length, 58 mm height, and 64 mm dia.

**Biology:** Not reported

**Fisheries/Aquaculture:** No information

**Countries where eaten**: It is an edible species in areas of its occurrence.

*Anadara globosa* (Reeve, 1844) (=*Scapharca globosa*)

**Common name(s):** Globose ark

**Distribution:** Western Pacific.

**Habitat:** Fine muddy-sand bottoms in bays and coastal lagoons; littoral and sublittoral to a depth of 20 m

**Description:** Shell of this species is rather thin but solid, strongly inflated, oval subquadrate in shape with rounded ventral margin, and almost as high as long. It is slightly inequivalve and left valve is overlapping the right valve on ventral and posterior margins. Cardinal area is moderately large. There are 32–36 radial ribs at each valve. Ribs are stout and flat. Periostracum is coarse and concentrically striated. There is no byssal gape. Color of outside of shell is white under the dark brown periostracum. Inner side is bluish white. Maximum length is 11.5 cm.

**Biology:** Not reported

**Fisheries/Aquaculture:** It is commercially exploited and cultivated in Venezuela

**Countries where eaten**: Venezuela

*Anadara inaequivalvis* (Bruguière, 1789) (*=Arca inaequivalvis*)

**Common name(s):** Inequivalve ark, blood cockle

**Distribution:** Indo-West Pacific, Mediterranean, and Black Sea

**Habitat:** Bays and coastal lagoons; from inshore brackish waters to 30 m depth on rocks; mud and sand with *Zostera nana* and *Cymodocea nodosa.*

**Description:** Shell of this species is thick and solid, inflated, inequilateral, roughly quadrate in shape with arcuate ventral margin and obliquely truncate posterior margin. It is slightly inequivalve and its left valve is distinctly overlapping the right valve ventrally and posteriorly. Cardinal area is rather long and narrow. There are 30–36 radial ribs at each valve. Ribs are as wide as the interstices and are granulated on left valve. Periostracum is well developed. Internal margins are with strong crenulations corresponding with the external radial ribs. There is no byssal gape. Color of outside of shell is white under the blackish brown periostracum. Inner side is whitish. Maximum length and height are 8 and 6.1 cm, respectively.

**Biology**

**Ecology:** Species eurythermal and euryhaline that can withstand extreme conditions. Comparative biometric studies in marine and lagoonal populations revealed that under extreme conditions of salinity and temperature (in lagoons), the shell is more fragile and the population density is limited. Maximum occurrence corresponds to areas with salinity 30‰ and sandy sea beds.

**Heavy metal contamination:** This species was found to have higher accumulation of metals in September. In comparison with the permissible limits set by the Thailand Ministry of Public Health and the Malaysia Food Regulations, mean values of As, Cd, Cu, Pb, and Zn were within acceptable limits, but the maximum values of Cd and Pb exceeded the limits. Regular monitoring of trace metals in this species is suggested for more definitive contamination determination (Pradit et al., 2016).

**Fisheries/Aquaculture:** It is commercially exploited and is cultivated in Japan and Philippines.

**Countries where eaten:** Malaysia, Thailand, Japan, and Italy.

*Anadara indica* (Gmelin, 1791) (=*Scapharca indica*)

**Common name(s):** Rudder ark

**Distribution:** Indo-West Pacific.

**Habitat:** Sand and mud bottoms of littoral and sublittoral area; depth range 0–25 m.

**Description**: Shell of this species is solid, not pearly, laterally compressed and subrectangular, inequilateral, striate, and twice as long as high. Left valve is larger and its edge projects beyond the right valve. There are 25–33 low and flat ribs. Dark brown or blackish coarse, hairy periostracum is present. Hinge area is long and narrow with taxodont teeth. Small anteriorly directed umbones are seen. Internal margins have crenulations corresponding with the external radial ribs. There is no byssal gape. Outside of shell is whitish and inner side is white, with a grayish tinge toward periphery. *Size*: Maximum shell length is 5 cm.

**Biology**

**Heavy metals:** Cu, Pb, and Zn content in the coastal sediment as well as the levels of *A. indica* and their relationship with water quality parameters have been studied. Smaller sized *A. indica* tends to have higher values of bioconcentration factor (BCF) for Cu and Zn compared to medium and larger sized samples. Furthermore, the BCF values of Cu, Pb, and Zn in *A. indica* were higher when the sediment had a lower concentration of organic matter (Takarina et al., 2013).

**Fisheries/Aquaculture:** It is commercially harvested and is a potential species for aquaculture in Japan and Philippines.

**Countries where eaten:** It is an edible species in Indonesia.

*Anadara kagoshimensis* (Tokunaga, 1906) (=*Arca subcrenata*)

**Common name(s):** Subcrenated ark shell, blood clam

**Distribution:** Northwest Pacific: China, Taiwan, and Japan

**Habitat:** Intertidal in sand and mud; depth range 0–30 m

**Description:** Shell of this species is slightly triangular or fan-shaped. Outer shell is ridgy, with existing or off brown hair. Top is prominent and curled inward. From the top to the ventral side there are 30–34 radial ribs. Inner shell surface is smooth and white. Edge is with the depression corresponding to the ridge of the outer side. Hinge has a row of toothlets. Shell size is 4–5 cm long and 3–4 cm high.

**Biology**

**Reproduction:** This species is characterized by a spawning period in summer. Monocyclic gametogenesis occurs throughout the year, and an inverse relationship between gonad development and glycogen content has been reported.

**Shell's chemical composition:** The shell contains large amounts of calcium carbonate and small amounts of calcium phosphate, magnesium, iron, silicates, sulfates, chlorides, and organic matter. The total calcium of the shell accounts for more than 93%.

**Fisheries/Aquaculture:** It is harvested on an intensive commercial basis in Japan. Commercial aquaculture of this species exists in Indonesia, China, Japan, and Vietnam.

**Countries where eaten:** Southern half of Japan

*Anadara satowi* (Dunker, 1882)

**Common name(s):** Ark shell

**Distribution:** It is native to the northwest Pacific, ranging from Hong Kong, China to Hokkaido, and Japan

**Habitat:** Muddy, subtidal environments and may occasionally attach to rocks

**Description:** Shell of this species is nearly rectangular. Shell coloration is white or light brown. Shell is thick and is strongly ribbed. Shell size—3.6 cm × 4.1 cm

**Biology:** Not reported

**Fisheries/Aquaculture:** It is fished in China and South Korea

**Countries where eaten:** China and South Korea

*Anadara similis* (Adams, 1852)

**Common name(s):** Brown ark, mangrove cockle

**Distribution:** Eastern Pacific: Mexico to Colombia; subtropical to tropical

**Habitat:** Most offshore in depth to 24 m

**Description:** This species is similar in size and appearance to *Anadara tuberculosa*, except that each valve has 40–44 ribs, and its periostracum which is without bristles. Dorsal margin is rounder and less angular. Blood of this species is red.

**Biology:** The average maturity length of this species has been reported to be 41.8 mm. Reproductive peaks were recorded in March (83.3%) and April (75.0%), but during all months of study reproductive activity was evident. Hermaphroditism has been recorded rarely.

**Fisheries/Aquaculture:** In Pacific coast of Columbia, it is harvested on a subsistence basis.

**Countries where eaten:** Colombia

*Anadara tuberculosa* (Sowerby, 1833)

**Common name(s):** Black ark, mangrove cockle, pustulose ark

**Distribution:** Pacific coast of Latin America

**Habitat:** Level mud sediments in mangrove swamps which occur along the main lands and islands of lagoons; below the base of the roots of the red mangrove, *Rhizophora mangle*

**Description:** Morphologically, this species is constituted of two valves of equal shape which are obliquely oval. Each valve has 34–37 radial ribs and a dark brown periostracum with bristles between the ribs. Dorsal margin of the valves is angular. Blood of this species is black. Maximum length for this species is 11 cm.

**Biology**

**Ecology:** This species can tolerate a wide range of temperature and salinity and maintain positive scope for growth. It can be grown under a wide variety of conditions, especially in tropical lagoons receiving fresh water runoff and shrimp ponds.

**Food and feeding:** It is a filter-feeding organism that feeds on benthic and tychoplanktonic diatoms and may also ingest some zooplanktonic organisms.

**Fisheries/Aquaculture:** In the Pacific coast of Columbia, this species is harvested on a subsistence basis. Commercial aquaculture exists for this species in Korea.

**Countries where eaten:** Columbia and Korea

*Arca imbricata* (Bruguière, 1789)

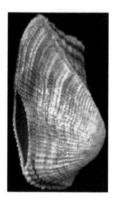

**Common name(s):** Mossy ark

**Distribution:** Atlantic Ocean

**Habitat:** Gulf; attaches to hard substrates up to 560 m depths

**Description:** Shell of this species is rectangular and elongate. Hinge area is long and straight. Radial ridges are crossing concentric growth lines, giving a beaded appearance to the shell. There is a strong ridge from beak to posterior margin. Anterior end is rounded and posterior end is more squarish with two pointed angles. Ventral margin of shell has a central flexure. Beak is somewhat pointed and there is an arched gap between beaks. Maximum length is 6 cm. Color of shell is reddish brown mixed with lighter streaks. Periostracum is hairy brown.

**Biology:** Not reported

**Fisheries/Aquaculture:** No information

**Countries where eaten**: S. E. USA to Brazil

*Arca noae* (Linnaeus, 1758)

**Common name(s):** Noah's arch shell

**Distribution**: Mediterranean and Adriatic Seas

**Habitat:** Lower part of the intertidal zone; depth range 0–60 m

**Description:** Shell of this species is shortened at the anterior end and elongated posteriorly. It is irregularly striped in brown and white and has fine sculptured ribs running from the umbones to the margin. Hinge is long and straight. Shell is attached strongly to the substrate by byssal threads. There are pallial eyes on the edges of the mantle, especially at the posterior end. Maximum length is 12 cm.

**Biology**

**Food and feeding:** Water is drawn into the shell mainly at the posterior end. Plankton and fine organic particles are filtered out as the water passes over the gills and inedible particles are rejected at the same time.

**Life span:** 25 years.

**Association:** This species often grows in association with *Modiolus barbatus*. Further, the shells are often heavily encrusted with epibionts.

**Fisheries/Aquaculture:** This species is fished commercially.

**Countries where eaten:** Montenegro Coast, Yugoslavia, Atlantic of Portugal in Angola, Mediterranean, and North West Africa. It is a Croatian cuisine.

*Arca pacifica* (Sowerby, 1833)

**Common name(s):** Not designated

**Distribution**: Baja California to Peru

**Habitat**: Intertidal zones; rocky ground; adhering to each other in large bunches; depth range 12–137 m depths

**Description:** Shell of this species is irregularly ribbed and is variable in shape, mostly with an expanded posterior portion. It is notched near the

posterior-dorsal margin. Narrow rounded V-shaped brown bands orna-
ment the whitish shell. Shell size: 13 cm length, 7 cm height, and 8 cm dia.

**Biology:** Not reported

**Fisheries/Aquaculture:** Potential species for aquaculture in Mexico

**Countries where eaten**: Mexico

*Arca zebra* (Swainson, 1833)

**Common name(s):** Turkey wing ark clam

**Distribution:** Atlantic coast of North America, ranging from North Carolina to the West Indies and Bermuda

**Habitat:** Gulf and bay; shell bottoms; reef areas

**Description:** Shell of this species is oblong, inflated, and almost rectangular shaped. Anterior end is short with straight or slightly curved margin,
angled at bottom. Posterior end is extended with concave indentation with
upper margin extending past lower margin. Dorsal (bottom) margin is
undulating. There are rounded radial ribs from beaks to margins, with fine
ridges between and across ribs. Beaks are slightly curved inward, widely

spaced, and area between is flat. Hinge is long and straight with numerous small teeth. Color of shell is yellow or white with reddish-brown wavy bands. Maximum size is 10 cm.

**Biology:** Shell of this species attaches itself to rocks or other hard substrates in shallow water with byssus threads. It is a plankton filter feeder and hermaphrodite.

**Fisheries/Aquaculture:** It is commercially harvested. Fishing of this species and extracting meat has an important socioeconomic impact in some communities in eastern Venezuela. Further, fishing is regulated since 1960 and is intended for the food industry where they are preserved, canned, and for the domestic market. It is also a potential species for aquaculture in Mexico.

**Countries where eaten**: Venezuela

***Barbatia barbata*** (Linnaeus, 1758)

**Common name(s):** Bearded ark

**Distribution:** Mediterranean and Black Sea: Turkey and Greece

**Habitat:** Rocky or coralligenous (coral-bearing) seabed; depth range 125–200 m

**Description:** Shape of the shell of this common species is quite variable. Usually, it is oblong and flattened, with many radial ribs cut by concentric lines. When alive, the shell has a characteristic hairy dark periostracum, covering the entire surface of the shell except for the apical part. Maximum length is 10 cm.

**Biology:** Not reported

**Fisheries/Aquaculture:** It is exploited for industrial purposes.

**Countries where eaten:** Portugal and from Mediterranean to Morocco

## *Barbatia foliata* (Forskal, 1775)

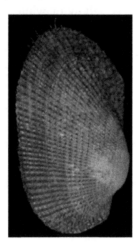

**Common name(s):** File ark, decussate ark, clothed ark

**Distribution:** Indo-Pacific: from East and South Africa, including Madagascar, the Red Sea, and the Persian Gulf, to eastern Polynesia; north to Japan and Hawaii, and south to Queensland

**Habitat:** Attached by byssus among rocks, underside of coral slabs, or nestling in crevices; littoral and sublittoral; occurs also in shallow areas of lagoons where it is attached to dead corals; depth range 0–20 m

**Description:** Shell of this species is equivalve, equilateral, and solid. There are 45–55 parallel rows. Radial ribs are raised into many small, scaly nodules. Below this region, dark brown bristles forming a border of the shell. Anterior beaks are present. Periostracum which has brown bristles has a concentric ridge just below the umbo region given an extrathickening to the anterior portion of the shell. Toward the posterior region 2–3 concentric lines are present. Byssal gap occupies almost in half of the shell length. Hinge is long. Cardinal area is short and deep with a narrow depression. Ligament area between umbo and hinge is narrow. Roughly equal anterior and posterior adductor scar is present. Pallial line is finely striate inside. Shell is white in color. Interior part is glossy white. Maximum length is 7.5 cm.

**Biology**: Not reported

**Fisheries/Aquaculture:** It is exploited commercially along with other clams for industrial purposes.

**Countries where eaten:** Southern half of Japan to Indo-Pacific

*Larkinia grandis* (Broderip & Sowerby, 1829) (=*Anadara grandis*)

**Common name(s):** Mangrove cockle

**Distribution:** From Bahia Magdalena, Baja California, Mexico, to Bahia de Tumbes, Peru

**Habitat:** Mangrove swamps

**Description:** Shell of this species is solid, massive, and heavy. Valves are almost square and convex with high umbos. They have 26 ribs separated by deep interspaces. Shell has a dark periostracum, and the blood of this species is red. White shell of this species is very large (11 cm long, 9 cm height, and 8 cm dia.).

**Biology:** It is very difficult to spawn the cockle under laboratory conditions. The need to investigate the possible role of temperature and salinity in this species reproduction biology is suggested.

**Fisheries/Aquaculture:** In the Pacific coast of Columbia, this species is harvested on a subsistence basis.

**Countries where eaten:** It is an edible species in areas of its occurrence.

*Larkinia multicostata* (Sowerby, 1833) (=*Anadara multicostata*)

**Common name(s):** Many-ribbed Ark

**Distribution:** Eastern Pacific: Newport bay, California; Baja California through Gulf of California, Panama, and Galapagos Islands

**Habitat:** Shallow waters; depth range 0–128 m

**Description:** Shell of this species is squarely rhomboid, solid, and equivalve. Sides are attenuated and angulated at upper part. Anterior side is shorter and posterior is angularly extended downward. Shell color is ivory-white, covered over with a brown horny epidermis which is little velvety between ribs. Shell has 6–30 ribs which are radial. These ribs are rather narrow, rounded, and smooth. Anterior ribs are slightly granulated and umbones are somewhat approximated. *Size of shell*: length is 6 cm; height 5 cm; and dia. 5 cm.

**Biology:** No information

**Fisheries/Aquaculture:** In the Pacific coast of Columbia, this species is harvested on a subsistence basis.

**Countries where eaten**: Columbia

*Lunarca ovalis* (Bruguière, 1789)

**Common name(s):** Blood ark clam

**Distribution:** Atlantic coast of North America, ranging from Massachusetts to the West Indies and Brazil

**Habitat:** Gulf and bay; muddy to sandy bottoms; depth range 0–68 m

**Description:** Shell of this species is round to circular to ovate and inflated. Left shell is slightly overlaps right shell. Color of shell is white and periostracum is furry and dark brown. There are strong radial ribs with weak groove in center of ribs. Beaks are curved anteriorly and they close together. Internal margins are deeply scalloped. Hinge line is slightly arched and all hinge teeth are present on posterior side of shell. Ligament on posterior side is very narrow. Maximum length of shell is 7.5 cm.

**Biology**

**Ecology:** It can survive at polyhaline to euyrhaline salinities (18–35 ppt).

**Blood:** It possesses red blood (hence its name) due to the red blood pigment hemoglobin and is one of the very few molluscs to have it. Most others possess the colorless pigments hemocyanin and hemerythrin.

**Food and feeding:** It is a suspension feeder, discreetly motile while feeding with its ctenidia, that is, the comb-like structure functioning as gill. It mainly depends on phytoplankton and detritus material for its nutrition.

**Reproduction:** Sexes are separate with low incidence of hermaphrodism. Males are dominant. Peak in gonadal pattern is in late spring to early summer and a minor peak in winter. Dribble spawning is the strategy to extend spawning period and increase reproductive success in this species. It is short lived and fast-growing species with recruitment occurring in coastal waters (Georgia) in the fall.

**Fisheries/Aquaculture:** A minor commercial exists for this species. Aquaculture of this species is in experimental stage.

**Countries where eaten:** In the USA, this species remains a potential fishery resource and food source.

*Senilia senilis* (Linnaeus, 1758)

**Common name(s):** Heavy African ark, cockle clam, bloody cockle, senile ark clam

**Distribution:** Eastern Atlantic: Mauritania, Sierra Leone, and Angola

**Habitat:** Intertidal

**Description:** This species has a large and massive shell with a smooth periastracum. Common length of shell is 10 cm.

**Biology:** No information

**Fisheries/Aquaculture:** It has been harvested by humans over thousands of years. In West Africa, this species is harvested on a subsistence basis.

**Countries where eaten**: This species has provided a source of food for humans since prehistoric times and is an important source of food for the people of Senegal.

***Tegillarca granosa*** (Linnaeus, 1758) (=*Anadara granosa*)

**Common name(s):** Blood cockle or blood clam, granular ark

**Distribution:** Indo-Pacific region: from the eastern coast of South Africa northward and eastward to Southeast Asia, Australia, Polynesia, and up to northern Japan.

**Habitat:** Lives in intertidal and shallow subtidal waters; muddy bottoms, mainly in protected bays and estuaries, or in mangroves; intertidal with silty bottoms with low salinity; depth range 0–20 m

**Description:** Shell of this species is equivalve, thick and solid, ovate, strongly inflated, slightly longer than high, and feebly inequilateral. Umbones are strongly protruding and cardinal area is rather large. There are 15–20 radial ribs with wide interstices at each valve. Ribs are stout and distinctly rugose, bearing regular, often rectangular nodules. Periostracum is rather thin and smooth. Internal margins are with strong crenulations corresponding with the external radial ribs. There is no byssal gape. *Color:* outside of shell is white under the yellowish brown periostracum. Inner side is white and is often tinged yellow toward the umbonal cavity. Adult size is about 5–6 cm long and 4–5 cm wide. Maximum length is 9 cm.

**Biology**

**Ecology:** This species, a shallow burrower, lives in an area of silty bottom with relatively low salinity. Its optimum temperature is ranging from 20 to 30°C.

**Blood:** Blood of this clam is red due to hemoglobin liquid inside the soft tissues.

**Food and feeding**: It is a filter feeder. Their feeding habit is related to the bottom feed where they live. Their important nutrient components are organic detritus, phytoplankton, and unicellular algae.

**Reproduction:** This species reproduces from August to February and begins to mature at the age of 1–2 years. One female can produce 5–23 lakhs of eggs.

**Fisheries/Aquaculture:** This species represents the most important commercial ark and is harvested on an intensive commercial basis in China, Vietnam, India, Malaysia, and Thailand. It is a potential species for aquaculture and is raised in the river estuaries of the neighboring Fujian Province as well. It is also cultivated China, Malaysia, Thailand, Japan, and Vietnam.

**Countries where eaten**: Japan and China. This species has provided a source of food for humans since prehistoric times. The meat of this bivalve is served steamed, boiled, roasted, or traditionally raw. Blood clams are considered one of the more delicious Chinese delicacies.

***Tegillarca nodifera*** (Martens, 1860)

**Common name(s):** Nodular ark

**Distribution:** Indo-West Pacific: from Myanmar to the Philippines; north to East China Sea and south to Malaysia

**Habitat:** Intertidal and shallow sublittoral waters; sand and mud bottoms; depth range 0–10 m

**Description:** Shell of this species is equivalve, moderately inflated and solid, oblong ovate, and distinctly longer than high. Umbones are moderately protruding and cardinal area is rather narrow. There are 19–23 radial ribs at each valve and these ribs are quite narrow and sharp, distinctly rugose, bearing regular rounded nodules on the top. Periostracum is rather thin and smooth. Internal margins are with strong crenulations corresponding with the external ribs. There is no byssal gap. *Color*: outside of shell is white under the medium brown periostracum. Inner side is milky white. Maximum length is 6 cm.

**Biology:** No information

**Fisheries/Aquaculture:** This species is commercially exploited and is cultivated in Thailand.

**Countries where eaten:** It is an edible species in areas of its occurrence.

## 3.5.2   PONDEROUS ARK CLAMS (NOETIIDAE)

These clams are more related to the ark clams. They are, however, differentiated from the ark clams by the presence of striations on the hinge ligament and on the placement of this ligament. Like the ark clams, however, their shells range from ovate to elongate, are inflated, and are brown and white with clear radial ribs. They usually grow to around 6 cm in length, with a maximum of 10 cm.

*Noetia ponderosa* (Say, 1822)

**Common name(s):** Ponderous ark clam

**Distribution:** Atlantic coast of North America, ranging from Virginia to Texas.

**Habitat:** Gulf and bay; sandy bottoms; depth range 0–11 m

**Description:** Shell of this species is ovate and inflated. Anterior end is rounded and posterior end is more angular with nearly straight upper margin. Margins are crenate (scalloped). There are flattened radial ribs (about 30) with fine concentric grooves crossing ribs and in between ribs.

Hinge margin is straight on top and is slightly curved on bottom. Hinge area is wide and beak is twisted toward posterior end. Posterior end is with ridge from beak to posterior margin. Posterior abductor muscle scar is with raised ridge. Color of shell is white. Periostracum is brown and hairy. Interior is white. Shell may reach a length of 7.5 cm.

**Biology:** Dimeric hemoglobins with immunological properties have been reported in this blood clam.

**Fisheries/Aquaculture:** It is fished and cultivated.

**Countries where eaten:** It is an edible species.

## 3.6 LUCINA CLAMS (BIVALVIA: LUCINIDAE)

The members of this family are found in muddy sand or gravel at or below low-tide mark. They have characteristically rounded shells with forward-facing projections. The valves are flattened and etched with concentric rings. Each valve has two cardinal and two plate-like lateral teeth. These molluscs do not have siphons but their foot is extremely long. Lucinids host their sulfur-oxidizing symbionts in specialized gill cells called bacteriocytes.

*Anodontia edentula* (Linnaeus, 1758)

**Common name(s):** Toothless clam, toothless lucine

**Distribution:** Indo-Pacific: from East and South Africa, including Madagascar and the Red Sea, to eastern Polynesia and Hawaii and from southern Australia to northern Japan.

**Habitat:** Often buried just under the surface of the sandy-muddy substrates near mangrove areas; occurs in mudflats; depth range 0–20 m

**Description**: Shell of this species is moderately thin and rounded in outline, very inflated, and globose. Anterodorsal margin is subhorizontal and is strongly rounded anteriorly. Posterodorsal margin is slightly convex and sloping posteriorly. Umbones are in front of midline of valves. There is a moderately large, flattish lunule which is slightly depressed near the umbones. Sunken ligament is in an oblique groove of posterodorsal margin. Outer surface of valves is dense, with irregular concentric growth lines. Periostracum is thin. There is an elongate adductor muscle, part of which is separate from the pallial line. Hinge is essentially toothless in adult. Outside of shell is dull white under the straw-colored periostracum. Interior is whitish. Maximum shell length is 9 cm.

**Biology:** These clams are chemosymbiotic and harbor sulfur-oxidizing bacteria in their gills from which they derive most of their nutrition. They have therefore lost their siphons and their ability to filter feed. They make connection with the outside world with their piston-like feet.

**Fisheries/Aquaculture:** It is commercially exploited for food. It is cultivated in Philippines.

**Countries where eaten**: It is an edible species.

*Codakia orbicularis* (Linnaeus, 1758)

**Common name(s):** Tiger lucine, Atlantic muck

**Distribution:** Western Atlantic Ocean; From Bermuda and Florida, throughout the Caribbean and southern Gulf of Mexico, to Brazil

**Habitat:** Gulf; sand or muddy bottoms; depth range 0–2 m

**Description:** Shell of this species is circular (but slightly longer than taller), compressed, and thick. Sculpture is of radial lines crossed by finer concentric threads, except for smooth surfaces of umbones. Lunule is deep, heart-shaped and is larger on right valve. Periostracum is thin. *Color of shell*: externally, it is white, and internally, it is white to pale lemon yellow, with pink margins. Periostracum is brownish. It has a maximum total length of 9 cm.

**Biology:** It has the ability to live in areas of high hydrogen sulfide concentrations and to derive nutrition from chemosynthetic bacteria within their tissue. Sulfur-oxidizing symbiont is hosted in the gills of this species. About 80% of the symbionts hosted by this species maintain respiratory activity throughout the year (Caro et al., 2007).

**Fisheries/Aquaculture:** It is commercially exploited.

**Countries where eaten:** It was a staple food of the Arawak Indians, the first inhabitants of the Bahamas. The raw meat of the clam is firm and sweet, but the gills are bitter when eaten raw. Often, it is prepared in soups and the entire clam may be eaten after steaming.

## 3.7   CARDITA CLAMS (FALSE COCKLES) (BIVALVIA: CARDITIDAE)

These clams have strong elongate to oval or round shells with strong radiating ribs that may be intersected by concentric scales or ribs. Margins are usually scalloped or toothed. They often attach to substrate using byssal threads and many of them brood their eggs inside their shells between the inner and outer branches of the gills. Cardita shells are edible.

*Cardites antiquatus* (Linnaeus, 1758)

**Common name(s):** Antique cardita, Greek seashell

**Distribution**: Mediterranean Sea: Greece.

**Habitat:** Muddy and sandy seabed; depth range 5–45 m

**Description:** Shell of this species is thick and sturdy and is characterized by sharp gnarled radial ribs. Coloration is brownish-tawny spots starting in the apical region and gradually extend to the edge. Umbo is generally white. Maximum size is 3 cm.

**Biology**: No information

**Fisheries/Aquaculture:** Not reported

**Countries where eaten:** People near Black Sea

*Cardites bicolor* (Lamarck, 1819)

**Common name(s):** Twotoned cardita

**Distribution:** Indo-Pacific: India, Sri Lanka, Myanmar

**Habitat:** Estuary; littoral and shallow sublittoral bottoms; also in sandy substratum

**Description:** Shell of this species is equivalve, often stout and inflated, trigonal ovate to trapezoidal in outline, and inequilateral. Umbones are generally anterior, prosogyrate, and prominent. Lunule is short and deep. Exterior is mostly with strong radial ribs. Ligament is external and is attached behind umbones on well-marked nymphs. Hinge plate is strong and is usually with two cardinal teeth which are unequal and often with fine transverse striations. Lateral teeth are more or less reduced or absent. There are two adductor muscle scars which are slightly inequal. Maximum length is 5.5 cm.

**Biology:** No information

**Fisheries/Aquaculture:** Subsistence fisheries exist for this species.

**Countries where eaten:** This mangrove-associated species is edible in India.

## 3.8   FILE SHELLS (BIVALVIA: LIMIDAE)

The Limidae are either attached by a byssus or are free-living animals. They occur in shallow to deep-water habitats, mostly sheltered in rock crevices, under stones and among marine growths. They are more or less buried or lying on the surface.

*Acesta rathbuni* (Bartsch, 1913)

**Common name(s):** Rathbun's giant lima, Rathbun's giant file shell

**Distribution**: Indo-West Pacific; limited to the seas, bordering the Philippines

**Habitat:** Offshore; depth range 0–400 m

**Description**: Shell of this species is thin-walled and fragile. Shell is rounded and slightly scalene, behind gaping. Periostracum is yellow-brown and is evenly covering the surface. Color of the outer surface is from white to creamy white with numerous small indistinct blur of pale pink color. This deep-water species are relatively rare. Maximum length is 21 cm.

**Biology:** No information

**Fisheries/Aquaculture:** Not reported

**Countries where eaten:** Philippines

## *Limaria hians* (Gmelin, 1791)

**Common name(s):** Flame shell, gaping file-shell

**Distribution:** English Channel to the Mediterranean

**Habitat:** Coarse bottoms; lower shore to about 100 m

**Description:** Shell of this species is thin, fragile, and obliquely oval along an anterior dorsal–posteriorventral axis. Anterior ears are distinctly more prominent. There is an elliptical, dorsal gape on anterior margin, and narrower gape along whole of posterior margin. Sculpture is of about 50 radiating ribs and coarse concentric lines. Ears, and parts of valves close to them, are with concentric lines only. Margin is serrated and growth stages are clear. Hinge line is straight and cardinal area is broadly lozenge-shaped. Color of shell is off-white. Maximum length is 2.5 cm.

**Biology:** This species builds protective nests. The nests are constructed by burrowing into a gravel substratum and consolidating the walls of the burrow with byssal threads. When disturbed in the nest, the animal performs locomotory movements which lead to enlargement of the nest. If the nest is broken the locomotory activities result in free swimming followed by attempts to burrow into the substratum to form a new nest. It is unlikely that this species swims freely in nature except when displaced from the nest.

**Fisheries/Aquaculture:** No information

**Countries where eaten:** It is an edible species in areas of its occurrence.

*Lima vulgaris* (Link, 1807)

**Common name(s):** Common file shell

**Distribution:** Indo-Pacific: from East and South Africa, to eastern Polynesia; north to Japan and south to Australia.

**Habitat:** Between tide marks and in subtidal waters; attached to rocks and underside of stones or coral slabs; also lives in sea grass species of *Posidonia oceanica*; depth range 0–50 m

**Description:** Shell of this species is thick with clear ridges which can close completely, with the animal retracted. Surface of the valves show 18–24 strong ribs covered with small scales. It is usually white or yellowish. Tentacles are long and are usually pale without markings. These tentacles are sticky and can break off if the animal is distressed. Animal can swim by "clapping" its valves. Maximum length is 9 cm.

**Biology:** Unattached animals can actively swim, with their long reddish pallial tentacles which are widely expanded beyond shell margins.

**Fisheries/Aquaculture:** It is commercially harvested

**Countries where eaten:** It is eaten in Montenegro Coast. It is a Croatian cuisine.

**Other uses:** In the Philippines, the shells of this species are used to make ornaments.

## 3.9 VENUS CLAMS AND THEIR ALLIES (VENEROIDA)

### 3.9.1 VENUS CLAMS (VENERIDAE)

This family has over 400 living species and it is one of the most colorful of the marine bivalve groups. Shell shape varies from circular to triangular, and from side view, it will appear as either ovate or cardioid. Other characteristics of the shell include highly finished porcelain-like shell; a complex tooth structure in the hinge and a well-developed sinus at the pallial line.

*Anomalocardia flexuosa* (Linnaeus, 1767)

**Common name(s):** Almeja venus

**Distribution**: Western Pacific: China and Hong Kong

**Habitat:** Intertidal in mud; lives shallowly buried in the sediment; depth range: 1–8 m

**Description:** Shell of this species is thick, triangular, moderately inflated, and variable in shape. Posterior end is commonly much produced with a pronounced umbonal ridge extending along the posterior slope from the beak to the posterior. Left and right valves are equal in size and contour (equivalve). Sculpture consists of very fine, radiating striations and of prominent, commarginal ribs which are developed anteriorly and over the posterior ridge as well as on the umbonal region. Pallial sinus is deep. Muscle scar of the anterior adductor muscle is smaller than that of the posterior adductor muscle. Internal margin of the shell is crenulated and shell is up to 4.5 cm maximum length.

**Biology**

**Food and feeding:** It feeds on suspended particles brought into the mantle cavity and trapped by its gills.

**Life span:** It rarely exceeds 2 years.

**Reproduction**: Sexes are separate. Fertilization is external and larval development is planktonic and rapid.

**Associations:** The following organisms are found attached to the shell of this species (epibionts): algae *Enteromorpha* sp., barnacles (Cirripedia), and eggs of the snail *Nassarius* sp. (Prosobranchia). This species may be found together with other molluscs, such as the bivalves *Chione intapurpurea, C. granulata* (Veneridae), *Eurytellina* sp. (Tellinidae), and the snail *Nassarius* sp. (Nassariidae).

**Fisheries/Aquaculture:** This species is harvested from the wild and not yet cultured.

**Countries where eaten:** It constitutes an important food item for coastal communities living along the Brazilian coast. It also has medicinal values.

*Austrovenus stutchburyi* (Wood, 1828)

**Common name(s):** New Zealand cockle, New Zealand little neck clam, tuangi cockle

**Distribution:** Southwest Pacific and Antarctic Indian Ocean: New Zealand and Kerguelen

**Habitat:** Harbors and estuaries; subtidal to intertidal zone, and when they are in the intertidal zone, they live between the low-tide mark and the mid-tide mark.

**Description:** This species is exceedingly variable in shell thickness, shape, inflation, height, and sculptural prominence. Shell is subcircular, weakly trigonal, or elongate-oval in outline, with umbo low-to-strongly protruding. Radial costae are low, wide, and flat-topped. These costae are becoming weak near ventral margin of large shells. Interior of ventral margin is complexly crenulate, with two orders of crenulating ridges. Hinge is wide, with long, prominent ligamental nymph. Right valve is with two thick, bifid cardinal teeth, thin posterior cardinal tooth, and short, knob-like anterior lateral tooth Left valve is with two thick, bifid cardinal teeth, weak posterior lateral ridge, and only anterior lateral socket. Adductor scars are deeply impressed and anterior is slightly narrower than posterior. Pallial sinus is short, narrow, and triangular. Maximum shell length is 9 cm.

**Biology**

**Reproduction:** Sexes are separate and the sex ratio is usually close to 1:1. Maturity appears to be primarily related to size rather than age and sexual maturity is occurring at a size of 1.8 cm shell length. Spawning extends over spring and summer, and fertilization is followed by a planktonic larval stage which is lasting for about 3 weeks.

**Bio-indicator:** This species has been reported to serve as a bio-indicator for lead in an estuarine environment.

**Chemical accumulation:** These cockles have been reported to bioaccumulate lipophilic compounds, including xenoestrogens, and recycle them back into the food chain when humans consume them. This may pose a health hazard to population groups who regularly consume cockles, or for those potentially more at risk from estrogenic chemicals such as men, infants, and the developing fetus.

**Fisheries/Aquaculture:** This species is noncommercially harvested in New Zealand for human consumption.

**Countries where eaten**: New Zealand

*Callista chione* (Linnaeus, 1758) (=*Cytherea chione*)

**Common name(s):** Smooth clam, smooth venus, brown venus, smooth callista

**Distribution**: Eastern Atlantic and the Mediterranean

**Habitat:** Sandy bottoms or with small pebbles in clean waters; depth range 0–25 m; burrows in clean sand, sublittoral and shallow shelf

**Description:** This species has a large and oval shell. Umbones are prominent and offset toward the anterior of the shell. In life, the shells are smooth and glossy with numerous, fine concentric lines and distinct growth bands. Sculpture is of fine concentric lines. Margin is smooth. Ligament is prominent, halfway to posterior margin. Pallial sinus extends to midline and ends in a point. Periostracum is thin and glossy. Outer surface of the shell is a light reddish-brown to chestnut color, often with darker streaks radiating from the umbones. Inner surface is of the shell is dull-white. Maximum length is 11 cm.

**Biology:** This species is an active suspension feeder.

**Fisheries/Aquaculture:** This species is commercially harvested in Spain, Italy, and Morocco.

**Countries where eaten:** This species is edible and different dishes are prepared throughout the Mediterranean in Spain, Italy, France, the Balkan, and the Magreb countries.

*Chamelea gallina* (Linnaeus, 1758)

**Common name(s):** Striped venus clam, chicken venus

**Distribution:** Eastern Atlantic coasts, from Norway and the British Isles, Portugal, Morocco, Madeira and the Canary Islands; Mediterranean Sea, Black Sea, and Adriatic Sea

**Description:** Shell of this species is solid and thick, with two equal sized valves. It is broadly triangular but asymmetrical, having a round anterior

margin but a somewhat elongated posterior. Periostracum is thin and ligament connecting the two valves is narrow. Lunule is short and heart-shaped
and light brown with fine radiating ridges. Shell is sculptured with about
15 concentric ridges. Color is whitish, cream, or pale yellow, sometimes
shiny, and usually with three red-brown radiating rays. Maximum length
is 5 cm and common length is 2.5–3.5 cm.

**Biology**

**Food:** It is an epibenthic suspension feeder, ingesting a variety of algae,
bacteria, and small detrital particles. Water enters the inhaling siphon and
passes through the gills where it is filtered to keep the nutriment and then
it comes out form the exhaling siphon.

**Reproduction**: Reproductive cycle begins in spring and goes on till the
end of summer. Sexes are separate and emitted gametes are fertilized in
the water and two planktonic larval forms follow one another. Afterward,
larvae settle on the bottom and, 1 month after birth, benthic life begins.

**Fisheries/Aquaculture:** It is commercially harvested in Mediterranean. It
is a potential species for aquaculture in USA, Canada, and Italy.

**Countries where eaten:** Southern-Europeans find these shellfish very
tasty and add them to all kinds of dishes.

*Chione californiensis* (Broderip, 1835)

**Common name(s):** California venus

**Distribution:** East Pacific

**Habitat**: Bay and embayment

**Description:** It is medium-sized edible saltwater clam. Shell is roundish,
heart-shaped and is striated longitudinally. Ribs are of an equal thickness
throughout. Margin is crenulated. Average size of animal is 1.5 cm.

**Biology:** An alternation of growth bands has been observed in this species. Wider dark bands appeared during summer and hyaline bands are formed in winter (Quezada, 2012).

**Fisheries/Aquaculture:** No information

**Countries where eaten:** It is eaten by the indigenous peoples of California.

*Chione cancellata* (Linnaeus, 1767)

**Common name(s):** Cross-barred venus

**Distribution:** Western Atlantic: from Honduras to southeast Brazil.

**Habitat:** Sand in shallow subtidal environments; often in seagrass beds; seamounts and knolls; depth range 2 m

**Description:** Shell of this species is thick and trigonal. Sculpture is of blade-like concentric ridges crossed by radial ribs. Interspaces between ribs are smaller than between ridges. Lunule is heart-shaped, dark. *Color*: Externally, it is white to light gray, sometimes with brown rays, and internally, it is white and frequently with blue-purple markings. Maximum length is 5 cm.

**Biology**

**Food and feeding:** The clam feeds by filtering plankton from the seawater, using a siphon system to draw water over a mucus collecting net.

**Reproduction:** These clams reproduce sexually. Sexes are separate and fertilization is external via broadcast spawning of gametes. Individuals reach maturity at around 1.5 cm shell length, at which time they become reproductive males or females at a ratio of approximately 1:1. Laboratory-based rearing studies reveal embryonic development from fertilization to straight-hinge D-stage veliger larvae in just 24 h at 25°C. At this time, the larvae possess a calcified shell, a poorly developed digestive tract, an apical sense organ, a single mantle fold, and a functional velum equipped with four ciliary bands. Lab-reared larvae settle out of the water column when individuals reach a size of about 170 μm. Larval duration is about 11 days from hatching to settlement.

**Association:** This species is offering a colonization site for epibiotic macroalgae and polydorid polychaetes.

**Fisheries/Aquaculture:** It is commercially harvested.

**Countries where eaten:** Incidentally, living cross-barred Venus clams are good human food. Tasty chowder is prepared from this species.

***Chionista fluctifraga*** (Sowerby, 1853) (=*Chione fluctifraga*)

**Common name(s):** Black clam, hard-shell cockle, smooth venus

**Distribution:** East Pacific from San Pedro into Mexico

**Habitat:** Bays in firm sand or sandy mud not too frequently disturbed by waves or strong tidal currents

**Description:** This species is relatively small, compact and rounded in outline, with firm heavy shells and short united siphons. Radiating ribs are most conspicuous, and more prominent than the concentric growth lines on the siphonate or posterior end of the shell. Pallial sinus is larger and sharply triangular. Concentric ribs are smooth and polished and are irregularly broken up by radial furrows that are not of even width. Shell is creamy to grayish white, stained dark-blue within, length, 5.1 cm; height, 4.9 cm; dia., 3.4 cm.

**Biology:** Due to their short siphons, they do not burrow deeply and the siphonate end of the shell may often be seen at the surface of the sand. In some cases, they were observed at the side of a burrow belonging to another species.

**Fisheries/Aquaculture:** No information

**Countries where eaten**: It is used in the restaurants for chowders and soups. The flavor is excellent, though because of the flatter and thicker shell the proportion of meat is not as large as in *Paphia*.

***Circe undatina*** (Lamarck, 1818)

**Common name(s):** Not designated

**Distribution**: Philippines

**Habitat:** Sandy beaches

**Description:** It is also of less width between the valves. Its maximum dia. is about 6 cm. It is yellowish white with black lines across the hinge margin and above the beak.

**Biology:** No information

**Fisheries/Aquaculture:** Not reported

**Countries where eaten:** It is eaten in Philippines. It is a good, clean food, and it has good marketability.

***Cyclina sinensis*** (Gmelin, 1791)

**Common name(s):** Venus clam, oriental cylcina, clams skin

**Distribution**: Indo-West Pacific

**Habitat:** Intertidal areas in sand and mud; depth range 0–50 m

**Description:** Shell of this species is nearly round, slightly thin, and strong. Shell surface is convex; the width of the height is of 2/3. Shell top is prominent and is slightly curved forward. Shell surface is without radiation ribs. Ligament is tan and is not protruding shell surface. Shell surface is light yellow, brown or dark purple range. Shell interior is white or pink. Maximum shell length is 5 cm.

**Biology**

**Ecology:** The activities of this species could positively influence sedimentary oxygen consumption and cause an increase in the fluxes of inorganic nutrients. These effects may substantially improve the primary productivity and water quality of earthen pond ecosystems (Nicholaus and Zheng, 2014).

**Fisheries/Aquaculture:** It is commercially harvested and is a potential species for aquaculture.

**Countries where eaten**: This species is eaten in southern half of Japan, Korea, China, and southeastern Asia.

***Dosinia anus*** (Philippi, 1848)

**Common name(s):** Ringed venus shell, ringed dosinia, biscuit clam

**Distribution:** Southwest Pacific: New Zealand

**Habitat:** Intertidal sand flats; depth range 5–10 m

**Description:** This species has a larger and rougher shell. Maximum length is variable between areas, ranging from 6 to 8 cm.

## Biology

**Reproduction:** The sexes are separate, and they are likely to be broadcast spawners with planktonic larvae. Spawning is likely to occur in the summer months and spat probably recruit to the deeper water of the outer region of the surf zone. Recruitment of surf clams is thought to be highly variable between years.

**Fisheries/Aquaculture:** Commercial fisheries exist for this species in New Zealand.

**Countries where eaten:** New Zealand

### *Dosinia concentrica* (Born, 1778) (=*Dosinia elegans*)

**Common name(s):** Elegant dosinia

**Distribution:** Atlantic coast of North America, ranging from North Carolina to Texas.

**Habitat:** Gulf; sand or sandy mud bottoms

**Description:** This bivalve is very compressed that is the two halves of the bivalve fit together rather flat, almost like a ladies' compact. Shell of this species is almost circular in shape, except for its small beaks, which point anteriorly. There are three teeth on each beak, left middle tooth, and right posterior tooth bifid (two lobes). It is a lovely white bivalve which may reach a diameter of 7.5 cm. Its surface is marked with numerous uniformly spaced concentric ridges that get wider toward the ventral margin. Pallial sinus is cone-shaped. Periostracum, or shell covering, is translucent, grayish, and varnish-like. Color of shell is off-white to light yellow. Common size is 7.6 cm. A length of about 1.5 cm is reached in a year, and an animal with the maximum size of 3 cm is 2–3 years old.

**Biology:** No information

**Fisheries/Aquaculture:** Not reported

**Countries where eaten**: Mexico

***Dosinia discus*** (Reeve, 1850)

**Common name(s):** Disk dosinia

**Distribution:** Western Central Atlantic and Western Central Pacific: North America and the Philippines.

**Habitat:** Gulf; sand or sandy mud bottoms

**Description:** Shells of this species is circular in shape, very slightly inflated and strong. Fine concentric ridges (about 20) are seen on shell. Beak is prominent, pointy, curved, and pointing anteriorly. Two large muscle scars are present on interior. Pallial sinus is cone shaped. Shell color is off-white. Maximum size is 7.6 cm.

**Biology:** Not reported

**Fisheries/Aquaculture:** No information

**Countries where eaten:** Gulf of Mexico

***Dosinia exoleta*** (Linnaeus, 1758)

**Common name(s):** Mature dosinia

**Distribution:** Eastern Atlantic and the Mediterranean

**Habitat:** Intertidal; deep burrower in mostly clean coarse sand and shell gravel; depth range 0–25 m

**Description:** Shell of this species is almost circular and inequilateral. Umbones are small but distinct, just anterior to midline. Dorsal margin is convex posteriorly and deeply concave immediately anterior to umbones, with a highly arched junction with anterior margin. Lunule is heart-shaped and is as broad as long. Sculpture is of numerous fine concentric ridges and surface is smooth. Growth stages are clear. Three cardinal teeth are seen in each valve. A short, rounded knob is present in front of the interior teeth of the left valve. Adductor scars and pallial line are distinct. Pallial sinus is deep, narrow, U-shaped and is extending anteriodorsally well into the anterior half of the shell. Color of shell is white, yellowish, or light brown, with irregular rays, streaks, or blotches of darker brown or pinkish brown. Inner surfaces are glossy white. Maximum length is 6 cm.

**Biology:** It is an active suspension feeder.

**Fisheries/Aquaculture:** It is commercially harvested.

**Countries where eaten:** Mediterranean, Norway, Senegal, and Gabon

*Dosinia lupines* (Linnaeus, 1758)

**Common name(s):** Smooth dosinia, smooth artemis, artemis shell

**Distribution:** Widely distributed on all coasts except the southern North Sea

**Habitat:** From the low intertidal to the outer shelf; deep burrower in fine sand, coarse sand, sandy mud, silty sand, and shell gravel; depth range 0–200 m

**Description:** Shell of this species is solid. Beaks are in front of midline. Outline of shell is suboval. Lunule is prominent but short. Curve of margin from lunule is low to anterior margin. Escutcheon is poorly defined. Evenly spaced concentric ridges and growth lines are seen. Margin of shell is smooth. Ligament is deeply inset and extends halfway to posterior margin. In lower valve, small protrusion is seen in front of anterior cardinal. Pallial sinus is very deep, extending toward top of adductor, narrowing toward end and rounded. Periostracum is thin, satin, and pale yellow. Color of shell is cream. Beige rays may extend from umbos. Maximum size is 3.8 cm. Common height is 2.3 cm and length is 2.4 cm.

**Biology:** No information

**Fisheries/Aquaculture:** Not reported

**Countries where eaten**: Mediterranean

*Gafrarium dispar* (Holten, 1802)

**Common name(s):** Transverse circe, discrepant venus

**Distribution:** Indo-Pacific: from East Africa, including the Red Sea to eastern Polynesia; north to Japan and south to Queensland.

**Habitat:** Intertidal areas in coarse sand and gravel; muddy bottoms; mangrove areas; depth range 0–20 m.

**Description:** Shell of this species is small, thick, solid and ovate-subquadrate in shape. It is covered with fine, well-marked concentric ridges and radial riblets. Umbones are low and rounded. Lunule is lanceolate and flattened. Hinge is with three cardinal teeth at each valve and three lateral teeth. Pallial sinus is very shallow. Outer shell surface is cream to buff-colored with irregular reddish brown patches or lines sometimes forming zigzag patterns. Maximum length is 3 cm and common length is 2 cm.

**Biology:** No information

**Fisheries/Aquaculture:** Not reported

**Countries where eaten:** This species is a commercially important edible bivalve which is caught and sold for human consumption.

*Gafrarium divaricatum* (Gmelin, 1791)

**Common name(s):** Forked venus

**Distribution**: Indo-West Pacific: from East and South Africa to the Philippines; north to Japan and south to Malaysia

**Habitat:** Intertidal areas in coarse sand and gravel; depth range 0–20 m

**Description:** Shell of this species is circular with a slight pointed tip, thick, and heavy. There are fine ribs which are parallel to the shell edge. Color of shell is white, usually with a pattern of thin dark lines perpendicular to the shell edges. Maximum length is 6 cm.

**Biology:** No information

**Fisheries/Aquaculture:** Not reported

**Countries where eaten**: Whole tissue of this species is used as a food source in coastal regions of India

## *Gafrarium pectinatum* (Linnaeus, 1758)

**Common name(s):** Comb venus and tumid venus

**Distribution:** Indo-Pacific and the Mediterranean Sea: from East and South Africa to eastern Polynesia; north to Japan and south to Queensland

**Habitat:** Intertidal and shallow water; muddy gravels and sands; depth range 0–30 m

**Description:** Shell of this species is medium sized and is slightly laterally compressed. Outline of shell is subovate and is distinctly longer than high. Sculpture is of nodulose radial ribs. Asymmetrical ribs are on the posterior slope where they change direction and are obliquely placed in relation to the central ribs. Divaricate ribs of the posterior slope are concentrically striae from the top to the bottom and concentric grooves continue into the symmetrical radial ribs near the ventral margin. Narrow heart-shaped lunule is seen. There is no pallial sinus. Inner margin is crenulate. Exteriorly shell is white to cream. Internally, it is white with a cream flush within the pallial line area, ligament stained with violet. Maximum length is 5 cm.

**Biology:** It lives for up to 3 years with low levels of recruitment taking place in two phases in spring and autumn. Gametogenic cycles confirm this picture. This species matures at a shell length about 2 cm after 1 year of life. Generally, it is a dioecious species with sex ratio about 1:1.

**Fisheries/Aquaculture:** It is a potential species for aquaculture in Chile.

**Countries where eaten:** It is an important food source in Fiji.

*Leukoma antiqua* (King, 1832)

**Common name(s):** King's littleneck

**Distribution:** Panama to Ecuador

**Habitat:** Deeply buried among rocks

**Description**: This species has a rounded ovate shell. Wavy concentric lamellae are well spread throughout the shell. Color is creamy white to light brown with rays and blotches of darker brown and white. Inner margins are crenulated except posteriorly. *Size of shell*: length 4.5 cm, height 3.9 cm, and dia. 2.9 cm.

**Biology:** No information

**Fisheries/Aquaculture:** It is commercially harvested and is a potential species for aquaculture in China and South Chile.

**Countries where eaten**: China and South Chile

*Leukoma staminea* (Conrad, 1837) (*=Protothaca staminea*)

**Common name(s):** Little neck clam, Pacific littleneck clam

**Distribution:** Northern Mexico to northern Alaska; Japan to Siberia

**Habitat:** Beaches; intertidal to 40 m; sand or gravel

**Description:** It is a roughly oval clam, with its beak nearer to the anterior end. Shell is sculptured with concentric lines and stronger radial ribs, forming beads as they cross. Radial ribs are better defined in the middle of the shell. Beaks are almost smooth. Ligament is sunken, and not elevated above the dorsal margin. Interior of the ventral margin is slightly crenulated. Siphons are united for their entire length. Adults range from 3.5 to 8.0 cm in length. Color of shell ranges from yellowish or gray in muddy environments to white with geometric blotches or lines on the open coast.

**Biology:** No information

**Fisheries/Aquaculture:** This species is extensively cultivated in Japan

**Countries where eaten:** Berring Sea to Mexico. It is a popular steamer clam

***Leukoma thaca*** (Molina, 1782) (=*Protothaca thaca*)

**Common name(s):** Greater littleneck, taca clam

**Distribution:** Southeast Pacific

**Habitat:** Infaunal burrower species in sandy substrates of the intertidal and subtidal zones; depth range 0–20 m.

**Description:** Shell of this species is oval-rounded and is crossed by concentric growth lines, with radial ribs present in the mid and posterior part. Periotracum is absent. Voluminous umbones are curved toward the anterior margin. Pallial sinus is long, reaching almost 1/2 the shell length.

Adductor muscles scars are oval and deep. Hinge is having three cardinal teeth in each valve: mid-one is in the left valve and the two posterior ones are in the right valve are bifid. Color of shell is dirty white externally, with some slight reddish tonalities. Maximum shell length is 8 cm. Common length varies from 5 to 7 cm.

**Biology:** This species shows an annual reproductive cycle with a short spawning season.

**Fisheries/Aquaculture:** The species is exploited commercially over all its distribution range. An artisanal (small-scale) fishery is developed for this species for local consumption. It is also extensively cultivated in India, Korea, China, and Taiwan.

**Countries where eaten:** It is eaten all over its distribution range.

*Macrocallista nimbosa* (Lightfoot, 1786)

**Common name(s):** Sunray venus

**Distribution:** Western Central Atlantic: North Carolina to Texas, USA and Mexico

**Habitat:** Gulf and bay, sandy bottoms; depth range 0–55 m

**Description:** This lovely shell is elongate, ovate, fairly thick, smooth, and shiny. Sculpture is almost completely absent, except for weak growth lines and radial riblets. Perioscracum is thin. Lunule which is purplish is oval.

External ligament is long. Brown, pinkish-gray, golden-tan, or lavender exterior is beautified with bands of darker colors which radiate from the beak (umbo) to the margin (periphery). Interior shell is white (or salmon-colored when the specimen is fresh). Maximum shell length is 15 cm.

## Biology

**Food and feeding:** It is suspension/filter feeder feeding mainly on phyto-plankton and detritus material for its nutrition.

**Fisheries/Aquaculture:** This species is commercially harvested in Florida.

**Countries where eaten**: Excellent clam chowder prepared from this species.

*Megapitaria aurantiaca* (Sowerby, 1831)

**Common name(s):** Chocolate clam

**Distribution:** Gulf of California to Ecuador

**Habitat**: Ecological zone of this species is unknown

**Description:** Overall shape of this species is subtrigonal and ovate. Posterior margin is more pointed than the anterior margin, but overall, the anterior, ventral, and posterior margins are rounded. Anterior dorsal margin is nearly straight, and the posterior dorsal margin is subconvex. Valves are equal in size and shape (equivalve), and there is no gape when they are closed. Sculpture consists of growth lines only. Lunule is spear-shaped, extends 2/3 of the anterior dorsal margin and is defined by a fine, shallow groove. Umbo is subanterior, and the beaks also point slightly to the anterior. Exterior is glossy, light brown-pink near the umbo and becomes more pink ventrally. Lunule may be partially covered with purple coloration, and the posterior margin may also be slightly purple. Beaks are typically white. Periostracum is red-brown and thin. Interior is white with purple coloration at the anterior and posterior of the hinge plate. Purple coloration may extend down the posterior dorsal margin. Anterior and posterior adductor muscle scars are similar in size and shape. Pallial sinus is deep. *Size*: length, 11 cm; height, 8.5 cm; dia., 6 cm.

**Biology**: No report

**Fisheries/Aquaculture:** Not known

**Countries where eaten**: West Central America and northwestern South America

***Megapitaria maculata*** (Linnaeus, 1758) (=*Macrocallista maculata*)

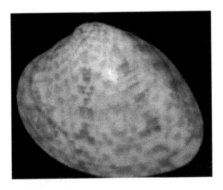

**Common name(s):** Calico clam

**Distribution:** Western Atlantic: Caribbean to Brazil

**Habitat:** Coarse sand, often near sea grass beds, in shallow waters; depth range 0–100 m

**Description:** Shell of this species is ovate and much less elongate than *Macrocallista nimbosa*. Surface is highly glossy. Sculpture is of very fine growth lines under glossy layer. Umbones are small. Lunule is small. *Color*: exterior is tan with irregular brown marks, sometimes arranged in radial bands. Internally, it is white. Maximum length of the shell is 7 cm.

**Biology:** Not known

**Fisheries/Aquaculture:** It is commercially harvested. The Calico clam has been protected from harvesting in Bermuda since the 1970s.

**Countries where eaten:** Bermuda and in areas of occurrence

## *Megapitaria squalida* (Sowerby, 1835)

**Common name(s):** Chocolate clam, Mexican chocolate clam

**Distribution:** Northwestern Mexico and Peru

**Habitat:** Lives buried in muddy and sandy bottoms; depth range 1–120 m

**Description:** Shell of this species is subelliptical, convex, and smooth and sculptured only with fine lines of growth. Periostracum is gray-brown and shiny. Overall shape is trigonal, ovate. Posterior margin is weakly pointed, and the anterior and ventral margins are rounded. Anterior dorsal margin is straight, and the posterior dorsal margin is subconvex. Valves are equal in size and shape (equivalve), and there is no gape when they are closed. Sculpture consists of growth lines only. Lunule is spear-shaped and extends half the anterior dorsal margin and is defined by a fine, shallow groove. Umbo is anterior, and the beaks also point slightly to the anterior. Exterior is typically light brown. There are typically two prominent, dark brown radial stripes. Periostracum is light brown or olive-brown. Interior is white, and there may faint purple coloration internally, especially along the some of the growth lines or outside of the pallial line. Anterior and posterior adductor muscle scars are similar in size and shape. Pallial sinus is deep. Maximum length is 14 cm.

### Biology

**Reproduction:** An unusual high incidence of hermaphroditism has been reported in this gonochoric species. This is largely due to the low population density resulting from overfishing. The species reproduces when the water temperature rises to a suitable point, and in regions where the temperature is always maintained at an optimal level, it reproduces throughout the whole year.

**Predators**: Gastropods, shore birds, and fish among others are its natural predators.

**Fisheries/Aquaculture:** It is commercially harvested. A positive and significant relationship has been observed between sea surface temperature and landing of this species.

**Countries where eaten:** This species is widely consumed by the population of several localities along the Pacific coast.

*Mercenaria campechiensis* (Gmelin, 1791)

**Common name(s):** Southern quahog, southern hard clam, cherry stone clam

**Distribution**: From the Gulf of St. Lawrence to the Gulf of Mexico

**Habitat:** Sandy or muddy shores between the tides in shallow water; depth range 0–40 m

**Description**: Heavy shell of this species is ovate, trigonal, and inflated with valves of equal size. Several concentric growth lines are present on the gray to whitish exterior. Three cardinal teeth (the middle one is split) are seen on each valve. In empty shells, the interior color is often porcelaneous white with purple marks occurring rarely. Two muscle scars are present on the interior surface, both attached to the pallial line. Maximum length is 17 cm.

**Biology**

**Locomotion:** This clam which tolerates high salinities uses its muscular foot to burrow into the sediment, where it is camouflaged from potential predators.

**Growth:** Most growth occurs in this species during the first 2–3 years of life.

**Reproduction:** It is a protandrous hermaphrodite, with about 98% of individuals beginning life as males and changing to females as they grow older. Like most bivalves, it reproduces via external fertilization, releasing gametes into the water column. Spawning generally occurs in warmer months. After fertilization, larvae pass through three main planktonic stages before developing into settled, juvenile clams. The first of these is the trochophore stage, which is formed 12–14 h following fertilization. After about 1 day, larvae enter the veliger stage. During this stage, the shell and internal organs develop. After 6–10 days, larvae develop the foot and are termed as pediveligers.

**Life span**: Maximum documented age for this species is about 22 years.

**Fisheries/Aquaculture:** Not known

**Countries where eaten**: This clam is an important food source in the areas of its occurrence especially in USA.

**Other uses:** The Indians used this shell for wampum by cutting small pieced beads of white and purple.

*Mercenaria mercenaria* (Linnaeus, 1758)

**Common name(s):** Northern quahog, northern hard clam, cherrystone, little-necked clam

**Distribution:** Gulf of St. Lawrence; east coast of the United States; Florida peninsula; Gulf Coast of Texas

**Habitat:** Intertidal species; estuaries or lagoons; muddy bottoms; depth range 0–10 m

**Description:** Shell of this species is large, extremely thick and heavy and is rather well inflated. It is Inequilateral with elevated beaks in the anterior part. Outline of shell is triangular and subovate. Sculpture is of numerous concentric lines which are widely spaced at the umbones and closely spaced near the margins. Color of shell is white or gray, sometimes with brown zigzag markings. Internally, it is white and is often stained violet. Maximum length is 15 cm.

**Biology**

**Ecology:** It is a euryhaline marine species and is sensitive to salinities below 12 ppt. It burrows shallowly in sediments of either mud or sand.

**Food and feeding:** Like other clams, it is a filter feeder. It largely depends on plankton and other microorganisms that are carried along the bottom by currents for food before and during spawning to provide sufficient energy to ripen the gonads. If the food supply is inadequate, spawning is diminished or nil.

**Growth and life span:** In this species, growth has been observed to cease after the age of 15 years, with annual growth at this age slowed to approximately 1 mm/year. The natural lifespan of this species is generally unknown; however, counts of growth rings indicate that this species, in the absence of predation or commercial exploitation, may live up to 40 years.

**Reproduction:** This species becomes reproductively active at approximately 1 year of age and will continue to produce broods throughout their lives. It is a protandric hermaphrodite, with the male line developing first. With increased age and size, approximately half of the males later change to females. The hard clam spawns during summer throughout its areas of occurrence. Water temperatures between 22 and 30°C were found to be favorable for these clams to spawn with maximum frequency. In the Indian River Lagoon, however, spawning occurs in the autumn after water temperatures drop below 23°C. Fertilized eggs become trochophore larvae within the first 12 h. Shells develop within 30 h. The veliger stage is reached in another 8–12 h. Veligers are planktonic for approximately 12–14 days before settling. With settlement, the velum disappears and use of the foot shifts from aiding in swimming to burrowing and crawling.

**Fisheries/Aquaculture:** This species is among the most commercially important species of invertebrate. It has the highest value of any fishery species in the Indian River Lagoon. It is commercially harvested and cultivated in Italy, Turkey, Spain, and France.

**Countries where eaten:** USA; Canadian Maritimes to the Gulf of Mexico

*Meretrix casta* (Gmelin, 1791)

**Common name(s):** Poker-chip venus

**Distribution:** Indo-Pacific oceans

**Habitat:** Estuarine

**Description:** Shell of this species is medium and solid, but less heavy than *Meretrtix meretrix*. Umbo is slightly anterior and beak is more attenuated. Lunule is well developed and it is about two-third or more of anterior dorsal margin. Shell is variable in shape and color. Anterior cardinal teeth are more strong on left valve. Pallial line is well impressed and pallial sinus is absent. Common length is 4 cm.

**Biology:** Not known

**Fisheries/Aquaculture:** It is a potential species for aquaculture in Indonesia and Sri Lanka.

**Countries where eaten:** It is edible in the areas of its occurrence. The meat of this species is used for the preparation of excellent recipes like cake and soup. Further, it has the potential to become a regular food item in the house hold diet with consumer acceptance.

*Meretrix lusoria* (Röding, 1798)

**Common name(s):** Common orient clam, Asian hard clam, poker chip venus

**Distribution:** Northwest Pacific: Asia from China, Korea, and Japan

**Habitat:** Estuaries and coastal waters; depth range 0–20 m

**Description:** This species has a triangular–oval shell, with a strongly rounded ventral margin, but fairly straight dorsal sides, peaking at the umbo. Umbo is slightly anterior of the midpoint of the shell and the anterior side of the shell drops more steeply than the posterior. Pallial sinus is shallow. Shell has fine concentric growth lines and a relatively smooth surface. Color of shell is highly variable, from white and brown concentric bands to fine reddish-brown or dark-brown mottling, forming dotted lines, and W-shaped markings. Maximum length is 10 cm.

**Biology:** No report

**Fisheries/Aquaculture:** It is heavily fished, and frequently raised in aquaculture in Asian waters. It is also extensively cultivated in USA and Canada. It is also commercially exploited for sushi, and its shells are traditionally used to make white stones.

**Countries where eaten:** This species serves as a popular seafood and is traditionally used as a Chinese remedy for liver disease and chronic hepatitis.

*Meretrix lyrata* (Sowerby, 1851)

**Common name(s):** Philippines seashell, white clam, Asian hard clam

**Distribution**: Indo-West Pacific: Vietnam, Taiwan, Philippines, and Southern China.

**Habitat:** Intertidal areas in sand and mud; tidal flats, estuaries and sandy beaches; depth range 0–20 m

**Description:** This species is easily recognized by its unique commarginal ribs, white shell, and the dark black streak covering one of the side edges. External shell is pearly white in color, while the interior is pale-yellow to off-white. Maximum length is 6 cm.

**Biology**: Not completely known

**Fisheries/Aquaculture:** Commercial fisheries exist for this species. It is heavily fished and frequently raised in aquaculture in Asian waters.

**Countries where eaten:** This edible saltwater clam is a commercially important species in East and Southeast Asia. In Vietnam, these hard clams are eaten boiled, steamed, or roasted. They are an important export item in the eastern and southern parts of Vietnam.

*Meretrix meretrix* (Linnaeus, 1758)

**Common name(s):** Asiatic hard clam, great clam

**Distribution:** Indo-West Pacific: from East Africa to the Philippines; north to Japan and south to Indonesia

**Habitat:** Intertidal areas in sand and mud; depth range 0–20 m

**Description:** Shell of this species is moderately large, heavy, ventricose, and compressed. Hinge is narrow and umbo is almost central. Beak is attenuated. Lunule is heart shaped. Sculpture is smooth and is of only growth lines. Anterior cardinal teeth are more strong on left valve. Muscle scars are well impressed; anterior adductor is smaller and semilunar; and posterior adductor scar is large with narrow anterior end. Pallial line is slightly arched posteriorly. Shell is variable in color. Maximum length is 7 cm.

**Biology:** It feeds on plankton and detritus and its life-span is about 8 years.

**Fisheries/Aquaculture:** It is fished commercially for human consumption, cultivated in China, and is a potential species for aquaculture in India.

**Countries where eaten**: India and China

***Paphia amabilis*** (Philippi, 1847)

**Common name(s):** True song buffy clam, short-necked clam, lovely venus

**Distribution:** Western Pacific: China, Taiwan and Japan; from Korea and Japan through Philippines and Malaysia, to Andaman Islands and Siam Gulf, and along Vietnamese and Chinese coasts to western Taiwan

**Habitat**: Sandy and muddy bottoms; depth range 10–70 m; from the subtidal zone to a depth of 90 m

**Description**: Shell shape of this species is oval to oblong. Hinged teeth are developed and hard. Exterior shell color is light reddish-brown and inside the surface is white. Shell length is about 3–7.5 cm.

**Biology**

**Reproduction:** The eggs are developed synchronously in a mature individual and are released at once. There are two spawn seasons each year. The period from fertilized egg to spat generally last for a period of 14–18 days (at temperature 28.5°C and salinity 30‰). It takes about 20 h for fertilized eggs to develop into early veliger (D-shape) larvae. In 3 days postfertilization, they become early umbo larvae, and in 8–13 days postfertilization, they turn into late umbo larvae with development of foot. Then in 18 days postfertilization, they become demersal spat through metamorphosis and finally in 33 days postfertilization, they grow into the juveniles.

**Fisheries/Aquaculture:** It is a promising species in aquaculture.

**Countries where eaten:** It is an important edible shellfish in southern Japan and China. This nutrient-rich clam has become a promising delicious species for consumption for common people of areas of its occurrence.

**Other uses:** The carotenoids of the meat of this species have an array of beneficial properties, such as antioxidative, antitumor, anticarcinogenic, and immune enhancement activities (Sri Kantha 1989).

*Paphia euglypta* (Philippi, 1847)

**Common name(s):** Crude cross rib clam, thick rib horizontal clams, well-carved venus

**Distribution:** Western Pacific

**Habitat:** Sand and mud; depth range 10–50 m

**Description:** This species has an oval shape, a light brown color, and it is plain. Ring veins are thick and parallel at the center of the shell. Common shell length is 7 cm.

**Biology:** The meat of this species is rich in amarouciaxanthin A, a carotenoid (Sri Kantha, 1989).

**Fisheries/Aquaculture:** No information

**Countries where eaten:** It is edible in areas of its occurrence.

**Other uses:** This species has pharmaceutical and biotechnological applications.

## *Paratapes textilis* (Gmelin, 1791) (=*Paphia textile*)

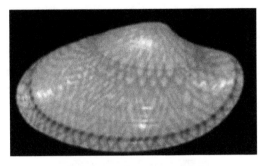

**Common name(s):** Textile venus clam, carpet clam

**Distribution:** Indo-West Pacific and the Mediterranean Sea: from eastern Africa to Papua New Guinea; north to the South China Sea and south to Indonesia

**Habitat:** Intertidal areas in sand and mud; offshore shelf bottoms; attached to rocks; depth range 0–20 m

**Description:** These shells are elongate, elliptical-ovate, and moderately inflated, with rounded margins. Umbones are well placed in the anterior part of shell. Dorsal margin is sloping to a narrowed posterior margin. Hinge is narrow, with three radiating cardinal teeth. Lunule is well defined and is slightly depressed. Surface is shining. Sculpture is smooth with growth lines only. Pallial sinus is round. Internal margin is smooth. Outer shell surface is smooth, glossy, pale yellowish-white, with pale purplish gray inverted V-shaped markings. Internally, the shell is white. Maximum shell length is 8 cm and common size is 5 cm.

### Biology

**Ecology:** It is filter feeder and it burrows in deep continental shelf waters bottom where it can dominate the molluscan fauna. Its affinities are with warm Indo-Pacific faunas and normal marine salinities.

**Reproduction:** It is dioecious but does not exhibit external sexual dimorphism. Spawning is year-round with 4 cm shell length size at first maturity. It exhibits negative allometric growth.

**Allergens:** Two major allergens, namely, a thermostable tropomyosin and a thermolabile allergen actin have been identified from this species (Yadzir et al., 2015).

**Fisheries/Aquaculture:** It is commercially harvested for food with other *Tapes* and *Paphia* species. *Paphia* fishery is a major source of livelihood, particularly in the coastal waters of Sorsogon, the Philippines.

**Countries where eaten:** It is one of the more popular edible shellfish in Malaysia. It is also an important commercial species in the central and southern Philippines. Its meat is consumed locally and abroad, while the empty shells are utilized for shell craft. *Paphia* fishery is a major source of livelihood, particularly in the coastal waters of Sorsogon.

***Paratapes undulatus*** (Born, 1778) (=*Paphia undulata*)

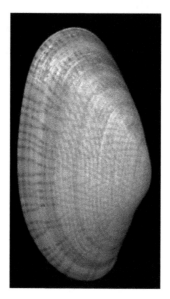

**Common name(s):** Undulate venus, undulated surf clam, short-necked clam, baby clam, nylon shell

**Distribution:** Indo-West Pacific: Red Sea to Papua New Guinea; north to Japan and south to New South Wales.

**Habitat:** Intertidal to sublittoral; shallow sandy seabed; coastal mudflats; depth range 0–50 m

**Description:** Shell of this species is suborbicular, with numerous thin elevated distant transverse ribs. Interstices are minutely striated longitudinally. Margin is crenulated. Maximum length is 6.5 cm.

**Biology**

**Growth:** The shell growth of this species is largely uncoupled to local environmental variables. Growth rates have been negatively correlated to seawater temperature and chlorophyll *a* levels and positively to salinity.

**Life span**: This species has a short lifespan of 3 years only.

**Fisheries/Aquaculture:** It is a popular shellfish resource harvested mainly for food. In Thailand, Malaysia, and Philippines, this species contributes to the regional or local fisheries. It is also a potential species for polyculture in Japan, Korea, and Taiwan.

**Countries where eaten:** This species is used as food in some Southeast Asian countries such as Thailand and Vietnam.

*Periglypta listeri* (Gray, 1838)

**Common name(s):** Princess venus clam

**Distribution:** Western Atlantic: North Carolina, southeastern and western Florida, including the Florida Keys, Texas, the West Indies, Caribbean Central America and northern South America

**Habitat:** Intertidal and shallow subtidal settings; coarse substrate and grassy areas; rocks, pilings, seawalls, seagrass beds, open sand, and lagoonal patch reefs; depth range 0–84 m.

**Description:** It is a large venerid, with thick walled, inflated, and trapezoid shell. External sculpture of shell is of erect commarginal ridges crenulated by underlying radial sculpture, and with posterior end vertically truncated. Exterior shell is cream colored, with scattered brown speckles, blotches, or flames. Interior is yellowish white with more-or-less strongly developed purplish brown stain around posterior adductor muscle scar, posterior margin, and above the pallial line. Anterior lateral hinge tooth is often with a purplish brown "hinge dot." A well-developed escutcheon and lunule are present. Hinge is with three cardinal teeth in each valve. Within the genus, this species is unique in having internal purplish brown coloration, and in the frequent presence of a purplish brown, "hinge dot" on the anterior lateral tooth. Maximum length is 10 cm.

**Biology:** No report

**Fisheries/Aquaculture:** Not known

**Countries where eaten:** It is an edible species in the areas of its occurrence.

*Periglypta puerpera* (Linnaeus, 1771)

**Common name(s):** Youthful venus and maiden's purse shell

**Distribution:** Indo-Pacific: from East Africa to Polynesia; north to Japan and south to Queensland and South Australia

**Habitat:** Intertidal in sand to subtidal; depth range 0–20 m

**Description**: Maximum length is 12.5 cm. No other information.

**Biology:** No information.

**Fisheries/Aquaculture:** This species is commercially harvested for food in Southeast Asia

**Countries where eaten**: Southeast Asia and Fiji

## *Pitar rudis* (Poli, 1795)

**Common name(s):** Rough pitar venus

**Distribution:** Bay of Biscay, Mediterranean, Aegean, Marmara, and Black Seas

**Habitat:** Shelly and muddy bottoms at a depth of about 200 m; muddy detritic areas

**Description:** This small shell has a similar shape to other clams, characterized by subtriangular rounded shape. Surface of this fragile shell is smooth, with light barely visible growth striae. Exterior of shell has light brown spots interrupted by concentric and radial white bands. *Shell size*: length 2.5 cm; height 2.1 cm; and width 1.4 cm.

**Biology**

**Feeding:** It is an active suspension feeder.

**Reproduction:** The eggs of this species are sewn into the water. The larvae swim in the water and are carried by currents over long distances. Then, they sink to the bottom.

**Association:** Accumulations of these bivalves form small biocenoses (interacting organisms living together) in the Crimea and the Caucasus coasts at a depth of about 25 m on the sandy and silty–sandy soil. Sometimes, these biocoenoses occur deeper about 50 m.

**Fisheries/Aquaculture:** No information

**Countries where eaten:** Mediterranean, Black Sea

***Polititapes aureus*** (Gmelin, 1791) (*=Venerupis aurea*)

**Common name(s):** Golden venus, golden carpet-shell, smooth butter clam

**Distribution**: Atlantic and Mediterranean regions

**Habitat:** Intertidal and sublittoral; sandy and rocky bottoms; shallow burrower in gravels and muds; depth range 10 to 90 m

**Description:** Shell of this species is solid. Outline of shell is subovate and is almost subtriangular. It is equivalve and inequilateral. Beaks are in front of midline. Dorsal margin is long and sloping to posterior. Lunule is heart shaped. Escutcheon is indistinct. Sculpture is consisting of thin concentric ridges and very fine radial lines. Margin is smooth. Ligament is extending from umbo 1/3rd the dorsal margin length. Hinge has three cardinals in each valve. Muscle scars are roughly equal: posterior is slightly fatter than anterior. Pallial sinus is just over a third of whole length of shell, rounded, but ends in a point. Periostracum is thin, smooth, and glossy. External shell is fawn/pale yellow/dirty-white/brown colored and may have red-brown and/or mid-brown zigzag/blotches/lines/streaks. Internal coloration is white sometimes with orange/purple/yellow spreading from umbo. Maximum size is 4.5 cm.

**Biology:** No information

**Fisheries/Aquaculture:** This species is commercially harvested. It is a potential species for aquaculture in Spain, Portugal, Ireland, Italy, and France.

**Countries where eaten:** It is an edible bivalve, though its quality inferior to other species. It is eaten in Black Sea, Mediterranean, Norway, and Mauritania.

*Polititapes rhomboides* (Pennant, 1777)

**Common name(s):** Not designated

**Distribution**: Northeast Atlantic: Portugal

**Habitat:** Burrows in deposits of gravel or coarse sand; depth range 0–25 m

**Description:** This species has a robust, thick shell which is equivalve and inequilateral. Beaks are seen in front of midline. Outline of shell is rhomboidal to subovate. Dorsal margin is long, gently sloping, and almost angular. Lunule is heart-shaped and is clearly defined. Escutcheon is poorly defined. Sculpture is consisting of thin concentric ridges and grooves. Margin is smooth. Ligament is extending from umbo halfway to posterior. Hinge has three cardinals in each valve. Muscle scars are roughly equal: posterior is slightly fatter than anterior. Pallial sinus extends 1/3rd of way to posterior and rounded. Periostracum is thin, smooth, and satin sheen. External shell is cream/yellow/pink-brown/pink-cream colored, but with reddish brown/mid-brown rays/blotches/zigzags. Sometimes, it has no pattern. Internal coloration is white sometimes with pink/orange spreading from umbo. Maximum length is 6.5 cm.

**Biology:** No information

**Fisheries/Aquaculture:** This species is commercially harvested and it is cultivated in Italy.

**Countries where eaten:** It is eaten in Mediterranean, Norway, and Morocco.

*Protapes gallus* (Gmelin, 1791) (=*Paphia malabarica*)

**Common name(s):** Short neck clam, yellow-foot clam, yellow clam

**Distribution:** Indian Ocean: India, Persian Gulf, Gulf of Aden, Pakistan, Sri Lanka, Myanmar, Malaysia, Indonesia, Philippines, China, and Australia

**Habitat:** Intertidal, continental shelf, sand and mud; Estuarine; shallow water; usually no deeper than 15 m

**Description:** Shell of this species is small, strong, thick walled, and trapezoidal oval. Crown is slightly shifted to the front edge. Sculpture is represented by concentric rounded edges almost the same size. Shell is light cream with a pattern of spots and light brown in color. Inner surface is white. Maximum length is 6 cm.

**Biology:** This species has peak spawning activity in November–December.

**Fisheries/Aquaculture:** It is a potential species for aquaculture.

**Countries where eaten:** It is edible in the areas of its occurrence.

*Ruditapes decussatus* (Linnaeus, 1758)

**Common name(s):** Grooved carpet shell

**Distribution**: Worldwide

**Habitat:** Lagoonal and coastal sites; lives burrowed in sand and silty mud; buried 15–20 cm deep in the sand from the middle of the intertidal zone to a depth of a few meters.

**Description:** Shell of this species is broadly oval to quadrate. Posterior hinge line is straight and posterior margin is truncated. Anterior hinge line is grading into the down-sloping anterior margin. Sculpture is of fine concentric striae and bolder radiating lines. Growth stages are clear. Lunule and escutcheon are poorly defined. Each valve is with three cardinal teeth: center one in left valve, and center and posterior in right are bifid. Pallial line and adductor scars are distinct. Pallial sinus is U-shaped, not extending beyond mid-line of shell, but reaching a point below the posterior part of the ligament. Lower limb of sinus is distinct from pallial line for the whole of its length. Inner surfaces are glossy white, often with yellow or orange tints, and with a bluish tinge along dorsal edge. Exterior is cream, yellowish, or light brown, often with darker markings.

**Biology**

**Food and feeding:** It filters water through its two siphons (one in and the other out) catching organic matter (detritus) and phytoplankton as food.

**Reproduction:** Its sexes are separate, although hermaphrodites can be found infrequently. Reproduction is external and takes places mainly during summer in the wild. In spring, clams can be artificially conditioned for hatching with higher temperature water and abundant food. The larvae swim freely for 10–15 days before settling as spat of about 0.5 mm on a sand and silty mud substrate.

**Predators:** Its main predators are shore crabs (*Carcinus maenas*), starfish (*Asteria rubens* and *Marthasterias glaciais*), gastropods (*Natica* sp.), and birds (*Larus* sp). An individual *Carcinus maenas* (6.5 cm width) can consume 5–6 clams/day.

**Fisheries/Aquaculture:** It is extensively cultivated in Thailand. It is also cultivated from the Atlantic coast of France, Spain, Portugal and in the Mediterranean basin. It is often grown with other bivalves such as *Venerupis pullastra*, *Venerupis rhomboideus*, *Venerupis aurea*, *Dosinia exoleta*, and *Tellina incarnata*.

**Countries where eaten:** Mediterranean, Norway, Congo (north of the Red Sea where the species has immigrated by the Suez Canal) and France. It is consumed fresh and canned.

**Other uses:** Due to its ecological and economic interest has been proposed as a bio-indicator.

*Ruditapes philippinarum* (Adams & Reeve, 1850) (*Venerupis philippinarum*)

**Common name(s):** Japanese carpet shell, Manila clam, Asari clam, Japanese littleneck

**Distribution:** Atlantic, Mediterranean, and Pacific Ocean: from India and Sri Lanka to Micronesia; north to Sakhalin, the Japan Sea, and Hawaii (introduced), and south to Indonesia.

**Habitat:** Marine and brackish waters (salinity range: 13.5–35.0 ppt); buries to only about 10 cm deep in a variety of substrates, from mud, sand, and gravel, high along intertidal areas; depth range 0–100 m.

**Description:** Shell of this species is solid, equivalve, and inequilateral. Beaks are located in the anterior half and are somewhat broadly oval in outline. Ligament is inset and not concealed. Lunule is elongate and heart-shaped. Sculpture is of radiating ribs and concentric grooves and the latter are becoming particularly sharp over the anterior and posterior parts of the shell, making the surface pronounced decussate. Growth stages are clear. Three cardinal teeth are present in each valve: center tooth is in left valve and center and posterior are in right and bifid. Pallial sinus is relatively deep though not extending beyond the center of the shell. Margin is smooth. Shell is extremely variable in color and pattern; exterior is white, yellow or light brown, sometimes with rays, steaks, blotches, or zigzags of a darker brown and is slightly polished; and inside of shell is polished white with an orange tint, sometimes with purple over a wide area below the umbones. Maximum length is 8.0 cm.

**Biology**

**Food and feeding:** It is a suspension feeder and its diet includes phytoplankton, benthic diatoms, and terrestrial organic matter.

**Reproduction:** These shells are strictly gonochoric and their gonads are represented by a diffused tissue closely linked to the digestive system. The period of reproduction varies, according to the geographical area. Spawning usually occurs between 20 and 25°C. A period of sexual rest is observed from late autumn to early winter. Gametogenes is in the wild lasts for about 5 months. In this species, spawning occurs throughout the year, peaking once. It is a broadcast spawner. Embryos develop into free-swimming trocophore larvae. After 2–4 weeks, it develops into a peliveliger with a formed foot to assist further with swimming, as well as byssal threads to help the clam attach itself onto the seafloor once it finds a suitable substrate to settle on. Once settled, it will stay in the substrate and continue to grow into a mature clam.

**Fisheries/Aquaculture:** It is commercially harvested for food. It is also extensively cultivated in Europe, Canada, Chile, and China.

**Countries where eaten:** Hokkaido, Japan to Korea; China: southern and northern sea; Canada, Japan, France, and Mediterranean

***Saxidomus giganteus*** (Deshayes, 1839)

**Common name(s):** Money shell, Washington butter clam

**Distribution:** East and northeast Pacific: USA and Canada

**Habitat:** Low intertidal; sheltered sand, sandy mud, and gravel beaches; burrows moderately deep (to 35 cm); depth range 0–50 m

**Description:** Shell of this species is heavy and thick with a thick, protruding ligament. It is only slightly longer than it is high, and the angle formed by the umbones is more than 110°. Sculpture can be nearly smooth or have fine commarginal ribs. Periostracum, if present, is not yellow or glossy. Adductor muscle scars are of nearly equal size. It has a true hinge plate with three cardinal teeth in each valve, but no chondrophore or byssal threads. Valves have a smooth but not glossy interior with a pallial sinus and a continuous pallial line. Shell gapes slightly at the posterior end. Outside of the shell may be white or may be stained by iron sulfide in anoxic sediments. Siphons are moderately long (about 4 cm). Maximum length is 15 cm.

## Biology

**Food and feeding:** Like most bivalves, it lives a sedentary lifestyle filter-feeding and consuming a diet of phytoplankton.

**Reproduction:** It spawns in the summer. In British Columbia, about half the clams are large enough to spawn by their third year. Larvae settle from the plankton after 4 weeks. This species may live 20 years or more.

**Predators:** Predators include the sea stars *Pycnopodia helianthoides* and *Evasterias troschelii*, the moon snail *Euspira lewisii*, Dungeness crabs *Cancer magister*, and sea otters.

**Association:** Pea crabs *Pinnixa littoralis*, *Pinnixa faba*, and *Fabia subquadrata* may be found in the mantle cavity. The crystalline style of this species may contain many large spirochaete bacteria (*Cristispira* sp).

**Fisheries/Aquaculture:** It has been extensively commercially harvested, especially for clam chowder. It is also extensively cultivated in the USA.

**Countries where eaten:** It is widely harvested for food and is eaten in areas of its occurrence. Great caution should be used before eating this species due to its saxitoxin accumulation.

**Other uses:** Indians formerly used the shells of this clam for money. The butter clam is also used for a variety of purposes, ranging from archaeological research to monitoring water pollution.

*Saxidomus nuttalli* (Conrad, 1837)

**Common name(s):** California butter clam

**Distribution**: East Pacific: northern Mexico to southern Oregon; native to the west coast of North America; extending from northern California to Baja California.

**Habitat**: Intertidal to 10 m

**Description**: Shell of this species is very similar to *S. giganteus*. *S. nuttalli* typically has prominent commarginal ribbing and a more elongate shape. It also frequently has purple staining on the posterior which is very rare in *S. giganteus*. Maximum size is 15 cm.

**Biology**: No information

**Fisheries/Aquaculture:** It is a commercially exploited species.

**Countries where eaten**: Edible in the areas of its occurrence.

**Other uses:** Shells of this species were used as currency by local native peoples (Chumash peoples on the central California coast).

***Sunetta scripta*** (Linnaeus, 1758)

**Common name(s):** Yellow clam

**Distribution:** Indo-Pacific: India, Sri Lanka, Indonesia, Myanmar, Philippines

**Habitat:** Estuarine; burrowing species

**Description:** Shell of this species is wedge shaped. Umbo is posteriorly placed, and beak is slightly in front of midlines. Anterior end is rounded and posterior is truncate. Sculpture is smooth with only growth lines. Shell is variable in color pattern. Exterior is white or lilac with brown violet zigzag lines or with deep violet patches. Inner ventral margin is crenulated. Maximum length is 3.8 cm.

**Biology:** No information

**Fisheries/Aquaculture:** It is fished in India for food.

**Countries where eaten:** India

*Tapes literatus* (Linnaeus, 1758)

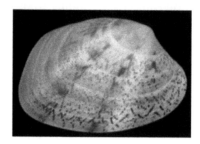

**Common name(s):** Lettered venus

**Distribution:** Indo-West Pacific: from East and Southeast Africa, to Melanesia; north to southern Japan and south to Queensland and New Caledonia.

**Habitat:** Fine sand and muddy sand near mangroves on tidal and shallow subtidal flats; depth range 0–20 m

**Description:** This species as an ovate-rhomboidal, moderately compressed, thin shell with almost parallel ventral and posterior dorsal margins. Fine concentric grooves which are strongest posteriorly are seen. Externally, it is creamy, with fine, brown, zigzag lines, and irregular blotches. Maximum length is 10.8 cm.

**Biology**: No information

**Fisheries/Aquaculture:** This species has a moderately important commercial fishery.

**Countries where eaten:** It forms an important food source in Fiji.

*Tawera elliptica* (Lamarck, 1818)

**Common name(s):** Gay's little venus

**Distribution:** Cosmopolitan; confined to southern South America

**Habitat:** Infaunal species found on sandy, muddy, gravel, and shell-covered substrates at depths of 2–80 m

**Description:** This species is characterized by an internal margin crenulated with three strong cardinal teeth in each valve and an internal reddish purple color. Maximum length is 2.8 cm.

**Biology**

**Reproduction:** In this species, the sex ratio is significantly different of 1:1. First sexual maturity occurs at 12 mm shell height in both sexes. A protracted gamete production with a principal spawning episode has been reported in this species. It is a gonochoric species with free-living larvae.

**Life span:** These clams can reach a maximum age of 15 years.

**Fisheries/Aquaculture:** This species is commercially exploited in some regions of Chile and limited artisanal fishery exists for it along the Argentinean coast

**Countries where eaten**: Chile

***Tivela mactroides*** (Born, 1778)

**Common name(s):** Trigonal tivela, trigonal clam

**Distribution:** Western Atlantic: Caribbean to Brazil

**Habitat:** Sand, from the intertidal to very shallow subtidal (2 m)

**Description:** Shell of this species is heavy, thick, inflated, and triangular. Umbones are central and prominent. Hinge is with three cardinal teeth. Smaller secondary teeth are also present. Lateral tooth is seen in left valve large. Lunula is large and escutcheon is absent. Exterior is whitish with brown tinges and rays. Maximum length is 3.4 cm.

## Biology

**Reproduction:** The species showed heterogeneous continuous gameto-genesis, with no resting period and the sex ratio did not differ significantly from 1:1. There was a high proportion of spawning in winter and spring; however, small releases of gametes may also have occurred during summer and autumn. As the reproductive cycle is heterogeneous and temporal seasonality in gametogenesis is not predictable, it is not possible to designate a closed fishing season for this population as a management strategy to avoid overexploitation.

**Fisheries/Aquaculture:** This species is commercially exploited. It is an important fishery resource for local artisanal fishermen in Brazil. A study on the population biology of this species in Caraguatatuba Bay, southeastern Brazil, revealed intense harvesting of this resource by both residents and tourists. Further, its fishery highlights key points for planning and implementing management measures, which will involve continuous monitoring of stocks, harvesting, and food safety.

**Countries where eaten:** Brazil

**Other uses:** It has also pharmaceutical values.

***Tivela stultorum*** (Mawe, 1823)

**Common name(s):** Pismo clam

**Distribution:** Eastern Pacific: Socorro Island, Mexico to Santa Cruz, California, USA

**Habitat:** Intertidal zone of relatively flat beaches exposed to open surf; depth range 7–150 m

**Description:** Shell of this species is strong, heavy, and generally smooth though sculptured with fine concentric growth lines. Beak is nearly central and ligament is elongate and is set in deep groove. Periostracum is shiny and greenish to brownish. Exterior shell is pale buff to dark chocolate, occasionally marked with brown or purple-brown bands. Maximum length is 15 cm.

**Biology**

**Food and feeding:** Due to its relatively short siphons, it typically buries itself 15 cm below the surface of the sand where it suspension feeds. Its main sources of nutrition are dinoflagellates, bacteria, zooplankton, gametes, and detritus.

**Reproduction:** Gametogenesis of this species begins in March or April and ripe gametes first appear in April and May. Gonadal development proceeds rapidly in June and July. Spawning begins in late July or early August and continues to the end of November. Discharge of gametes normally takes place when there is a rise in water temperature. After a short free-swimming larval stage, the young settle in the sand. They are at this stage protected from wave action by the presence of byssus which is anchoring them to sand grains. They grow rapidly until about November.

**Predators:** Its predators include crabs, snails, gulls, sea otters, sharks, rays, and some fishes.

**Fisheries/Aquaculture:** It is commercially exploited. The clam is relatively important for human use as a food item and fishing restrictions have been implemented to protect its populations.

**Countries where eaten:** This is the most important edible bivalve of California.

*Venerupis aspera* (Quoy & Gaimard, 1835)

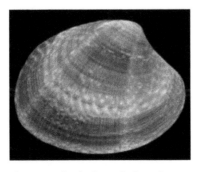

**Common name(s):** Short-necked clam, baby clam

**Distribution:** Japan, China

**Habitat:** Muddy shores

**Description:** It grows to a maximum size of 3.4 cm. No other information is available.

**Biology:** No information

**Fisheries/Aquaculture:** It is widely cultivated in China.

**Countries where eaten:** China

***Venerupis corrugata*** (Gmelin, 1791) (=*Venerupis pullastra*)

**Common name(s):** Pullet carpet shell, corrugated venus

**Distribution**: Eastern Atlantic and the Mediterranean Sea: Mauritania to Angola; eastern Atlantic Ocean

**Habitat:** Intertidal and sublittoral; buried in the sediment on the sea bed in shallow parts; depth range 0–65 m

**Description:** Shell of this species is solid, equivalve, and inequilateral. Beaks are located in the anterior half. Shells are somewhat broadly oval in outline. Ligament is inset and is not concealed. A thick brown elliptical arched body is extending almost half-way back to the posterior margin. Lunule is elongate and heart-shaped, with light and dark-brown fine radiating ridges. Escutcheon is reduced to a mere border of the posterior region of the ligament. On the surface light lines parallel to the border of the shell are seen. Growth stages are clear. Pallial sinus is relatively deep though not extending beyond the center of the shell. Margin is smooth. Exterior is gray to cream with bands of darker color. Siphons are attached along the whole length. Maximum length is 5 cm.

## Biology

**Food and feeding:** This species extends its siphons to the surface of the sediment in which it is buried and draws in water through one of them and expels it through the other. During this process, the water is passing through the gills and phytoplankton and other organic food particles (detritus) are filtered out for its nutrition.

**Reproduction:** Individuals clams are either male or female, and breeding takes place mostly in summer by the release of gametes into the water. The resulting larvae drift with the currents as part of the plankton for about 2 weeks before settling on the seabed, undergoing metamorphosis and becoming juveniles known as spat.

**Fisheries/Aquaculture:** This species is commercially harvested for human consumption in Spain and other parts of Western Europe. It is extensively cultivated in China, Korea, Japan, USA, Spain, Portugal, France, and Italy for human consumption.

**Countries where eaten:** China, Korea, Japan, USA, Spain, Portugal, France, and Italy. These shellfish are very tasty.

***Venerupis largillierti*** (Philippi, 1847) (=*Ruditapes largillierti*)

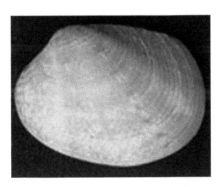

**Common name(s):** Oblong venus shell

**Distribution:** North, South, Stewart, Chathams, Auckland, and Campbell Islands

**Habitat:** Burrowing in muddy and sandy substrates in large, enclosed bays; depth range 5–20 m

**Description:** This species has a moderately large shell which is elongated and subrectangular, thick and solid, with smooth ventral margin.

Umbones are located at anterior fifth of length. Dorsal and ventral margins are curved and subparallel. Anterior and posterior margins are nearly straight to lightly convex, converging toward dorsum. External sculpture is of many low, wide, almost smooth, rather irregular, weakly anastomosing, commarginal costae. Lunule is concave, almost smooth, long and narrow, and weakly differentiated. Hinge is with three narrow, cardinal teeth in each valve; left posterior one is lamellar; right anterior one is low, lamellar; and the other two in each valve are medially grooved. Pallial sinus is deep and oval. *Size*: length 6.5 cm, width 7.0 cm, and height 5.5 cm.

**Biology:** Fertilized eggs develop into trochophore larvae by 24 h at 20°C and D veligers appear within 48 h. Development to pediveliger stage takes between 11 and 16 days at 20°C, and metamorphosis to spat in 16 and 19 days.

**Fisheries/Aquaculture:** The aquaculture potential of this subtidal venerid clam warrants further investigation.

**Countries where eaten:** This species is eaten in the areas of its occurrence. It has limited use to people and the seafood industry.

***Venus casina*** (Linnaeus, 1758)

**Common name(s):** Ridged venus

**Distribution:** Eastern Atlantic and the Mediterranean: Norway to Senegal and Benin

**Habitat:** Intertidal; variety of substrates from muddy sand to muddy gravels; depth range 5–40 m

**Description:** Shell of this species is solid, equivalve, and inequilateral. Beaks are present in front of the midline. Approximately, shell is circular in outline, with a slight truncation on the posterior margin below the escutcheon. Ligament is extending almost half-way to the posterior margin. Lunule is well defined, heart-shaped, and light brown with very fine radiating ridges. Shell sculpture is of prominent concentric ridges interspersed with small ridges, and between the ridges and troughs there are fine concentric striae. Growth stages are not clear. There are three cardinal teeth in each valve. Pallial sinus is small and triangular. Inner margins of the valves are crenulate except below the escutcheon. Exterior of the shell is dirty white to pale fawn in color, sometimes with red-brown rays. Interior of shell is white. Periostracum is rich chestnut-brown. Maximum shell length is 5.1 cm.

**Biology:** It is a suspension feeder.

**Fisheries/Aquaculture:** No information

**Countries where eaten**: Mediterranean

***Venus nux*** (Gmelin, 1791)

**Common name(s):** Saltwater clam

**Distribution:** Mediterranean Sea and the Atlantic Portugal and Spain

**Habitat:** Circalittoral and bathyal soft bottoms; depth range 40–800 m

**Description:** The shells are globular, with dense, concentric ribs on the surface. Maximum shell length is 4.5 cm.

**Biology:** No information

**Fisheries/Aquaculture:** It is commercially harvested and is the most abundant mollusc caught in the fish trawl (42%) in southern Portugal.

**Countries where eaten:** Spain

*Venus verrucosa* (Linnaeus, 1758)

**Common name(s):** Warty venus

**Distribution**: Eastern Atlantic and the Mediterranean

**Habitat:** Burrows in mud and sand; subtidally to 155 m

**Description:** Shell of this species is oval, tumid, and rugose. Sculpture is of numerous prominent, concentric ridges. Growth stages are indistinct. Lunule is deeply impressed, oval, with fine striations, and deep brown or chestnut. Escutcheon is elongate and is unequally developed. There are three cardinal teeth in each valve, with a small lateral tooth in front of anterior cardinal of left valve, and a corresponding pit in right. Posterior adductor scar is often tinted pinkish brown, chestnut, or light purple. Pallial line is faint, and sinus small and triangular, extending no further forward than inner edge of posterior adductor scar. Inner shell margin is very finely crenulate. Exterior is off-white to light brown and is darkest about umbones and anteriodorsal margin. Periostracum is deeper brown. Inner surfaces are white and very glossy. Maximum shell dia. is 6 cm and length is 7 cm.

**Biology:** No information

**Fisheries/Aquaculture:** It is commercially harvested.

**Countries where eaten:** Western Adriatic, Western France, Montenegro Coast, Yugoslavia, British Isles, Angola, and South Africa. It is an expensive delicacy in France. It is eaten either raw or baked in the oven with garlic butter.

## 3.9.2   COCKLES (CARDIIDAE)

The family Cardiidae is one of the largest and best known of bivalves. The cockle shell is strong, compact, and heart-shaped. Its siphons which bear light receptors are short, and the foot is also well developed. Shell features are completely symmetrical and equal sized valves; prominent umbones; strong radial ribs, which in some species carry spines; equal size muscle scars; no pallial sinus; and two cardinal teeth in each valve.

*Acanthocardia aculeata* (Linnaeus, 1758)

**Common name(s):** European spiny cockle

**Distribution:** Mediterranean Sea and in North East Atlantic

**Habitat:** Buried in sandy bottoms, from 10 m deep.

**Description:** Shell of this species is robust, broadly oval, with a heart-shaped profile, equivalve and inflated, with crenulated margins. Surface shows 20–22 prominent radial ribs, with rows of sharp spines, especially at sides. Hinge is with two cardinal teeth. Right valve is with two anterior teeth and one rear, and left with only one in every position. Basic coloration is usually pale brown and interior is white. Maximum size is 11.5 cm.

**Biology:** These animals are phytoplankton feeders.

**Fisheries/Aquaculture:** It is commercially harvested

**Countries where eaten:** This species is eaten in the south from Norway to Morocco; Mediterranean and Italy.

## *Acanthocardia echinata* (Linnaeus, 1758)

**Common name(s):** European prickly cockle

**Distribution:** Eastern Atlantic and the Mediterranean Sea: from Iceland, Norway, Sweden, south to Morocco, Madeira Island and Canary Islands, and the Mediterranean.

**Habitat:** Fine sand; depth range 0–619 m

**Description**: Shell of this species is brittle and obliquely oval. Anterior hinge line is sloping gently to convex anterior margin and posterior hinge line is more steeply inclined. Posterior margin is only slightly convex. There are 18–23 ribs, each with a sharp central keel and is regularly produced into erect, sharp spines. Concentric sculptures are of numerous wavy ridges which are pronounced between ribs. Growth stages are clear. There are two cardinal teeth in each valve; right valve is with two anterior and one posterior lateral teeth and left valve is with single anterior and posterior laterals, the anterior being longer, thicker, and more prominent than posterior one. External sculpture is visible on inner surfaces as grooves which extend the whole depth of shell. Adductor scars and pallial line are indistinct. Maximum length is 7.5 cm.

**Biology:** This is a sessile-burrower and suspension feeder. It feeds on phytoplankton and suspended organic materials.

**Fisheries/Aquaculture:** This species is commercially harvested.

**Countries where eaten**: North from Norway and Iceland to Morocco and the Canary Islands; and Mediterranean and France.

*Acanthocardia paucicostata* (Sowerby, 1839)

**Common name(s):** Poorly ribbed cockle

**Distribution**: Northeast Atlantic and the Mediterranean

**Habitat:** Beaches; depth range 0–25 m

**Description**: Shell of this species is very light, with very few extended radial ribs (16) and separated by wide channels. Exterior of the shell has tangential brown bands on a lighter background, and with the brown uniform height. Maximum shell length is 3.6 cm.

**Biology:** No information

**Fisheries/Aquaculture:** No report

**Countries where eaten**: Black Sea, Bay of Biscay in Morocco and Mediterranean

*Acanthocardia spinosa* (Solander, 1786)

**Common name(s):** Sand cockle

**Distribution:** Mediterranean Sea

**Habitat:** Sand and mud, from low waters to 120 m.

**Description:** Shell of this species is robust, round with a heart-shaped profile, equivalve and inflated, with crenulated margins. Surface shows thick narrowly spaced radial ribs, with rows of pronounced thorny hooks. Basic external coloration is usually pale brown and interior is white. Maximum length is 9.5 cm.

**Biology:** Like almost all bivalves, these molluscs are phytoplankton feeders.

**Fisheries/Aquaculture:** No information

**Countries where eaten:** Mediterranean and Italy

*Acanthocardia tuberculata* (Linnaeus, 1758)

**Common name(s):** Tuberculate cockle, Moroccan cockle, knotted cockle, rough cockle

**Distribution**: Northeast Atlantic and the Mediterranean

**Habitat:** Sandy and gravel substrates; depth range 0–100 m

**Description:** Shell of this species is thick and strong and is approximately rhombic in shape. Hinge line is sloping gently from umbones on each side. Posterior margin is scarcely convex and almost straight. Anterior margin is more strongly convex. Sculpture is of 18–20 bold ribs and fine concentric grooves and ridges. Dorsally, each rib has a central keel, bearing short pointed spines; and ventrally the keel is obscured and spines appear separate. Growth stages are clear. There are two cardinal teeth in each valve; right valve is with two anterior and one posterior lateral teeth; and left valve is with single anterior and posterior laterals. On the inner surface, the pallial line and adductor scars are distinct. External sculpture is visible as grooves extending from ventral margin to pallial line, fading rapidly beyond it and smooth beneath umbones. Coloration of the shell is off-white, yellow, or light brown, often in concentric bands of different shades, and frequently darker about umbones. Periostracum is thin and yellowish. Inner surface is white and glossy ventrally. Maximum shell length is 9 cm.

**Biology:** The animal has long gills, like siphons. The foot is big, long, and cylindrical. Mantle edge is pointed and like siphons presents a dotted corresponding to photoreceptor bodies. It is an active suspension-feeder and deposit-feeder.

**Fisheries/Aquaculture:** It is commercially harvested

**Countries where eaten**: From Great Britain to Morocco and the Canary Islands; Mediterranean and France. It is a Croatian cuisine.

***Clinocardium nuttallii*** (Conrad, 1837) (=*Cardium corbis*)

**Common name(s):** Pacific cockle, Nuttall's cockle, basket cockle, heart cockle

**Distribution:** Northeast Pacific and Caribbean Sea; Alaska to California; Japan to Venezuela

**Habitat:** Intertidal zone; sheltered areas; sand-gravel substrate; eelgrass beds; depth range 0–50 m

**Description:** This species has a thick shell and is just as high or slightly higher (dorsal to ventral distance) than long (anterior to posterior end). It has about 30 distinct radial ribs covering the entire valve. Its profile from the side is heart shaped. Its siphons are short. When buried in the sand, the siphon edges appear white with white hairs radiating from their tips. Color of shell is whitish and is irregularly spotted or stained with yellowish-brown, orange, or pale purple. Maximum shell length is 14 cm.

**Biology**

**Growth:** Growth rings may be prominent, especially in the northern parts of its range, as this species nearly ceases feeding in winter. Yearly growth lines are much less prominent farther south, but tidal cycle growth lines can often be seen. Mantle margin has tiny tentacles with tiny eyes.

**Reproduction:** This species is a simultaneous hermaphrodite. In Puget Sound, these animals mature in their second year and spawn in July and August. It lives for 15–19 years in Alaska.

**Predators:** Predators of this species include *Pycnopodia helianthoides*, *Pisaster brevispinus*, *Cancer magister*, and gulls. It has a strong escape response to Pycnopodia—rapidly extending its foot and jumping away.

**PSP:** This species may be a source of paralytic shellfish poisoning (PSP) for humans.

**Association:** This species may contain small pea crabs such as *Pinnixa faba* inside its mantle cavity.

**Fisheries/Aquaculture:** It is commercially harvested and is a potential aquaculture species.

**Countries where eaten:** It is used by the indigenous peoples of California and the Pacific Northwest as food.

***Dallocardia quadragenaria*** (Conrad, 1837) (=*Cardium quadragenarium*)

**Common name(s):** Spiny pricklycockle, forty-ribbed cockle

**Distribution:** From Santa Barbara southward.

**Habitat:** It is an off shore (deep water) species.

**Description:** This species is characterized by its large size and spiny ribs. Ribs are usually more than 40. When it is older, newer parts are set with yellow teeth or short horns. Strong crenulations are seen on the edges of valves which are also yellow.

**Biology:** No information

**Fisheries/Aquaculture:** Not reported

**Countries where eaten:** Though this species is of large size, it is too seldom taken to be of economic importance and is included here merely for the sake of completeness.

***Dinocardium robustum*** (Lightfoot, 1786)

**Common name(s):** Atlantic giant cockle, Vanhyning's heart cockle

**Distribution:** Western Central Atlantic: USA and Mexico

**Habitat:** Shallow waters along the coast including beaches, bays, and estuaries

**Description:** Shell of this species is very large for family, inflated, and is obliquely ovate. Sculpture is of about 32–36 rounded, smooth radial ribs. Pallial line is simple. Margins are crenulated. Umbones are rounded. *Color*: pale tan to yellowish brown, mottled irregularly with red-brown. Posterior slope is mahogany brown. Interior is salmon pink. Maximum length is 11.9 cm.

**Biology:** Not known

**Fisheries/Aquaculture:** It is commercially harvested.

**Countries where eaten:** It is edible in areas of its occurrence.

*Laevicardium crassum* (Gmelin, 1791)

**Common name(s):** Norway cockle, Norwegian egg cockle, smooth cockle

**Distribution:** Northeast Atlantic and the Mediterranean

**Habitat:** Sublittoral zone; substratum with generally coarse sand and gravel, often overlain with pebbles, cobbles and dead shell, depth range 9–200 m

**Description:** Shape of the shell of this species is oval and elongated. It has a white-to-light-yellow shell exterior with occasional dark markings. Surface of shell is smooth. Valves are with 40–50 nonprojecting or faint ribs around the center. Shell margin is crenulated. Exterior is cream, light yellow to fawn, with blotches of pink, chestnut, or brown. Periostracum is greenish, thin. Inner surfaces are glossy white tinted with pink and are darkest close to margin. Maximum length is 7.5 cm.

**Biology**

**Feeding:** It is a suspension feeder

**Behavior:** It is a sessile-burrower. If it wants to escape, it is just like an underwater kangaroo. It can jump large distances on the sea floor by pressing out water.

**Fisheries/Aquaculture:** Not reported

**Countries where eaten**: Black Sea

*Laevicardium elatum* (Sowerby, 1833) (=*Cardium elatum*)

**Common name(s):** Giant egg cockle, giant Pacific cockle, yellow cardinal cockle

**Distribution:** East Pacific

**Habitat:** Shallow water in mud; intertidal to 6 m deep

**Description:** Shell of this species is ovate to trigonal and its height is greater than length. There are 45–48 low, smooth radial ribs. Anterior and ventral margins are finely crenulated within. Posterior margin is smooth within. Periostracum is thick, dehiscent, tan to dark brown, and silky. Exterior color is yellow to brown and interior is white. Maximum length is 19 cm.

**Biology**

**Reproduction:** It is a functional hermaphrodite with male and female follicles intermingled in the gonads. The development and spawning of male and female gametes are synchronous and spawning is normally related to high food availability.

**Fisheries/Aquaculture:** No information

**Countries where eaten**: It is edible in areas of its occurrence.

## *Laevicardium oblongum* (Gmelin, 1791)

**Common name(s):** Oblong egg cockle

**Distribution:** Mediterranean Sea: Greece

**Habitat:** Sandy or muddy bottoms ; depth range 3–400 m.

**Description:** Shell of this species has an oval shape and is elongated. There are numerous ribs which are thin, flattened, and more pronounced toward the ventral margin. Umbonal region is even smoother. Color of shell exterior is uniform light brown, with some mottling more accentuated toward the apex. Interior of the valve is white porcellaneous. Periostracum is thin shiny and brown. Maximum length is 7 cm.

**Biology:** No information

**Fisheries/Aquaculture:** No information

**Countries where eaten:** Black Sea, Straits of Gibraltar, and Mediterranean

## *Monodacna colorata* (Eichwald, 1829)

**Common name(s):** Colored egg cockle, colored lagoon cockle

**Distribution:** Northeast Atlantic and the Black Sea

**Habitat:** Brackish-water in the lagoons and the mouth of the rivers

**Description:** This species has oval shell with rounded dorsal edge and evenly rounded ventral edge. There are up to 28 wide ribs which are divided by narrow spaces. Shell is red or brownish-yellow in color. *Shell size*: length 2.7 cm and dia. 3.0 cm.

**Biology**

**Ecology:** It is well adapted to seawater with salinities of 1–10 ppt. It best develops at 5 ppt, while older life stages dwell and reproduce at 8–9 ppt.

**Food and predators:** This species filter feeds on planktonic green algae, diatoms, and detritus and is preyed upon by bottom dwelling fish.

**Fisheries/Aquaculture:** No information

**Countries where eaten:** It is edible in areas of its occurrence.

*Cerastoderma edule* (Linnaeus, 1758)

**Common name(s):** Common cockle

**Distribution:** Northeast Atlantic and the Mediterranean

**Habitat:** Brackish; shallow coastal and estuary areas; subtidal zone; depth range 0–50 m

**Description:** Shell of this species is broadly oval. There are 24 broad ribs and closely spaced concentric ridges. Ridges on ribs may be developed as flattened, scale-like spines. Growth stages are distinct. Each valve is with two small cardinal teeth; right valve is with two anterior and two posterior laterals. Adductor scars and pallial line are distinct. In larger specimens, posterior adductor and areas around it may be tinted chestnut or light purple. External sculpture is visible as grooves extending from ventral margin, fading rapidly beyond pallial line. Color of exterior shell is cream, light yellowish, or pale brown. Periostracum is yellowish or greenish. Inner surfaces are dull white, brown and thin. Maximum length is 10 cm.

**Biology:** It moves by spitting water.

**Fisheries/Aquaculture:** It is an important species for the fishing industry and is commercially fished in the United Kingdom, Ireland, and France. Previously, the greatest catch was from the Netherlands, but now fisheries restrictions have been put in place due to environmental concerns. This edible saltwater clam is extensively cultivated in Philippines, Bangladesh, United Kingdom, the Netherlands, and Portugal.

**Countries where eaten:** This cockle is cooked and eaten in several countries (including the United Kingdom, France, Germany, Ireland, Japan, Portugal, and Spain).

***Cerastoderma glaucum*** (Bruguière, 1789) (=*Cerastoderma lamarcki*)

**Common name(s):** Lagoon cockle

**Distribution:** Coasts of Europe and North Africa, including the Mediterranean and Black Seas and the Caspian Lake, and the low-salinity Baltic Sea.

**Habitat:** Shallow burrows in saline lagoons, or sometimes on lower shores in estuaries.

**Description:** Shell of this species is rather thin and brittle. It is markedly eccentric and its anterior margin is shallowly convex, forming sharp junction with hinge line. Posterior margin is continuous with hinge line, sloping steeply from umbones. There are 24 ribs, with closely spaced concentric corrugations, forming small, flattened scales on ribs. Growth stages are distinct. Each valve is with two small cardinal teeth; right valve is with two anterior and two posterior laterals, and left valve is with single anterior and posterior laterals. On inner surfaces ribs are visible as distinct grooves throughout whole of shell, extending beneath umbones. Posterior margin is not crenulated. Shell is off-white, yellowish, or greenish, with darker brown areas, overlain by a greenish-brown periostracum. Inner surfaces are dull white or brown. Maximum length is 5 cm.

## Biology

**Ecology:** It is a euryhaline species living in salinities of 4–100‰. It is a substrate-specialist slow-burrowing species and it shows a clear preference for the sediment where it occurs naturally.

**Reproduction:** It spawns in May–July, and the planktonic larval phase takes 11–30 days. The life span of the settled cockle is typically 2–5 years.

**Fisheries/Aquaculture:** It is commercially harvested.

**Countries where eaten:** All Europe, Spain, and Black Sea

**Other uses:** Its cultivation also helps as a food source for fish and/or as an indicator of the environmental conditions.

*Fragum unedo* (Linnaeus, 1758)

**Common name(s):** Strawberry cockle

**Distribution:** Indo-West Pacific

**Habitat:** Shallow areas within the lagoon

**Description:** This species is characterized by prominent red scales on the ribs. Shell is solid and inequilateral. Umbonal keel is rounded and is not sharply angulate. Posterodorsal wing is fairly broad. Length of posterodorsal and anterodorsal margins is approximately equal. Posterior and ventral margins meet almost at right angles. Sculpture is of strong radial ribs. Mean number of ribs 27, and it ranges from 23 to 31. Ribs are broad, flat with irregular, transverse red scales. Lunule is narrow and smooth. Dorsal margin in lunule is raised. Posterior lateral teeth are only slightly closer to cardinals than anterior laterals. Exterior of shell is white with dark red scales, sometimes with pale brown blotches. Interior is white. Measurements of largest shell: length 5 cm, height 6 cm, and width 5 cm.

**Biology:** No information

**Fisheries/Aquaculture:** Not reported

**Countries where eaten**: It is consumed in Indonesia.

*Fulvia mutica* (Reeve, 1844) (=*Cardium annae*)

**Common name(s):** Egg cockle, Japanese cockle

**Distribution:** Northwest Pacific; Mutsu Bay Kyushu; Korean Peninsula; China

**Habitat:** Sandy mud; depth range 10–50 m

**Description**: Shell of this species is subcircular and inflated. It is almost equilateral, thin, white and is profusely marked with pale flesh-pink zigzag streaks. Beaks are fleshy-brown. Sculpture is of fine radial riblets. Anteriorly, there are 16 very regular riblets. Posteriorly riblets are coarser and less regular. Right valve has two contiguous cardinal teeth. In left valve, cardinals diverge. Common length is 2.7 cm and dia. is 1.6 cm.

**Biology**

**Reproduction:** It is a hermaphrodite species. Seasonal variations in the concentrations of estradiol-17β (E2) and testosterone (T) have been investigated in the gonad of this species. Both steroids exhibited a similar annual pattern that correlated with the reproductive cycle. These results suggest that E2 and T could play a role as endogenous modulators in the reproductive cycle of this species (Liu et al., 2008).

**Microbe:** A novel gram-staining-negative, nonspore-forming, nonflagellated, nonmotile, aerobic, saffron-colored, rod-shaped bacterium, *Bizionia fulviae* sp. nov. has been isolated from the intestinal tract of this species (Kim et al., 2013). The ecological role and physiological potential of this novel strain are yet to be investigated.

**Fisheries/Aquaculture:** It is a potential species for aquaculture in United Kingdom, Netherlands, and France.

**Countries where eaten:** It is edible in areas of its occurrence

*Hypanis plicata* (Eichwald, 1829)

**Common name(s):** Folded lagoon cockle

**Distribution:** Black Sea

**Habitat:** Lagoons and brackish water lake; soft ground or in dense silt

**Description:** Shell of this species is turtle oval-rectangular, thick-walled and is with a few prominent peaks, located closer to the front edge. Its surface is covered with about 40 narrow, triangular, sharp edges. Intercostal spaces are wide. Castle area is narrow and sometimes on the right wing, there is a rudimentary cardinal tooth. Mantle sinus is narrow and it reaches almost to the middle of the length of the shell. Color of shell is painting white. Shell length, 3.8 cm; height 2.5 cm; and width 1.6 cm.

**Biology:** No information

**Fisheries/Aquaculture:** Not reported

**Countries where eaten:** Black Sea

### 3.9.3   BEAN CLAMS OR WEDGE SHELLS (DONACIDAE)

The Donacidae are prolific filter feeders and are an important part of coastal food chains. The animals of this family are sensitive to coastal industries such as dam-building and dredging. Members of this family have asymmetric, elongated, compressed shells. The two siphons are short, but are completely divided and the foot is large. They are vigorous burrowers.

***Donax cuneatus*** (Linnaeus, 1758)

**Common name(s):** Bean clam, cuneate donax, wedge clam, cuneate wedge shell

**Distribution:** Indo-West Pacific

**Habitat:** Intertidal; shallowly buried in sand of exposed sandy beaches; migrating up and down between the high and low tide marks

**Description:** Shell of this species is solid, compressed, and distinctly inequilateral and is trigonal-ovate in outline. Shell surface is smooth with fine concentric lines. Radial sculpture is distinct at posterior end, forming reticulate scale-like sculpture. Color of shell is variable. Exterior is white, cream, brown, and gray and is usually with radial bands of varying thickness and prominence. Interior is white to purple and is often with tinted yellow, brown, or purple blotches. Maximum shell length is 4 cm.

**Biology:** Not reported

**Fisheries/Aquaculture:** This species is known for its local artisanal exploitation. It is a potential species for aquaculture in Canada and USA.

**Countries where eaten:** It is edible in areas of its occurrence.

*Donax deltoides* (Lamarck, 1818)

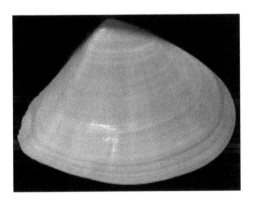

**Common name(s):** Goolwa cockle, Coorong cockle, surf clam, common pipi

**Distribution:** Indo-West Pacific: from Andaman and Nicobar Islands to Indonesia and throughout Australia

**Habitat:** Intertidal and shallow subtidal waters; surf beaches

**Description:** Shell surface is smooth. Sculpture is of fine concentric striae with stronger radial striae posteriorly. Color of shell is variable. Exterior is white, green, brown, purple, mauve, yellow, or orange. Interior color is equally varia ble. Maximum shell length is 8 cm.

**Biology**

**Behavior:** They make shallow burrows, just below the surface of the sand. They can quickly right themselves if washed around by the surf.

**Feeding:** They are suspension feeders and feed by filtering phytoplankton from the water. They need heavy surf to live, as the surf concentrates the phytoplankton they feed on and increases the oxygen in the water.

**Reproduction:** They mature at around 1 year of age and live to about 5 years. They are dioecious serial broadcast spawners, with spawning taking place over a long period of time peaking in the spring. Larvae drift as plankton for about 8 weeks.

**Fisheries/Aquaculture:** It is commercially harvested. This species has been considered as a potential species for aquaculture.

**Countries where eaten:** This species is eaten as soups and chowders in Australia. It is also sold as bait for recreational fishers.

*Donax faba* (Gmelin, 1791)

**Common name(s):** Pacific bean donax

**Distribution:** Indo-West Pacific: New Caledonia; north to southern Japan and south to New South Wales

**Habitat:** Intertidal areas; sandy beaches, in all but the most sheltered areas, often in dense populations; most common in wave-beaten areas

**Description:** Shell of this species is thick, compressed, inequilateral, and trigonal-ovate in outline. Shell surface is smooth with fine concentric lines that are usually more pronounced and appear as ridges on the posterior side. Color of shell is variable. The species is highly polymorphic for shell color; 14 morphs are present. Highly polymorphic for shell color and 14 morphs have been reported. Exterior is white, cream, brown, green, gray, purple; often with one or more radial bands and random maculations. Interior is white, often with tinted yellow or with purplish blotches, to purple, usually with whitish blotches and/or radial bands. Maximum length is 2.5 cm.

**Biology:** It is migrating up and down the beach with the tide and is frequently preyed upon by crabs and birds during migration.

**Fisheries/Aquaculture:** It is a potential species for aquaculture in Myanmar.

**Countries where eaten:** Mauritius

***Donax fossor*** (Say, 1822)

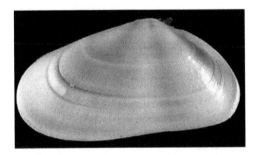

**Common name(s):** Northern coquina, fossor coquina

**Distribution:** North America

**Habitat:** Exposed beaches; depth range 0–1 m

**Description:** This small clam can be almost white in color, or yellow, orange, grayish, or purplish. Darker shells may be rayed with purple on the inside, but these rays are usually only rather faintly visible on the outside. Maximum shell length is 1.8 cm.

**Biology**

**Association:** This species often has the hydroid *Lovenella gracilis* growing on the posterior tip of the shell. The *Donax–Lovenella* relationship is a candidate for a case of mutualism because of the reduced predation rates by moon snails on clams with a hydroid colony.

**Fisheries/Aquaculture:** No information

**Countries where eaten:** USA

## *Donax semistriatus* (Poli, 1795)

**Common name(s):** Half-striated donax

**Distribution:** Northeast Atlantic and the Mediterranean

**Habitat:** Shallow water in the sandy beaches fine well sorted clean sand; depth range 0–25 m

**Description:** Valves of this species have a streak end while the front part is completely smooth and is covered with a thin, shiny periostracum. It is more brightly colored. Maximum shell length is 4 cm

**Biology:** Not reported

**Fisheries/Aquaculture:** No information

**Countries where eaten:** Black Sea and south of Portugal in Morocco

## *Donax trunculus* (Linnaeus, 1758)

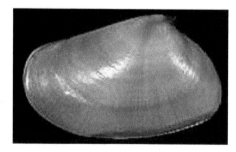

**Common name(s):** Abrupt wedge shell, wedge clam, coquina clam, donax clam

**Distribution**: Western Europe and north-western Africa

**Habitat:** Shallowest two meters along the coastline; forming colonies

**Description:** Shell of this species is bright, oblique shaped, very thin and is slightly flattened with a smooth surface. If examined microscopically shallow concentric grooves are seen on the surface. Front is rounded and the back is shorter. Anterior dorsal margin is straight and inclined. Ventral margin is slightly convex. It is light colored and covered by a bright brown periostracum. Exterior is yellowish or brownish-purple and whitish inside. Maximum length is 5 cm.

**Biology**

**Food and feeding**: It is a suspension feeder, feeding on suspended particles in seawater.

**Reproduction:** The sexual dimorphism of this species is largely gonadal and both sexes are equally represented in populations.

**Predators:** It is a sought after prey by carnivorous gastropods

**Fisheries/Aquaculture:** It is commercially harvested for food and is a potential species for aquaculture in India.

**Countries where eaten:** India and in other areas of its occurrence, namely, Black Sea, France, and Mediterranean. This species is appreciated for its delicate taste and tender meat. It may be used in several recipes especially in first courses. Like clams, these molluscs are a low source of calorie, fat and omega-3 fatty acids. The protein content is not high. They are a good source of minerals also.

**Donax variabilis** (Say, 1822)

**Common name(s):** Variable coquina

**Distribution**: Warm water species; Western Atlantic: North America, from New York to Texas and Mexico, throughout the Gulf of Mexico; the Sea of Marmara and the Mediterranean

**Habitat:** Brackish-water; intertidal zone of sandy beaches depth range 0–11 m

**Description:** Coquinas have small, long, triangular-shaped shells. Exterior of shell can have any one of a wide range of possible colors, from almost white, through yellow, pink, orange, red, purple, to brownish and blueish, with or without the presence of darker rays. Maximum length is 1.9 cm.

**Biology**

**Behavior:** Coquinas typically live in close proximity of each other, sometimes in colonies. These animals migrate up and down the beach with the tides. They actively migrates by jumping out of the sand and riding specific waves.

**Food and feeding:** Coquinas are filter feeders, feeding primarily on phytoplankton, algae, detritus, bacteria, and other small particles suspended in the surf as the waves ebb and flow. Feeding is performed through their short siphons.

**Reproduction:** Coquina clams are dioecious (male and female) broadcast spawners. Eggs and sperm are released synchronously into the water for external fertilization.

**Fisheries/Aquaculture:** This species has subsistence fisheries.

**Countries where eaten:** The coquina is edible in USA and is used to make broth.

**Other uses:** The colorful shells of this species are used for crafts.

***Donax variegatus*** (Gmelin, 1791)

**Common name(s):** Smooth donax

**Distribution:** Atlantic from the British Isles to Morocco; Mediterranean and Black Sea

**Habitat:** Intertidal; depth range 25–25 m.

**Description:** Shell of this species is thin, solid, equivalve, and inequilateral. Beaks are behind midline. Shell outline is narrowly ovate/elongate, and wedge-shaped. Posterior region is half the length of the anterior region. Posterior margin is more narrowly rounded than anterior. Sculpture is of concentric growth lines and fine lines. Margin is finely serrated but smooth to the touch. Hinge has two cardinals in each valve. Pallial sinus is deep and extends just past the midline. It is confluent with the pallial line for some of its length. Cruciform muscle scars are not visible. Periostracum is thin and satin. Color of exterior shell is white/pale yellow/olive green sometimes with pale cream splotches. Maximum length is 4 cm.

**Biology:** It is an active suspension feeder.

**Fisheries/Aquaculture:** No information

**Countries where eaten**: Black Sea

*Donax venustus* (Poli, 1795)

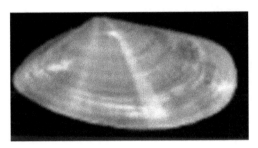

**Common name(s):** Lovely donax

**Distribution:** Eastern Atlantic and the Mediterranean

**Habitat**: Brackish water; commonly found along beaches; fine well sorted sand and in areas influenced by outflows from estuaries; depth range 7–9 m

**Description:** This species has a smooth and shiny shell, except on the back where concentric grooves are appreciated. Posterior ventral margin has crenulated area. Common length of the shell is 3 cm.

**Biology:** It is an active suspension feeder.

**Fisheries/Aquaculture:** No information

**Countries where eaten**: Black Sea

*Iphigenia laevigata* (Gmelin, 1791)

**Common name(s):** Smooth false donax

**Distribution:** Eastern Central Atlantic: Mauritania to Congo; Florida and Southern California

**Habitat:** Offshore in sand; exposed sandy beaches; mid-tide down to low water at the surface or only slightly buried.

**Description:** This is the smallest of the economic species. Valves are heavy and strong, deeply arched and marked by indistinct radiating lines and concentric growth lines of varying degrees of distinctness. Posterior or siphonate end is abruptly truncated or cut off, forming nearly a right angle at the umbos which thus come to lie near the siphonate end of the shell. Margins of the valves are heavily crenulated. Color varies from white to blue and purple and is often conspicuously striped. Common length is 7 cm.

**Biology:** No report

**Fisheries/Aquaculture:** No information

**Countries where eaten:** It is used entire in the making of soup; and its flavor is said to be excellent. It is eaten in areas of its occurrence.


### 3.9.4   GIANT CLAMS (TRIDACNIDAE)

True giants among the clams of this family weigh in at 150 kg and 1 m in length. These large molluscs are highly specialized in both their structure and nutrition. The shell remains attached by a large byssus that, contrary to other clams, appears to emerge from a gap on the dorsal side, near the hinge. The hinge and umbones have migrated 180° to the ventral position with respect to the internal organs. One adductor muscle has disappeared, and the mantle—actually, enlarged fleshy siphons—fills the entire fluted opening of the shell. These tridacnids harbor the zooxanthellae and thus they play a key role in coral reef formation by providing food for many invertebrate organisms.

***Hippopus hippopus*** (Linnaeus, 1758)

**Common name(s):** Bear paw clam, horse's hoof clam, strawberry clam

**Distribution**: Indo-West Pacific

**Habitat:** Intertidal areas on corals; ocean floor in sandy and rubble areas; seaweed beds depth range 0–30 m

**Description:** Valves of this species are thick, heavy, and triangular in shape, often covered with reddish spots and obscured by encrustations. Mantle is a deep yellow-green, irregularly mottled at the periphery and in the center. Maximum length is 40 cm.

**Biology**

**Nutrition:** Tridacnids derive their nutrition from uptake of dissolved matter through their epidermis and from their symbiotic zooxanthella, *Symbiodinium microadriatiacum*.

**Reproduction:** This clam reaches maturity at about 3–5 years old, when it measures about 15 cm. Remarkably, all horse's hoof clams spend the first 2–3 years of life as males, but then develop reproductive organs that enable them to produce both sperm and eggs. This clam spawns in June, releasing sperm and then eggs into the surrounding waters, where fertilization takes place. The fertilized eggs quickly develop into a swimming larval stage before settling on the substrate, where they develop into the adult clams and are unable to move from their position on the coral reef.

**Association:** This clam has a special relationship with photosynthetic algae, which live inside the clam's fleshy mantle tissue and provide it with the majority of its nutrition. The algae use sunlight, like plants, to produce sugars, which are then released into the bloodstream of the clam. In return, the algae may receive some nutrition from the clam's waste products.

**Fisheries/Aquaculture:** It is commercially harvested. This species is extensively cultivated in Portugal.

**Countries where eaten:** Philippines, Cook Islands, and Indo-Pacific region.

**Other uses:** Shells of this species are collected for the shell craft industry. This hardy species is prized for its very colorful and decorative shell. It is used in shell craft industry and as an aquarium species.

*Hippopus porcellanus* (Rosewater, 1982)

**Common name(s):** China clam

**Distribution:** Western Central Pacific: Restricted in the tropical western Pacific.

**Habitat:** Shallow waters

**Description:** The name China clam of this species is largely due to its porcelain like appearance. It is the most rare of the giant clam species. Its shell is thinner and smoother than that of *Hippopus hippopus*. It is usually devoid of pigmentation and is more semicircular in profile. Mantle is similar to that of *H. hippopus*, except that prominent papillae line the margins of the incurrent siphon. Inhalant siphon of this species has large, branching tentacles. Maximum length is 40 cm.

**Biology**

**Ecology:** Giant clams are also quite sensitive to chemicals or toxic substances dissolved in the water. High pH, high salinity, and high temperatures can also cause problems in both wild and aquarium environments.

**Behavior:** Young specimens often byssally attached to coral heads and mature specimens lack a byssus and lay unattached on the substrate.

**Reproduction:** These animals are broadcast spawners. Embryos develop into free-swimming trochophore larvae, succeeded by the bivalve veliger, resembling a miniature clam.

**Fisheries/Aquaculture:** It is commercially harvested and is a potential species for aquaculture

**Countries where eaten:** Areas of its occurrence, especially in Philippines.

**Other uses:** Shells of this species are collected for the shell craft industry. This hardy species is prized for its very colorful and decorative shell. It is used in shell craft industry and as an aquarium species.

*Tridacna crocea* (Lamarck, 1819)

**Common name(s):** Boring clam, crocus clam, crocea clam, saffron-colored clam

**Distribution:** Indo-West Pacific: from Andaman Islands to Fiji Islands; north to Japan and south to New Caledonia and Queensland.

**Habitat:** Intertidal areas on corals; depth range 0–20 m.

**Description:** Valves of this species are grayish white, often fringed with orange or yellow both inside and out. They are triangularly ovate in shape. Mantle coloration is predominantly blue but shows great variability. Maximum length is 15 cm.

**Biology**

**Polymorphism:** In this species, eight morphs have been identified, but there also existed considerable intramorph variation. Morphs with brown and/or and green pigments dominate the larger clam size category (>8 cm).

**Pigment:** The photosynthetic pigment, chlorophyll $c$, has been isolated from the associated zooxanthellae of this species.

**Fisheries/Aquaculture:** This species is commercially harvested for food. It is also a potential species for mariculture.

**Countries where eaten:** Indo-Pacific; Philippines

**Other uses:** It is a potential species for aquarium trade.

*Tridacna derasa* (Röding, 1798)

**Common name(s):** Southern giant clam, Derasa clam, smooth giant clam.

**Distribution:** Indo-West Pacific: Australia, Cocos Islands, Fiji, Indonesia, New Caledonia, Palau, Papua New Guinea, Philippines, Solomon Islands, Tonga, and Vietnam

**Habitat:** Intertidal areas on corals; shallow waters; outer edges of reefs; depth range 0–20 m.

**Description:** This species is the second largest of the species among the clams. It is characterized by a low primary and radial sculpture, variable shape, massive umbonal area, and smooth white shell. Maximum shell length is 60 cm.

**Biology**

**Food and feeding:** As in other tridacnids.

**Reproduction:** The breeding season of the southern giant clam usually occurs in spring and summer, although they may be induced to spawn through the year

**Conservation:** It is a vulnerable species as per the IUCN Red List.

**Fisheries/Aquaculture:** It is commercially harvested for food. It is also extensively cultivated in Japan. Though it is a potential species for mariculture, this is probably the hardest of the tridacnids in captivity.

**Countries where eaten:** Indo-Pacific Cook Islands and Fiji

**Other uses:** It is also locally exploited for its shell.

*Tridacna gigas* (Linnaeus, 1758)

 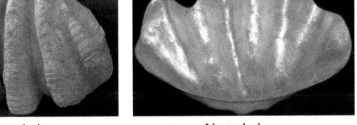

Dorsal view          Ventral view

**Common name(s):** Giant clam, gigas Clam

**Distribution:** South Pacific and Indian Oceans

**Habitat:** Reef flats and shallow lagoons to a depth of up to 20 m.

**Description**: This is the true giant clam, world's largest clam and is the most endangered of the species. Shell is extremely thick and lacks bony plates. When viewed from above, each valve has four to five inward facing triangular projections. Mantle is quite visible between the two shells and is a golden brown, yellow or green, although there may be an abundance of small blue-green circles giving a beautiful iridescent color. A number of pale or clear spots are seen on the mantle and these help in allowing sunlight to filter in through the mantle. Mantle is completely fused with the exception of two holes (siphons). Gills are visible through the inhalant siphon, while the exhalant siphon is tube-like and is capable of expelling a large volume of water during spawning. It has a maximum shell length of 140 cm and weight of 300 kg.

**Biology**

**Nutrition:** Giant clams have an inhalant siphon, which they use to draw in seawater that is then filtered for their planktonic food. However, the majority of the clam's nutrients are obtained by a mutually beneficial relationship with minute algae known as zooxanthellae as in other tridacnids.

**Reproduction:** It reaches sexual maturity at age 9–10 years. Adult giant clams are completely sessile and are unable to move from the coral reef. They reproduce by expelling sperm and eggs into the sea where fertilization occurs. The fertilized eggs quickly enter a swimming stage (trochophores), before entering a planktonic stage. During this stage, the larvae (veligers) inhabit the open ocean for 1 week, before settling in the substrate.

**Life span**: The average lifespan of this species in the wild is more than 100 years.

**Association:** Giant clams provide protection for a species of pea crab (*Xanthasia murigera*). A single pair will often be found living within the cavity of the clam.

**Fisheries/Aquaculture:** It is commercially harvested and is extensively cultivated in Australia, Philippines, Papua-New Guinea, Fiji, and Solomon.

**Countries where eaten**: Philippines to Micronesia. The adductor muscle is considered a delicacy in Japan.

*Tridacna maxima* (Röding, 1798)

**Common name(s):** Maxima clam, small giant clam

**Distribution:** Indo-Pacific region: East Africa, India, China, Australia, Southeast Asia, the Red Sea and the islands of the Pacific.

**Habitat:** Surface of reefs or sand or partly embedded in coral

**Description:** As an adult, this species has a large shell that adheres to a rock by its byssus. When open, the bright blue, green or brown mantle is exposed and obscures the edges of the shell with its prominent and distinctively furrowed edges. Attractive colors of the small giant clam are the result of pigment cells, which have a crystalline structure inside. These are thought to protect the clam from the effects of intense sunlight, or to enhance photosynthesis, carried out by the tiny algae living within. Maximum shell length is 20 cm.

**Biology**

**Food and feeding:** This sessile mollusc attaches itself to rocks or dead coral and siphons water through its body, filtering it for phytoplankton, as well as extracting oxygen with its gills. However, it does not need to filter-feed as much as other clams since it obtains most of the nutrients it requires from tiny photosynthetic algae known as zooxanthellae as in other tridacnids.

**Reproduction:** The tiny fertilized egg of this clam hatches within 12 h, becoming a free-swimming larva. This larva then develops into a more developed larva which is capable of filter-feeding. At the third larval stage, a foot develops, allowing the larva to alternately swim and rest on the substrate. After 8–10 days, the larva metamorphoses into a juvenile clam, at that time, it can acquire its symbiotic zooxanthellae. The juvenile matures into a male clam after 2–3 years, becoming a hermaphrodite when larger. Reproduction is stimulated largely by the lunar cycle, the time of day, and the presence of other eggs and sperm in the water. Hermaphroditic clams release their sperm first followed later by their eggs, thereby avoiding self-fertilization.

**Fisheries/Aquaculture:** It is collected for food and is a potential species for aquaculture.

**Countries where eaten:** Philippines to Micronesia and Fiji

**Other uses**: It is also a potential species for aquarium trade.

*Tridacna mbalavuana* (Ladd, 1934)

**Common name(s):** Devil's clam, tevora clam

**Distribution**: Northeastern barrier reef of New Caledonia

**Habitat:** Fairly in the deeper water; depth range 20–30 m

**Description:** The "devil clam" is so named because of its appearance with contrasting dark and bright colors. It is different from other clams in that the mantle does not overhang the edge of the shell. It has an unusual mantle color and texture. It is one of the largest clams in the world with a maximum size of 50 cm in length.

**Biology**

**Behavior and feeding:** Studies on the devil clam have shown that it has the incredible ability to live at great depths because of its ability to increase its photosynthetic efficiency at great depths, while being dependent on the symbiotic zooxanthellae which occur on their exposed mantle. Like other clams, these clams are filter feeders and have particularly been able to obtain nutrients at great depths and low light intensity. They are also morphologically adapted to life in deep sea and low light intensity (which the zooxanthellae would need for photosynthesis).

**Conservation and threats:** It is a vulnerable species in the IUCN Red List of Threatened Species. Devil clams are naturally rare. Overharvesting for consumption and for the marine aquarium trade because of their role as potential biofilters in aquarium tanks) is the major possible threats to this species.

**Fisheries/Aquaculture:** No information

**Countries where eaten**: Fiji and Tonga

***Tridacna rosewateri*** (Sirenko & Scarlato, 1991)

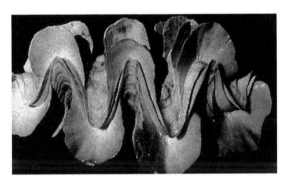

**Common name(s):** Bénitier De Rosewater (French)

**Distribution:** Indo-West Pacific: Mauritius and the Philippines

**Habitat:** Not reported

**Description:** Valves of this species are moderately thin, with large widely spaced scales on primary radial folds. Dorsal valve margins are with 4–5 very large triangular interdigitating processes. Maximum shell length is 20 cm.

**Biology:** It is a vulnerable species.

**Fisheries/Aquaculture:** No information

**Countries where eaten**: Mauritânea

## *Tridacna squamosa* (Lamarck, 1819)

**Common name(s):** Fluted giant clam, scaled clam, squamosa clam

**Distribution**: South Pacific and Indian Oceans

**Habitat:** Intertidal areas on corals; outer reef slope; lagoon; sandy bottoms; rubble or in reef pockets; anchored amidst Acropora corals; depth range 0–20 m

**Description**: This species is distinguished by the large, leaf-like fluted edges on its shell called "scutes" and a byssal opening that is small compared to those of other species of Tridacnidae. Normal coloration of the mantle ranges from browns and purples to greens and yellows arranged in elongated linear or spot-like patterns. Its maximum shell length is 45 cm.

## Biology

**Nutrition:** Sessile in adulthood, this clam's mantle tissues act as a habitat for the symbiotic single-celled dinoflagellate algae (zooxanthellae) from which it gets a major portion of its nutrition. By day, the clam spreads out its mantle tissue so that the algae could receive the sunlight they need to photosynthesize.

**Fisheries/Aquaculture:** This species is commercially harvested for food. It is a potential species for aquaculture.

**Countries where eaten**: It is an important food source in Fiji. It is also eaten in Red Sea and eastern Africa, eastern Polynesia (with the exception of Hawaii), and Philippines.

**Other uses:** It is also an aquarium species.

## 3.9.5 SURF CLAMS, TROUGH CLAMS OR DUCK CLAMS (MACTRIDAE)

Surf clams burrow in rocks, and their shell is consequently quite strongly constructed. Shell is equivalve and ovate or trigonal to transversely elongated. Shell surface is smooth, with concentric growth lines, and covered with a thin periostracum. Resilium (part of the hinge ligament inside the edges of the valves) is large and situated in a chondrophore, the spoon-shaped pit on the interior of the bivalve shell that contains the internal ligament. Lateral teeth are usually present, and the cardinal teeth are weakened. Mantle line has a posterior sinus. Mactridae represent generally species of secondary importance in the harvest of edible bivalves, although a few species are quite regularly fished locally.

***Eastonia rugosa*** (Helbling, 1778)

**Common name(s):** Rugose mactra clam

**Distribution:** Western Mediterranean and the Atlantic Ocean from Portugal to Dakar (West Africa).

**Habitat:** Sandy and muddy bottoms at depths between 0.5 and 35 m

**Description:** Solid shell of this species is wide oval with a center blunt umbo. Rear is more pointed than the front side. Sculpture is composed of irregular radial ribs which are separated from each other by a variable space between ribs. Ribs are lacking in front and rear of the shell.

In addition to irregular growth lines, virtually there is no concentric sculpture. On the inside, the bottom edge is crenulated. There are two large muscular impressions. There is a very striking deep resilium. Shells have a white or pink color. Common size: length: 6.0 cm; height: 4.5 cm.

**Biology**

**Food:** The animals filter food (plankton) from the seawater.

**Fisheries/Aquaculture:** No information

**Countries where eaten:** Straits of Gibraltar, Mediterranean, and Spain

*Lutraria angustior* (Philippi, 1844)

**Common name(s):** Narrow otter shell

**Distribution:** Northeast Atlantic and the Mediterranean

**Habitat:** Mixed soft substrata; depth range 5–25 m

**Description:** Shell of this species is elongate and somewhat quadrate. Anterior hinge line is sloping more steeply than posterior. Gaping is at both ends. Umbones are anterior to midline. Sculpture which is of numerous fine grooves is developing as fine ridges close to the margin. Growth stages are clear. Right valve is with two cardinal teeth and a single, poorly developed, posterior lateral. Left valve with anterior two cardinal teeth forming a solid, forked structure, third thin and indistinct and has single, thin, anterior and posterior laterals. Adductor scars and pallial line are distinct. Anterior adductor scar is elongate and irregular. Lower edge of pallial sinus is largely fused with pallial line. Color of exterior shell is dull white or yellowish. Periostracum is pale yellowish brown. Inner surfaces are white and glossy. Maximum length of shell is 10 cm.

**Biology:** No information

**Fisheries/Aquaculture:** Not reported

**Countries where eaten:** Mediterranean and Spain

*Lutraria lutraria* (Linnaeus, 1758)

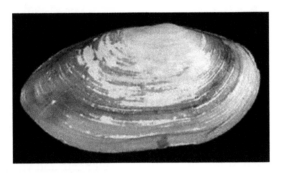

**Common name(s):** Otter shell

**Distribution:** From Norway south to the Mediterranean and West Africa

**Habitat:** Burrows in mixed soft substrata from the lower shore to about 100 m

**Description:** Shell of this species is elongate and elliptical. Gaping is at each end. Umbones are anterior to midline. Sculpture is of fine concentric lines, developing as fine ridges anteriorly and posteriorly. Growth stages are clear. Right valve is with two thin cardinal teeth which are anterior to chondrophore, and a single, poorly developed posterior lateral tooth. Left valve is with two cardinals forming a solid, forked structure; a third, very fine posterior cardinal, and very thin, single anterior and posterior laterals. Adductor scars and pallial line are distinct. Anterior adductor is larger and more elongate than posterior. Lower edge of pallial sinus is distinct from pallial line. Color of exterior shell is white or yellowish, often tinted pink or purple, and glossy. Periostracum is olive, thick, and brittle. Inner surfaces are dull white. Maximum length is 13 cm.

**Biology:** This species spawns in late spring and continued to do so through the summer. Animals become spent in August and September.

**Fisheries/Aquaculture:** No information

**Countries where eaten**: Mediterranean and Norway

*Lutraria oblonga* (Gmelin, 1791)

**Common name(s):** Oblong otter shell

**Distribution**: Indo-Pacific and Mediterranean Sea

**Habitat:** Burrows in mixed soft substrata from the lower shore to about 100 m

**Description:** Shell of this species is more elongate and valves are moderately thick. Umbo lies rather far forward. Ventral edge is fairly evenly curved. Dorsal edge behind umbo is curved upward. Anterior end is evenly rounded and posterior is obliquely truncate. Posterior gape is much wider than other species. Valves are covered by a dark-brown periostracum. Outer surface is rather irregular and is marked with concentric growth lines. Maximum length is 15 cm.

**Biology:** No information

**Fisheries/Aquaculture:** Not reported

**Countries where eaten**: Black Sea and Mediterranean

*Mactra chinensis* (Philippi, 1846)

**Common name(s):** Sunray surf clam, hen clam

**Distribution:** Northwest Pacific: Vietnam, Korea, Mainland China, and Taiwan

**Habitat:** Sandy substrates in shallow marine habitats; depth range 0–10 m

**Description:** Shell of this species is oval and beaks are slightly in front of middle. It is moderately thin and pure white inside. External shell is well painted with flesh-colored rays on a whitish background and is covered with a very thin, yellow cuticle toward margins. Anterior and posterior dorsal areas are closely and deeply radially sulcate; lower part of anterior half is concentrically irregularly sulcate; and rest of the surface is smooth. Common length of shell is 6 cm and dia. is 3 cm.

**Biology:** A carbamoylase, which catalyzes hydrolysis of the carbamoyl (or *N*-sulfocarbamoyl) moiety of paralytic shellfish toxins, has been purified from the digestive glands of this species (Lin et al., 2004).

**Fisheries/Aquaculture:** It is a potential species for aquaculture in Australia, Philippines, Papua-New Guinea, Fiji, and Solomon.

**Countries where eaten**: It is edible in areas of its occurrence.

**Other uses:** A carotenoid mactraxanthin has been isolated from this species. This compound may find use in food and feed additives (Sri Kantha, 1989).

*Mactra cuneata* (Gmelin, 1791)

**Common name(s):** Wedge trough shell, Maki wedge clam

**Distribution**: Indo-West Pacific: from East Africa to Melanesia; north to southern Japan and south to Queensland and New Caledonia

**Habitat:** Intertidal areas in sand; muddy-sand bottoms; littoral to sublittoral; depth range 0–50 m

**Description:** Shell of this species is equilaterally triangular in outline and solid. Shell sculpture consists of very fine concentric striae. Anterior dorsal margin is slightly convex, and the posterior dorsal margin is almost straight. Anterior end is rounded and is connected by a curved ventral edge to the posterior end which is more pointed than the anterior. Umbones are prominent, inflated, and rounded and placed near the mid line of the shell. Lunule and escutcheon are large. Anterior and posterior muscle scars are of the same size. Pallial sinus is shallow, about as deep as it is high, and the pallial line is very thick. Left valve has one anterior and one posterior lateral tooth, and the right valve has two anterior and two posterior lateral teeth. Shell exterior is bluish white to purplish gray, with reddish-brown rays of varying thickness. Shell interior is light brown to dark reddish-brown, and the areas near the ventral edge are always marked white. Maximum length is 4 cm.

**Biology**: No information

**Fisheries/Aquaculture:** Subsistence fisheries exist for this species

**Countries where eaten:** In areas of its occurrence

*Mactra glauca* (Born, 1778)

**Common name(s):** Glaous trough shell, five shilling shell, gray rough shell

**Distribution:** Northeast Atlantic and the Mediterranean

**Habitat:** Continental shelf; coarse gravelly sand; very fine sand; depth range 25–25 m

**Description:** Shell of this species is large, broadly triangular, thin and glossy, and equivalve. Left valve has two teeth conjoined to form an inverted V shape. Surface has fine concentric lines. Color of exterior shell is creamy white with radiating pale brown rays; and interior is pale brown. Periostracum is with brown, satin texture. Maximum length is 11.5 cm and breadth is 7.5 cm.

**Biology:** No information

**Fisheries/Aquaculture:** Not reported

**Countries where eaten:** Black Sea and Mediterranean

*Mactra mera* (Reeve, 1854)

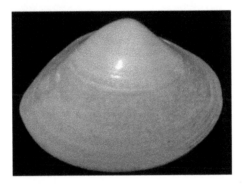

**Common name(s):** Plain trough shell

**Distribution**: Western Pacific: from southern Japan to the Philippines and Indonesia

**Habitat:** Intertidal areas; depth range 0–30 m

**Description:** Shell of this species is trigonal to subtrigonal in outline and is almost equilateral in some specimens. Shell is solid, and the exterior of adult specimens is covered with a thin brown periostracum that is thicker nearer the ventral margins and somewhat lacking near the umbones. Shell sculpture consists of fine concentric striae which become prominent at the lunule. Anterior and posterior dorsal margins are almost straight to mildly convex. Anterior end is rounded and the posterior end is pointed. Umbones are rounded, pointing inward, and placed near the midline. Lunule and escutcheon are large. Adductor muscle scars are large and rounded. Pallial sinus is rounded and moderately shallow. Hinge plate is whitish. Left valve has one anterior and one posterior lateral tooth, and the right valve has two anterior and two posterior lateral teeth. Color of the shell is purplish brown, with lighter and darker radial and concentric bands. Shell interior is purplish brown and somewhat glossy. Maximum length is 9 cm.

**Biology:** No information

**Fisheries/Aquaculture:** Commercial fisheries exist for this species.

**Countries where eaten**: Areas of its occurrence

*Raeta plicatella* (Lamarck, 1818)

**Common name(s):** Channeled duck clam

**Distribution:** North America.

**Habitat:** Gulf; sandy bottoms of outer surf zones; depth range 0–11 m

**Description:** Very fragile shell of this species is somewhat triangular and moderately inflated. Posterior end is narrow and angular and anterior end is broad and rounded. Prominent, widely spaced concentric ridges are present in outer and inner shell. Hinge has a spooned-out depression for the inner ligament. Posterior end has a gap between the two shells. Beak is pointed and points toward the narrow posterior end. Pallial sinus is fairly well developed. Color of exterior shell is white. Periostracum is red-brown. Maximum length is 7.5 cm.

**Biology:** Not reported

**Fisheries/Aquaculture:** No information

**Countries where eaten:** Mexico

*Simomactra planulata* (Conrad, 1837)

Image not available

**Common name(s):** Flat surfclam

**Distribution:** East Pacific

**Habitat**: No information

**Description:** Shell of this species is subequilateral. Pallial sinus is shallow and narrow. Anterior lateral tooth is short in left valve and is not aligned with cardinal.

**Biology:** Not reported

**Fisheries/Aquaculture:** No information

**Countries where eaten**: Alaska

*Mactra quadrangularis* Reeve, 1854 (=*Mactra veneriformis*)

**Common name(s):** Venus mactra, corners clams

**Distribution**: Western Pacific

**Habitat:** Intertidal in sand and mud; depth range 0–20 m

**Description:** Shell of this species is thick and slightly square shaped. Growth lines are thick and there is an uneven formation of concentric rings. Hinge is wide. Inner surface of the shell is white. Common length is 4 cm.

**Biology:** Gonads of both sexes develop in March, and animals with mature gonads (ripe phase) are available from spring to summer (April–August). These clams spawn several times a year in Korean coast.

**Fisheries/Aquaculture:** It is extensively cultivated in Palau, Indonesia.

**Countries where eaten**: In areas of its occurrence

*Mactra stultorum* (Linnaeus, 1758)

**Common name(s):** Rayed trough-shell

**Distribution**: Northeast Atlantic and the Mediterranean ; Black Sea, the Mediterranean coasts, and the west coast of Europe, from Norway to the Iberian Peninsula, and south to Senegal.

**Habitat:** Brackish water; sessile-burrower; sandy (rarely soft) bottoms; beaches; depth range 5–12 m

**Description:** Shell of this species is thin, brittle, and oval. Umbones are just anterior to midline. Sculpture is of very fine concentric lines and growth stages are clear. Shell margin is prominent at hinge line. Lunule and escutcheon are absent. Right valve has two cardinal teeth and left valve is with three cardinal teeth. Chondrophore is triangular. Adductor scars and pallial line are indistinct. Pallial sinus is broad and rounded, not extending far into shell. Exterior shell is white; tinted purple about the umbones and is with light brown rays of varying width radiating from umbones. Periostracum is light brown and thin. Inner surfaces are glossy, white, and tinted purple. Maximum shell length is 6.4 cm.

**Biology**

**Food and feeding:** It is an active suspension feeder and it mainly depends on phytoplankton and detritus material for nutrition.

**Association:** Symbiotic crude-oil degrading bacteria have been isolated from this species. The isolated strains belong to: *Alcanivorax dieselolei* strain BHA25, *Idiomarina baltica* strain BHA28, *Alcanivorax dieselolei* strain BHA30, *Alcanivorax* sp. strain BHA32 and *Vibrio azureus* strain

BHA36. These strains were also found to degrade: 64%, 63%, 71%, 58%, and 75% of crude oil, respectively (Bayat et al., 2016).

**Fisheries/Aquaculture:** No information

**Countries where eaten:** It is eaten in areas of its occurrence. It is an important source of food for humans, fish, and seabirds.

*Mactra violacea* (Chemnitz, 1782)

**Common name(s):** Violet trough shell

**Distribution:** Western Central Pacific; Eastern Indian Ocean: Indo-West Pacific, from Eastern Indian Ocean to Indonesia and the Philippines

**Habitat:** Shallow subtidal levels

**Description:** Shell of this species is inequilaterally triangular, thin, glossy, and covered with a thin grayish-gold periostracum which is more prominent nearer the ventral margin and lacking toward the umbones. Shell sculpture consists of very fine concentric striae, which tend to be more regular and prominent dorsally and less so toward the ventral margin. Anterior dorsal margin is slightly concave, and the anterior end is broad and round. Posterior dorsal margin is straight to slightly convex and terminates in a pointed but truncated, gaping edge. Umbones are placed near the midline and are high and pointed, bending inward. Lunule and escutcheon are large but their borders are not well defined. Fine, irregular radial grooves emit from the umbonal cavity toward the ventral edge. Muscle scars are small and elliptical, and more rounded in the posterior. Pallial sinus is rounded but shallower than that of *Mactra mera*. Hinge line is white. Anterior cardinal teeth in both valves do not reach the ventral edge of the hinge plate. Left valve has one anterior and one posterior lateral tooth, and the right valve has two anterior and two posterior lateral teeth. Dorsal most anterior lateral tooth in the right valve is reduced to a short ridge. Color of the exterior shell ranges from deep purple to white. Shell interior is of the same color as the exterior, but less glossy. Maximum length is 9.5 cm.

**Biology:** No information

**Fisheries/Aquaculture:** This species has subsistence fisheries.

**Countries where eaten:** Areas of its occurrence

*Mactrellona alata* (Spengler 1802)

**Common name(s):** Winged surfclam, Caribbean winged mactra, duck clam

**Distribution:** Tropical eastern Pacific; Western Atlantic and Eastern Pacific: Caribbean to southeast Brazil

**Habitat:** Infaunal; shallow subtidal sand

**Description:** Shell of this species is thin, triangular, inflated, and light. Posterior slope is flattened and bound by characteristically elevated "keel-like" ridge. Hinge is with anterior lateral teeth short. Umbones are prominent twisted inward. Periostracum is thin and flaky when dry. Color of exterior shell is white and periostracum is yellowish. Maximum length is 10 cm.

**Biology:** No information

**Fisheries/Aquaculture:** Not reported

**Countries where eaten:** Brazil

*Rangia cuneata* (Sowerby, 1832)

**Common name(s):** Atlantic Rangia

**Distribution:** Native to the Gulf of Mexico and introduced to the NW Atlantic

**Habitat:** Low salinity (5–15 ppm) estuarine habitats

**Description:** Valves of this species are thick and heavy, with a strong, rather smooth pale brown periostracum. Shells are equivalve, but inequilateral with the prominent umbo curved anteriorly. An external ligament is absent or invisible, but the dark brown internal ligament lies in a deep, triangular pit immediately below and behind the beaks. Both valves have two cardinal teeth, forming an inverted V-shaped projection. Upper surface of the long posterior lateral teeth is serrated. Inside of the shell is glossy white, with a distinct, small pallial sinus, reaching to a point halfway below the posterior lateral. Pallial line is tenuous. Maximum length is 9.4 cm.

**Biology**

**Feeding:** It is a nonselective filter-feeder turning large quantities of plant detritus and phytoplankton into clam biomass.

**Reproduction:** In this species, there are two spawning periods ranging from March–May and late summer–November in Louisiana. Spawning may also be continuous.

**Remarks:** As a biofouling species, it may cause problems in industrial cooling water systems.

**Fisheries/Aquaculture:** No information

**Countries where eaten:** Gulf Coast and Southern Atlantic

## *Spisula aequilateralis* (Deshayes, 1854)

**Common name(s):** Triangle shell

**Distribution:** Endemic to New Zealand

**Habitat:** Low water mark of sandy beaches with great surf; depth range 0–8 m

**Description:** Shell of this species is heavy and solid, inflated, and triangular. Four of these shells, placed together with their beaks pointing inward, make a circle. Coloration of exterior shell is creamy-white under a thin yellowish periostracum and the beak is often purplish. Maximum shell length is 7.4 cm and height is 6 cm.

**Biology**

**Reproduction:** The sexes are separate, they are broadcast spawners. Nothing is known of their larval life.

**Fisheries/Aquaculture:** No information

**Countries where eaten:** The meat of this species is used as the traditional food of the Maori.

## *Spisula elliptica* (Brown, 1827)

**Common name(s):** Not designated

**Distribution:** Whole North Sea; northward to the Barents Sea

**Habitat:** Mixed soft substrata; offshore to about 100 m

**Description:** Shell of this species is thin and elongate oval. Umbones are close to midline. Sculpture is of fine concentric lines and grooves and growth stages are clear. Lunule and escutcheon are poorly defined, the former perhaps most distinct. Right valve is with two separate, but closely spaced cardinal teeth and paired, elongate anterior and posterior laterals. Left valve is with three cardinal teeth. Interlocking surfaces of lateral teeth are serrated. Chondrophore is posterior to cardinal teeth. Adductor scars and pallial line are distinct. Pallial sinus is oval, extending to a point below and beyond midline of posterior lateral teeth. Color of exterior shell is dull white with greenish or grayish brown periostracum. Inner surfaces are glossy and white. Maximum shell length is 3 cm.

**Biology:** No information

**Fisheries/Aquaculture:** Not reported

**Countries where eaten:** Northern Atlantic

*Mactromeris polynyma* (Stimpson, 1860) (=*Spisula polynyma*)

**Common name(s):** Stimpson's surf clam, Arctic surfclam

**Distribution:** Northeast Pacific; Arctic Sea, from Rhode Island (western Atlantic) to Alaska and Puget Sound (east Pacific); Japan

**Habitat:** Low-tide line to around 200 m depth

**Description:** Shell of this species is oval in shape and is beaking very near the middle of the valve. Anterior end of the shell is smaller than the elliptical posterior end. Worn shells have coarse, concentric, wide growth lines. Chondrophore is not set off by a plate. Ligament is partially external. Exterior shell is chalky, dirty-white, and periastrcum is varnish-like, yellowish brown. Maximum shell length is 16 cm.

**Biology:** It is an active burrower and is preyed by gastropods, sea stars, and also by sea otters.

**Fisheries/Aquaculture:** Commercial fisheries exist for this species. It is a potential species for aquaculture in Korea.

**Countries where eaten:** Alaska; North America's northern Atlantic coastline

***Spisula sachalinensis*** (Schrenck, 1862) (=*Mactra sachalinensis*)

**Common name(s):** Sakhalin surf clam

**Distribution:** Northwest Pacific: Japan

**Habitat:** Near wave-cut zone on sorted medium-grained sands of the upper sublittoral; depth range 18–36 m

**Description:** Shell of this species is triangularly rounded and massive. Surface of the valves is smooth, with thin lines rise and small concentric wrinkles. Exterior of the shell is brown or yellow-brown and periostracum is green-brown. On inner part, well-pronounced cardinal and noticeably elongated lateral lock-forming teeth are seen. Largest specimen is up to 13 cm long.

**Biology:** Digs into sand to length of shell, whose hind part usually protrudes over floor surface.

**Fisheries/Aquaculture:** There is a commercial fishery for this species.

**Countries where eaten:** It is a commercial species. The massive leg and most of the body are used as food in areas of its occurrence.

**Commercial uses:** This species produces an enzyme called endo-1,3-beta-D-glucanase which may find use in enological practices during the aging process of wine.

*Spisula solidissima* (Dillwyn, 1817)

**Common name(s):** Bar clam, hen clam, skimmer, sea clam, Atlantic surf clam

**Distribution:** Western Atlantic Ocean

**Habitat:** Lived buried in coarse or fine sand; offshore as well as in the low intertidal and surf zones; depth range 0–70 m

**Description:** Shell of this species is oval in shape, strong, and smoothish, except for small, irregular growth lines. Lateral teeth bear very tiny, saw-tooth ridges. Chondrophore is not set off by a plate. Ligament is partially external. Color of exterior shell is yellowish white with a thin yellowish brown periostracum. Maximum shell length is 20 cm.

**Biology**

**Feeding:** It is an infaunal suspension-feeding bivalve. These clams use their siphons to pull in and then filter fine particles of organic matter and plankton from the surrounding seawater. Like almost all clams, they are filter feeders.

**Reproduction:** Full sexual maturity is in the second year of life (4.5–8.5 cm shell length) and spawning occurs usually in summer. Rapid growth occurs to about age 7 and maximum age of 37 years is obtained at a shell length of 22.6 cm.

**Predators:** Sea stars and gastropods

**Fisheries/Aquaculture:** There is a commercial fisheries for this species and is exploited commercially as a food item. It is also a potential species for aquaculture in Korea and Japan.

**Countries where eaten:** Labrador to Gulf of Mexico. The meat of the clam is used as "strips," chowder, and sushi. Further, the foot of this clam is commercially valuable because it is cut into long strips which are breaded and fried and served as clam strips.

**Other uses:** Shell of this species is used as a decorative dish or ashtray

*Spisula subtruncata* (da Costa, 1778)

**Common name(s):** Cut trough shells

**Distribution:** Whole North Sea; from Norway to the Mediterranean and Canary Isles

**Habitat:** Burrows in muddy or silty sand, from the lower shore into the shallow sublittoral

**Description:** Shell of this species is thick, strong and subtriangular but distinctly asymmetrical. Umbones are close to midline and posterior hinge line and margin are sloping more steeply than anterior end. Posterior end is appearing slightly drawn out. Sculpture is of fine concentric lines and grooves and growth stages are clear. Lunule and escutcheon are broad and elongate. Right valve is with two short cardinal teeth and paired, elongate anterior and posterior laterals. Left valve is with three cardinal teeth. Chondrophore is posterior to cardinal teeth. Adductor scars and pallial line are distinct. Pallial sinus is broad and rounded, extending to a point below and behind the midline of the posterior lateral teeth. Color of exterior shell is dull white to cream and periostracum is grayish brown. Inner surfaces are glossy and white. Maximum shell length is 3 cm.

**Biology:** Concentrations of trace metals, namely, mercury, cadmium, lead, copper, zinc, and chromium have been studied in this species collected from the Belgian Continental Shelf. Average concentrations (in mg/kg wet weight) of these elements were found to be 0.02 for mercury, 0.08 for cadmium, 0.38 for lead, 2.5 for copper, 12.7 for zinc, and 0.47 for chromium. These values were, however, well below legally authorized limits (Vyncke et al., 1999).

**Fisheries/Aquaculture:** No information

**Countries where eaten:** Black Sea and Mediterranean

*Tresus capax* (Gould, 1850)

**Common name(s):** Fat gaper clam, horse clam, Alaskan gaper, summer clam, otter clam

**Distribution:** Kodiak Island, Alaska to central California

**Habitat:** Quiet bays; middle and low Intertidal and subtidal to 30 m; prefer sand, mud, and gravel substrates, burrowing in mud and clay, to depths of up to 1 m

**Description:** These clams are also commonly called gapers because their shells are flared around the siphon and do not completely close, rather like geoduck clams. This species has two valves which are similar to one another and two adductor muscle scars that are similar in size. It has a true hinge plate with teeth and a socket-like chondrophore in both valves. A pallial sinus and continuous pallial line are seen and there are no radial ribs. *Tresus capax* has a shell often over 10 cm long (up to 20 cm) and with a wide gape (over 1/4 the width of the shell) at the posterior end. Umbones are near the end of the anterior third of the shell. Periostracum, where it is still attached, is a dark brown or black. Exterior shell is chalky white or yellow. Shell is about 1.5 times as long as high. Maximum shell length is 2 cm.

## Biology

**Food and feeding:** These clams have extremely long siphons which extend up from their deep location to the surface of the mud. The incurrent and excurrent siphons are fused together but the internal channels are separate. It burrows in the mud and feeds on particles brought through its siphons.

**Life span:** It has a life-span of around 15 years.

**Reproduction:** In this species, spawning occurs usually in late winter. In Washington, it reaches sexual maturity in 3–4 years.

**Predators:** Predators of this species include the moon snail *Euspira lewisii*, the crab *Cancer magister*, and the sea stars *Pycnopodia helianthoides* and *Pisaster brevispinus*. Fusitron oregonensis will attack it if it gets the chance.

**Association:** Unlike other members of the *Tresus* genus, this species has a fringe of tissue known as the visceral skirt, which is commonly harbored by crabs. Commensals of this species include the crabs *Pinnixa faba* and *Pinnixa littoralis*, in the mantle cavity.

**Fisheries/Aquaculture:** This is a sport species, commercially harvested in British Columbia. It is harvested intertidally by recreational fishers and harvested subtidally by geoduck divers.

**Countries where eaten:** This clam is often eaten by man, especially by native Americans. It is an important human food source. The meat is of good flavor and makes excellent chowder.

*Tresus nuttallii* (Conrad, 1837)

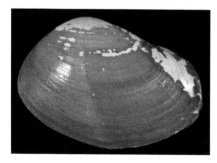

**Common name(s):** Pacific gaper, horse clam

**Distribution:** Eastern Pacific: from Alaska to Baja California, Mexico

**Habitat:** Intertidal zones; prefer sand, mud, and gravel substrates; depth range 0–50 m

**Description:** Shell of this species is oval (i.e., longer and narrower) and chalky-white or yellow with patches of brown periostracum. It has larger siphonal plates. These clams are also commonly called "gapers" because their shells are flared around the siphon and do not completely close, rather like geoduck clams. Maximum shell length is 22.5 cm and weight is 1.8 kg.

**Biology:** It feeds on phytoplankton and detritus and it is a summer spawner.

**Fisheries/Aquaculture:** Commercial fisheries exist for this species. It is harvested intertidally by recreational fishers and is harvested subtidally in Baja California by geoduck divers.

**Countries where eaten:** Kodiak Island; Alaska to southern Baja California

## 3.9.6   RAZOR SHELLS (SOLENIDAE)

Shell of the members of this family is elongate, equivalve, and inequilateral. Umbones are present at anterior end. Gaping is anteriorly and posteriorly. Ligament is external. One or two peg-like cardinal teeth are seen in each valve. Lateral teeth are present or absent. Adductor scars are unequal; anterior is typically elongate and posterior is small. Pallial sinus is present. These are filter-feeding animals and are adapted to swift and deep-burrowing in soft bottoms with their powerful foot. They are capable of swelling at the end and protruding anteriorly through the narrowly elongate shell. The intertidal species of Solenidae are sometimes actively exploited for food, notably in the Philippines, Indonesia, and Malaysia.

*Ensis directus* (Conrad, 1843)

**Common name(s):** Atlantic jackknife clam, bamboo clam, American jackknife clam

**Distribution:** Eastern Central Pacific and Atlantic Ocean

**Habitat:** Littoral and sublittoral sediment; sand and mud; intertidal or subtidal zones in bays and estuaries; brackish water; depth range 0–260 m

**Description:** It is characterized by an elongated (razor-shaped) shell, with both valves connected by a hinge in the anterior end. Hinge has few very small "teeth" and an elastic ligament. Shell shape is as slightly curved to almost straight. Dorsal and ventral margins are almost parallel. Anterior end is rounded while the posterior end is truncate. Maximum length is 26 cm.

## Biology

**Behavior:** Because of its streamlined shell and strong foot, this species can burrow in wet sand very quickly and is also able to swim.

**Feeding:** These animals filter feed on plankton and detritus.

**Reproduction**: This species is dioecious, but males and females cannot be distinguished externally. Males release sperm into the water, and the sperm enters the females through the inhalant siphon. Eggs are fertilized in the interior of the gill, and newly fertilized zygotes develop into larvae, which are then released in the water. The zygotes develop into trochophores, translucent free-swimming larvae, which develop further into lightly shelled "D" veliger larvae that become more rotund with age to form the umbonate stage. The final larval stage is a pediveliger, which is capable of swimming and crawling with a "foot." The planktonic stage lasts 10–27 days, resulting in its dispersal by currents. After the larval stage, young individuals settle onto sandy or muddy habitats and begin their development into adults. Most animals achieve their sexual maturity in their first year. Reproduction occurs between May and September, based on rising water temperatures.

**Predators:** These include birds, such as the ring-billed gull (*Larus delawarensis*) in North America and the Eurasian oystercatcher (*Haematopus ostralegus*) in Europe, and the nemertean worm *Cerebratulus lacteus*.

**Life span:** This species is relatively long-lived for a bivalve, with an estimated maximum age of up to 20 years.

**Fisheries/Aquaculture:** It is commercially exploited for food.

**Countries where eaten**: It is widely regarded as a delicacy in coastal Massachusetts. This species also forms an important food source for seagulls.

*Ensis ensis* (Linnaeus, 1758)

**Common name(s):** Sword razor, razor shell

**Distribution:** Northeast Atlantic and Mediterranean

**Habitat:** Estuarine and inshore areas in substrates of silt to very fine sand; brackish water; depth range 0–200 m.

**Description:** Valves of this species are parallel-sided, narrow, and curved. Valves are thin and rather brittle. Edges are parallel and are tapering slightly toward the posterior and are off-white with transverse bands of brown. Shell is sculpted with fine comarginal lines and the annual growth lines can be seen. Inner side of the shell is white with a purplish sheen. Periastracum is olive green and the foot is reddish. Posterior adductor muscle is positioned about one and a half times its own length from the pallial sinus, and that the muscle that retracts the foot is posterior to the insertion point of the ligature. Maximum shell length is 13 cm.

**Biology**

**Feeding:** It is an active suspension feeder and a microvore that feeds mainly on organic detritus. It has two short siphons that normally extend up to the surface of the sand. It draws in water through one siphon and expels it through the other. By this means it both respires and extracts food particles from the water at the same time.

**Reproduction:** This species matures at about 3 years old and may live for 10. Reproduction takes place in the spring and the larvae are pelagic and form part of the zooplankton. After about a month, they settle out and burrow into the substrate.

**Association**: This species is often found living in association with other burrowing animals including the sea potato, *Echinocardium cordatum*, and other bivalve molluscs, namely, *Tellina fabula* and *Chamelea gallina*.

**Fisheries/Aquaculture:** Commercial fisheries exist for this species in Italy.

**Countries where eaten**: Mediterranean, Norway, Baltic, and Morocco

*Ensis macha* (Molina, 1782)

**Common name(s):** Razor clam, Navaja

**Distribution:** Southern Pacific and Atlantic: Chile and Argentina

**Habitat:** Buried in shallow subtidal soft bottoms, specifically in coarse sandy sediments or silty sands at depths of 2–50 m.

**Description:** Shells of this species are narrow and edges are parallel and arcuate with smooth surface. Leading edge is rounded, while the latter is slightly truncated. Umbones are near the leading edge. Externally, the periostracum is thin, yellowish to greenish brown. Hinge has three cardinal teeth, two in the left valve and one on the right valve. Maximum shell length is 20 cm.

**Biology**

**Fisheries/Aquaculture:** It is commercially harvested for food. It is also one of the world's economically most significant species, with a total catch of 6000 t in Chile in 1999.

**Countries where eaten:** Chile

**Remarks:** Occasionally intake of this species may lead to allergies such as generalized pruritus and facial angioedema in humans (Jiménez et al., 2005).

*Ensis magnus* (Schumacher, 1817) (=*Ensis arcuatus*)

**Common name(s):** Arched razor shell, sword razor

**Distribution:** Northeast Atlantic and the Mediterranean Sea

**Habitat:** Brackish water; subtidal region in soft bottoms of fine to medium-grained sand; areas influenced by estuarine outflows

**Description:** Shell of this species is thin and brittle, and elongate. Dorsal margin is almost straight and ventral margin is distinctly curved. Shell is deepest at midline. Anterior and posterior margins are obliquely truncate. Sculpture is of fine lines following the growth stages. Left valve is with two, projecting, peg-like cardinal teeth and two elongate, posterior laterals, one above the other. Right valve is with one cardinal and a single, elongate, posterior lateral. Adductor scars and pallial line are distinct. Anterior adductor is elongate, dorsal and is extending to behind ligament; and posterior is small, round and is separated from pallial sinus. Color of shell is white or cream, with reddish or purplish brown streaks and blotches. Periostracum is glossy, light olive or yellow-brown and is darker gray-green along posterior dorsal region. Inner surfaces are white, tinted with pink or pale purple. Maximum length is 15 cm.

**Biology:** No information

**Fisheries/Aquaculture:** It is commercially harvested for food.

**Countries where eaten:** Spain and Norway

*Ensis minor* (Chenu, 1843)

**Common name(s):** Minor jackknife clam

**Distribution:** Western Central Atlantic and the Mediterranean Sea: USA

**Habitat:** Gulf and bay; edges of bays and lagoons

**Description:** It is recognizable by its smaller size, obliquely truncated anterior end, and the anterior pallial scar running parallel and very close to the anterior edge. Shell is rather broad, straight, or slightly curved. Edges are parallel or nearly so. Posterior end is of approximately the same breadth as the anterior end and is often squarely truncate. Posterior aperture is narrow and is more or less compressed in outline. Sculpture is of fine concentric lines, following the growth increments. Left valve is with two projecting, peg-like cardinal teeth of about equal size and two elongate, posterior laterals. Right valve is with a single cardinal and a single, elongate posterior lateral. Anterior adductor scar is elongate, extending to behind ligament. Posterior scar is small and round. Color of exterior shell is white or cream, with pale purple, pink, or reddish streaks. Periostracum is glossy, light olive or yellow-brown, and gray-green along posterior dorsal region. Inner surfaces are white, often tinted blue or purplish. Maximum length is 17 cm and breadth is 2.5 cm.

**Biology:** Not reported

**Fisheries/Aquaculture:** No information

**Countries where eaten:** Mediterranean, Norway, Baltic, and Morocco

## *Ensis siliqua* (Linnaeus, 1758)

**Common name(s):** Pod razor, razor clam, giant razor, sword razor shell

**Distribution:** Northeast Atlantic and the Mediterranean: from Norwegian Sea and the Baltic, south to the Iberian Peninsula, into the Mediterranean, and along the Atlantic coast of Morocco

**Habitat:** Brackish-water; depth range 0–59 m

**Description:** Shell of this species is thin, brittle, and elongate. Dorsal and ventral margins are parallel, and practically straight. Anterior and posterior margins are obliquely truncate, with rounded corners. Sculpture is of fine concentric lines, following the growth increments. Left valve is with two projecting, peg-like cardinal teeth and two elongate, posterior laterals situated one above the other. Right valve is with a single cardinal and a single, elongate posterior lateral. Adductor scars and pallial line are clear. Anterior adductor scar is elongate, dorsal and is extending to behind ligament. Posterior scar is small and round and is separated from pallial sinus. Color of exterior shell is white or cream, with pale purple, pink, or reddish streaks. Periostracum is glossy, light olive or yellow-brown, and gray-green along posterior dorsal region. Inner surfaces are white, often tinted blue or purplish. Maximum length of shell is 20 cm.

**Biology**

**Nutrition:** It is a sessile-burrower and suspension feeder depending mainly on phytoplankton and detritus material for nutrition.

**Fisheries/Aquaculture:** It is fished commercially, especially in Spain, Ireland, and Scotland.

**Countries where eaten:** Mediterranean and Northern Atlantic areas

### 3.9.7 JACKKNIFE CLAMS (PHARIDAE)

Shell of the members of this family is very elongate and is gaping at both ends. Dorsal margin is slightly to very concave and ventral margin is convex. Umbones are terminal. Left valve is with two horizontal teeth separated by a groove and two vertical teeth. Right valve is with one horizontal tooth and one vertical tooth. Anterior adductor scar is very elongate.

***Pharus legumen*** (Linnaeus, 1758)

**Common name(s):** Bean solen, bean razor clam

**Distribution:** Atlantic Ocean and the Mediterranean

**Habitat:** Buried in the sand on the coast; from the lower shore into the shallow sublittoral; depth range 5–25 m

**Description:** Shell of this species is thin and brittle, elongate and is about four times as long as deep. Anterior and posterior margins are rounded and gaping is at both ends. Anterior end is distinctly tapered. Umbones are low and indistinct, sited about one-third of length from anterior margin. Ligament is immediately posterior to umbones, situated in middle third of shell. Sculpture is of numerous fine concentric lines, with a group of very fine striae radiating from umbones to middle third of ventral margin. Growth stages are visible but not conspicuous. Right valve is with a single cardinal tooth. Left valve is with two elongate, closely spaced, cardinal teeth. Adductor scars and pallial line are distinct. Pallial sinus is short, and quadrate. Cruciform muscle scars are unclear. Color of exterior shell is white or light brown, with a glossy, light olive or yellow periostracum. Posterodorsal section of shell is fawn. Inner surfaces are white and glossy. Maximum shell length is 12.7 cm.

**Biology:** It is a sessile-burrower, suspension feeder

**Fisheries/Aquaculture:** It is commercially harvested in Italy.

**Countries where eaten**: Adriatic, Mediterranean, British Isles, Morocco, and Norway to Senegal.

*Siliqua patula* (Dixon, 1789)

**Common name(s):** Pacific razor clam, Northern razor clam

**Distribution:** Pacific Northwest: from Pismo Beach in southern California to the Aleutian Islands in Alaska

**Habitat:** Open sandy beaches

**Description:** This species has elongate shells which are thin, flat, and smooth. It is covered with a heavy, glossy, yellowish periostracum. A prominent rib is extending from the umbo to the margin on the inside of the valve. Foot is large and powerful and is never pigmented. Siphons are rather short and united except at tips. Umbos are nearer to anterior than posterior end. Maximum shell length is 14.5 cm.

**Biology:** No information

**Fisheries/Aquaculture:** It is a potential species for aquaculture in USA.

**Countries where eaten:** Bering Sea in Alaska to as far south as San Luis Obispo; France and Great Britain; USA and Canada

*Siliqua radiata* (Linnaeus, 1758)

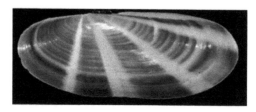

**Common name(s):** Sunset shell, sunset siliqua, purple clam

**Distribution**: Western Pacific: China; Indian Ocean; India, South African coast, Sri Lanka, Myanmar, Indonesia, and Philippines

**Habitat:** Intertidal in sand; shallow muddy areas; estuarine; depth range 0–50 m.

**Description:** Shell of this species is thin, elongate compressed, elliptical, and gaping at both the ends. Umbo is anterior and beaks are slightly in front of midline. Anterior side is short and is excavated near umbone. Posterior is elongated. Dorsal margin are almost straight and ventral margin is convex, arched and is sloping upward toward the end of both sides. Internal ribs are almost straight. Shell is brown to lilac with four white radial rays. Maximum shell length is 9 cm.

**Biology:** Sudden population eruption of this species was recorded during monsoon months (June–August 2008) along Versova beach in Mumbai, India. The average density was estimated as 14 per square meter and the biomass was 153 g/m². The proportion of meat was 60%. Analysis of its meat showed high amount of heavy metals namely copper, cadmium, and arsenic (Sujit et al., 2010).

**Fisheries/Aquaculture:** It is a potential species for aquaculture in Chile and Peru.

**Countries where eaten**: Areas of its occurrence

*Sinonovacula constricta* (Lamarck, 1818)

**Common name(s):** Razor, Jack knife clam, Chinese razor clam, Asian marine razor-clam

**Distribution:** Indo-Pacific (West Pacific); Japan and China

**Habitat:** Brackish mudflats along sheltered bays and estuaries, and seaward of mangrove forests; intertidal areas with flat, muddy or sandy-muddy substrates; depth range 0–10 m

**Description:** Shells of this species are elongate, narrow, and rather fragile shell with bluntly rounded gaping ends. Surface is smooth with concentric striations. Small beaks are seen near center. Periostracum is yellowish. Maximum shell length is 10 cm.

**Biology**

**Ecology:** This species has wide tolerances of temperature and salinity.

**Behavior:** This species is deep-burrowing and sedentary, moving vertically in the upper substrate layer with the help of its well-developed foot. During high tide the clam is usually seen in the upper part of its vertical hole with its siphons extending beyond the sand/water interface. At low tide, they retreat deep down their burrow for protection. The depth and diameter of the hole depends on the size of the clam, texture of the seabed and season. It has been observed that the depth is generally 5–8 times the length of the shell. These clams are unlike other bivalves, such as the scallop, that frequently move over long distances.

**Food and feeding:** Although razor clams are filter feeders, they are passively taking food organisms from the surrounding seawater by means of their filter organs. Observations have shown that the major food organisms in their stomach are benthic diatoms such as *Nitzschia* sp., *Coscinodiscus* sp., etc., while few spiny phytoplankton species have also been recorded.

**Reproduction:** It is a dioecious mollusc; however, the two sexes can only be distinguished during the spawning season when the reproductive organs are ripe. The male gonad usually appears milky-white with a smooth surface, while the female organ has a rough surface, granular in appearance and beige in color. The spawning season of this species is closely related to water temperature and therefore specimens from different localities spawn at different times. Adult and fully mature individuals can spawn 3–4 times in one spawning season, at intervals of about 2 weeks. A razor clam has been reported to lay an average of 193,000 eggs per spawning.

**Fisheries/Aquaculture:** It is commercially exploited. It is a potential species for aquaculture in Sri Lanka and India. It is an excellent aquaculture species in muddy coastal intertidal areas because its culture is low cost. Further, it has a short life cycle, gives high yields, is a valued commodity, its management is simple, and production is relatively stable. Further, these clams tend to dwell in one spot throughout their life. Thus, it is easy to culture them in open sea and no special facilities are required to prevent them from escaping. It has been extensively harvested from enhanced mudflat environments in China for nearly 500 years, especially in Fujian and Zhejiang. More recently, it has been widely cultured in Shandong.

**Countries where eaten:** It is economically important and is commonly used as fresh, dried, or canned food in Japan and China. It is also widely used as a health food and medicine in China, Japan, and Korea.

***Solen grandis*** (Dunker, 1862)

**Common name(s):** Grand razor shell

**Distribution**: Western Pacific: from Thailand to the Philippines; north to Korea and Japan

**Habitat:** Intertidal areas in sand; depth range 0–20 m

**Description:** Shell of this solen is very thin and bamboo shaped. Its outside is marked with pale rosy concentric bands. Periostracum is greenish-yellow. Maximum shell is length 15 cm, and common length is 12 cm.

**Biology:** No report

**Fisheries/Aquaculture:** No information

**Countries where eaten**: China: Southern and Northern sea. It is occasionally marketed in Thailand and the Philippines. The meat of this solen is chubby and tasty, full of amino acid and succinic acid, so delicious and is deemed as the best among the solen group. The solen in Nantong is served in the menu name of "fried Razor Clam" and "Quick Fried Solen."

**Other uses:** A novel fibrous protein, solenin, has been isolated from the ligament of this species. This natural fibrous protein will provide a new template for fabricating of bioinspired materials (Meng et al., 2014).

*Solen marginatus* (Pulteney, 1799)

**Common name(s):** Grooved Razor Shell, European razor clam

**Distribution:** Northeast Atlantic and the Mediterranean Sea

**Habitat:** Fine well-sorted sand; near shore shelf depths

**Description:** Shell of this species is thin, brittle, and elongate. Dorsal and ventral margins are straight and parallel. Anterior end of shell is obliquely truncate, with a distinct dorsoventral groove close to margin. Sculpture is of fine concentric lines which are distinct along growth stages. A single cardinal tooth is present in each valve and there are no laterals. Anterior adductor scar is almost as long as ligament; and posterior is small and round, often close to the dorsal edge of the pallial sinus. Color of exterior shell is white or yellow and periostracum is glossy, light olive or brown. Inner surfaces are white or yellowish. Maximum shell length is 12 cm.

**Biology:** In this species, 1-year-old juveniles measured 38.5 mm and commercial size was reached with 3-year-old individuals (80.1 mm).

**Fisheries/Aquaculture:** It is mainly harvested in Italy as bait for sport fishing and occasionally for human consumption. There is ample scope for its commercial mariculture.

**Countries where eaten**: Mediterranean, Black Sea, France, Norway, Baltic, and Senegal

*Solen regularis* Dunker, 1862

Image not available

**Common name(s):** Razor clam

**Distribution:** Malaysia

**Habitat:** Mudflats

**Description**: This species has a pair of elongated thin shells which are nearly straight and subcylindrical with parallel margins and terminal beaks and they are gaping at both ends. It has a large and powerful foot which is particularly useful as a feeding and burrowing mechanism.

**Biology**

**Reproduction:** The male gonads of this species appear beige in color while female gonads are rewhitish. It has five stages of reproductive cycle, namely, (1) gonadal development, (2) maturation, (3) spawning, (4) spent, and (5) resting period. Spawning period is from end of March–April to September.

**Fisheries/Aquaculture:** It is a potential species for aquaculture in future.

**Countries where eaten**: Areas of its occurrence. It is a highly priced bivalve collected as food source and is preferred by consumers.

***Solen sicarius*** (Gould, 1850)

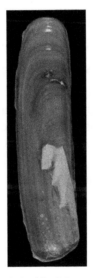

**Common name(s):** Sickle jack knife, blunt razor shell

**Distribution:** Northeast Pacific: USA and Canada

**Habitat:** Buried in sandy-muddy substrate, in relatively sheltered areas along the intertidal zone to depths of 55 m; at times on eelgrass

**Description:** Compared to most clamshells of this species are relatively long and narrow, slightly curved, and blunted on the ends. Hinge is close to one end. Periostracum is shiny with color ranging from greenish yellow to brown. It can reach a maximum length of 13 cm.

**Biology:** These animals are very active diggers and can quickly dig deeper into the mud if threatened. This evasive clam also has the ability to "leap" several centimeters to avoid potential predators. It prefers the higher salinity waters.

**Fisheries/Aquaculture:** No information

**Countries where eaten**: This species was nowhere found in abundance and though edible, it cannot be considered as of economic importance.

**Other uses**: Pearl production has been reported in this species.

***Solen strictus*** (Gould, 1861)

**Common name(s):** Gould's razor shell

**Distribution**: Northwest Pacific: China, Taiwan, and South Korea

**Habitat:** Intertidal zone to shallow sea; sand and mud; prefer a depth about 20–50 cm.

**Description**: Shell of this species is thin and fragile, elongated, rectangular and straight. Both ends of the shell cannot be completely closed. Top of the shell deflects toward the front, where the front opening is an ax-footed extension, and the rear opening is out of the water pipe. Shell surface is milky white with fine hinge teeth. Common size is 10 cm.

**Biology:** No information

**Fisheries/Aquaculture:** Not reported

**Countries where eaten:** China and Taiwan where it is of most edible economic value.

## 3.9.8   OCEAN QUAHOG (ARCTICIDAE)

Shell of this family is equivalve and inequilateral. Umbones are anterior to midline. Periostracum is thick and conspicuous. Ligament is external. Cardinal teeth and lateral teeth are present. Pallial line is without a pallial sinus. The only living species in the family is *Arctica islandica*.

*Arctica islandica* (Linnaeus, 1767)

**Common name(s):** Ocean quahog

**Distribution:** Northern Atlantic and the Mediterranean

**Habitat:** Dense beds over level bottoms, just below the surface of the sediment which ranges from medium to fine grain sand; prefers depths of 25–61 m; depth range 0–482 m

**Description:** Shell of this species is thick and strong, broadly oval and is with prominent umbones. Periostracum is typically brittle and peeling readily in largest shells. Anterior hinge line is strongly curved and lunule is ill-defined. Sculpture is of concentric lines and few irregular grooves and growth stages are distinct. Right valve is with three prominent cardinal teeth and a single posterior lateral tooth. Left valve is with three cardinals. Inner surfaces are smooth and glossy. Adductor scars are distinct and posterior one is slightly larger than anterior one. Ventral margin is crenulate. Color of exterior shell is dull white with a thick periostracum which is deep greenish brown to black in large specimens. Maximum shell length is 12 cm.

**Biology**

**Food and feeding:** Adult ocean quahogs are suspension feeders on phytoplankton, using their relatively short siphons which are extended above the surface of the substrate to pump in water.

**Reproduction:** Its earliest age of maturity is 7 years for both sexes, and maturity is reached at 49 mm shell length. Spawning of this species is protracted, lasting from spring to fall. Extended spawning period is from May through December, with several peaks during this time. Multiple annual spawnings may also occur at the individual and population levels. The eggs and larvae of ocean quahogs are planktonic.

**Life span:** It may reach a maximum age of 225 years.

**Fisheries/Aquaculture:** Commercial fisheries exist for this species. It is a potential species for aquaculture in Italy, Spain, and France.

**Countries where eaten:** Norway; Newfoundland to North Carolina. Quahogs are a favorite of New Englanders for chowders, clam sauces.

### 3.9.9   TELLIN CLAMS (TELLINIDAE)

These shells live fairly deep in soft sediments in shallow seas and respire using their long siphons that reach up to the surface of the sediment. Tellinids have rounded or oval, elongated shells which are much flattened. The two valves are connected by a large external ligament. The two separate siphons are exceptionally long, sometimes several times the length of the shell. These siphons have a characteristic cruciform muscle at their base. Tellins are collected for food by coastal populations in many areas, and their delicate and often colorful shells are frequently used to make decorative items.

*Austromacoma constricta* (Bruguière, 1792)

**Common name(s):** Constricted macoma

**Distribution:** Western Central Atlantic: Belize

**Habitat:** Bay; sand or muddy bottoms

**Description:** Shell of this species is oval and is slightly inflated. Anterior end is broadly rounded and posterior end is angulate and slightly truncate with weak notch on lower margin. Beak is small, pointed, and central. Posterior end is about the same length as anterior end. Fine concentric growth rings are seen. Weak ridge is from beak to posterior end. Pallial sinus is doubles back on pallial line and nearly touches anterior muscle scar before turning toward posterior scar. Periostracum is thin and brown. Color of exterior shell is white to brown and its maximum size is 6.4 cm.

**Biology:** No information

**Fisheries/Aquaculture:** Not reported

**Countries where eaten:** Mexico

*Limecola balthica* (Linnaeus, 1758)

**Common name(s):** Baltic macoma, Baltic clam, Baltic tellin

**Distribution**: Northwest Pacific: Northern Alaska to southern California

**Habitat**: Lives buried to 20 cm below surface in sandy or muddy bays; and in estuaries from intertidal to a depth of 40 m

**Description:** Shells of this species are smooth, relatively flat, oval or somewhat trigonal in shape. Concentric growth rings which are indicating the age of the specimen are often clearly visible. Shell color is polymorphic, varying between individuals and between localities. Often most specimens are white, sometimes most are pink, and also yellow and orange shells may occur. Maximum shell length is 4 cm.

**Biology:** No information

**Fisheries/Aquaculture:** Not reported

**Countries where eaten**: Western Pacific USA (Puget Sound and San Francisco Bay)

**Other uses:** These shells are known for their lovely colors. They are often prized.

*Macoma nasuta* (Conrad, 1837)

**Common name(s):** Bent-nosed clam, bent-nose macoma

**Distribution:** Northeast Pacific: Alaska, Canada, Baja California Sur

**Habitat:** Low intertidal to 50 m; buried in mud flats; gravel, sand, or muddy clay; depth range 0–50 m

**Description:** This clam has a rounded shell with neither valve very flat nor very inflated, and the anterior and posterior ends are shaped differently. There are two adductor muscle scars of similar size on each valve. Umbones are near the middle of the dorsal side. It has no radial ribs. Hinge has a true hinge plate with two cardinal teeth on both valves. Hinge ligament is mostly external. Valves have a pallial sinus and a continuous pallial line. Valves gape only slightly, if at all, at the posterior end. Shells are chalky white and usually unstained, with some grayish-brown periostracum. Maximum length is 7.5 cm.

**Biology**

**Ecology:** This species is very hardy and can be found in areas that have very poor circulation and can live in very soft, silty mud.

**Behavior:** This common clam lies on its left side instead of vertically, at a depth of 10–20 cm. It rocks back and forth while digging.

**Food and feeding:** The siphons extend out the right side and up to the surface. The siphons are used to suck debris from the surface of the sediment like a vacuum cleaner. The clams digest mainly diatoms and some flagellates from the sediments. They ingest large quantities of sediment but reject 97% of it, producing copious pseudofeces.

**Reproduction:** The species spawns in early summer in Oregon.

**Predators:** These include the moon snail *Polinices lewisii*.

**Association**: The pea crab *Pinnixa littoralis* or *Pinnixa faba* may live in the mantle cavity as may the Nemertean worm *Malacobdella grossa*.

**Fisheries/Aquaculture:** Not reported

**Countries where eaten**: California. This clam is an important food of the coastal Indian tribes and to Chinese immigrants in San Francisco but is little used commercially today because of the debris that is usually in the gut.

***Rexithaerus secta*** (Conrad, 1837) (=*Macoma secta*)

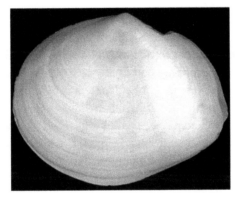

**Common name(s):** White-sand macoma, butterfly clam, white sand clam

**Distribution**: Northeast Pacific

**Habitat:** Buried to a depth of 50 cm in sand or sandy mud; intertidal to a depth of 50 m

**Description:** Shell of this species is white in color and glossy. Periostracum can often be seen along the edge of the shell. Shell is angled sharply downward at the posterior end. Hinge is visible but is relatively short. Maximum length is 10 cm.

**Biology:** These clams, which are deposit feeders, have white siphons. As in other Macoma clams, the siphons are separate.

**Fisheries/Aquaculture:** No information

**Countries where eaten**: California

*Megangulus bodegensis* (Hinds, 1845) (=*Tellina bodegensis*)

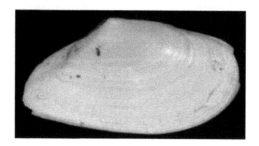

**Common name(s):** Bodega tellin, bodega clam

**Distribution:** Southern Alaska to Bahia Magdalena, Baja California, Mexico

**Habitat:** Exposed sandy shores; intertidal zone to depths of 100 m

**Description:** As of other species of Family Tellinidae, outside shell has no radial ribs and only fine concentric growth lines are seen. Shell is about twice as long as high, and the anterior end of the shell is broader and more rounded than the pointed posterior end. Hinge ligament is external. Shell valves close together or nearly so on the posterior end. Inside the clam, it has two adductor muscle scars of almost equal size, a pallial sinus, and two cardinal hinge teeth on each valve. It has no chondrophores. Shell is white outside and may have yellow or pink tinge inside. Maximum length of the shell is 6 cm.

**Biology:** No report

**Fisheries/Aquaculture:** No information

**Countries where eaten:** California

*Macomangulus tenuis* (da Costa, 1778)

**Common name(s):** Thin tellin, petel tellin

**Distribution:** Off the coasts of northwest Europe and in the Mediterranean Sea; buried in sandy sediments.

**Habitat:** Fine sand generally from about the middle of the intertidal zone to a depth of a few meters.

**Description:** Shell of this species is brittle, somewhat flattened, inequilateral, and slightly inequivalve. Right valve is a little bigger than the left. Beaks are just behind the midline directed inward and backward, almost touching. Almost oval in outline with the posterior region which is somewhat attenuated and curved slightly to the right. Ligament is a prominent green-brown arched band extending backward from the beaks one-third the way to the posterior margin. Sculpture is of concentric lines grouped into growth stages which may be accentuated by color bandings. Right valve is with two cardinal teeth. Left valve has two cardinals. Cruciform muscle scars are sometimes distinct. Pallial sinus is very deep. Margin is smooth. Color of shell is white, pink, rose, orange, or yellow. Periostracum is transparent and glossy. Inside of shell is colored like the outside but normally fainter. Maximum length is 1.9 cm.

**Biology**

**Behavior:** It burrows in clean sand and has a large foot and two long siphons which it extends to the surface of the sediment. Water and food particles are drawn down to the mollusc through one siphon while water is expelled through the other. At the low tide level, it may be the most abundant animals in the benthos and may attain a density of 3000 individuals per square meter.

**Food and feeding:** It is both a deposit and a filter feeder. It usually lies on its left side when feeding. When the tide is out it descends to a depth of about 10 cm but ascends to nearer the sediment surface when feeding.

**Reproduction:** Individuals are either male or female and gametes are liberated into the water table during the summer. The larvae are free swimming and form part of the zooplankton for a period. The life-span is up to 5 years.

**Fisheries/Aquaculture:** No information

**Countries where eaten:** In France, it is served with garlic, parsley, and olive oil.

## 3.9.10   BASKET CLAMS (CORBICULIDAE)

The members of this family are of moderate-sized clams, often tinged or colored with violet on the interior. The name "corbicula" (little basket) is due to their characteristic concentrically ribbed shells and general shape. They are ovoviviparous, brooding internally fertilized eggs in brood pouches.

***Corbicula japonica*** (Prime, 1864)

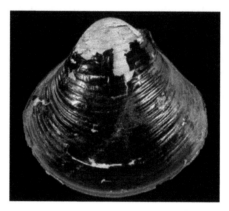

**Common name(s):** Japanese corbicula

**Distribution:** Northwest Pacific: from south of Japan to the south of Sakhalin

**Habitat:** Brackish water; subtidal areas of estuaries in mud; depth range 0–6 m

**Description:** Shell of this species is rounded, trigonal, and well inflated, with thick dark brown or black periostracum. Beak is high. Outer surface is ornamented with more or less regularly spaced commarginal ribs which are more distantly spaced in young shell. Young shell is with yellowish green radial streaks. Shell color inside is violet and sinus is absent. Maximum shell length is 6 cm.

**Biology:** No information

**Fisheries/aquaculture:** It is commercially harvested for food.

**Countries where eaten:** Very popular in Japan. In Primorye, this mollusc is landed commercially with subsequent export abroad.

**Other uses:** This species is known for its vitamin and carotenoid contents. In Lake Togo (Tottori prefecture, Japan), the vitamin B12 content of this species remained high (17.3–22.5 μg/100 g wet weight. The total carotenoid content of this species was found to be 5.3 mg/100 g in the edible part (wet weight). These carotenids may be of great use in food formulations (Maoka et al., 2005).

***Polymesoda bengalensis*** (Lamarck, 1818)

**Common name(s):** Bengali Geloina, mangrove clam, marsh clam

**Distribution:** Indo-West Pacific: from India to Thailand and Celebes and possibly also from the Philippines to northern Australia.

**Habitat:** Freshwater; marine; buried in stiff mud in mangroves; tolerates long periods of low tide

**Description:** Important character of this family is the heterodon type of tooth at the hinge of the shell. Shell is without ventral posterior margin. There is a close relationship between the length and height with thick shells. Maximum length of the shell is 9.5 cm.

**Biology:** No information

**Fisheries/Aquaculture:** Not reported

**Countries where eaten:** Vietnam

**Other uses:** This species is economically very important. In India, huge quantities of these shells are collected from the different parts of the Sundarbans and brought to the shell factories where they are powdered by crushing for use as calcium supplement in poultry feed. Further, the shells are also used in making handicrafts (Do et al., 2012).

*Geloina erosa* (Lightfoot, 1786) (=*Polymesoda erosa*)

**Common name(s):** Common geloina

**Distribution:** Indo-West Pacific: from India to Vanuatu; north to southern islands of Japan, and south to Queensland and New Caledonia

**Habitat:** Intertidal areas; fresh and brackish waters of mangrove swamps, estuaries, and larger rivers

**Description:** Shell of this species is subrhomboidal-ovate in outline. Shell is without ventral posterior margin. Maximum length is 10.5 cm.

**Biology:** It is highly tolerant to surface desiccation of its habitat and can survive by aerial respiration at the posterior mantle margin for a period of a few days and feed from subterranean water by means of water exchange through a narrow anterior gape of valves.

**Fisheries/Aquaculture:** It is widely collected as food in Asia.

**Countries where eaten:** Malaysia

**Other uses:** The hydrolysate of this species has the potential to be used in food formulations for human consumption (Normah and Noorasma, 2015).

*Geloina expansa* (Mousson, 1849) (=*Polymesoda expansa*)

**Common name(s):** Broad geloina, marsh clam

**Distribution:** Indo-West Pacific: from India to Vanuatu; north to Viet Nam and south to eastern Java

**Habitat:** Brackish to almost fresh water areas of mangrove swamps; buried in the stiff mud

**Description:** Shell of this species is expanded posteriorly in outline. Maximum length is 10 cm.

**Biology:** Its diurnal rhythm of activity and inactivity strongly depending on the tides and rainfall. It can survive during drought periods by aerial respiration at the posterior mantle margins. It also has the ability to resume filter-feeding rapidly when inundated.

**Fisheries/Aquaculture:** It is widely collected as food in Asia

**Countries where eaten:** Malaysia

## 3.9.11  JEWEL BOX CLAMS (CHAMIDAE)

These bivalves superficially resemble oysters as most species are cemented to hard substrates such as rocks or other shells. The shells are thick, with a somewhat round outline but irregularly shaped. They are usually covered with short spikes or leaf-like structures. The ligament is external, while the hinge thick and arched with a few large curved teeth and corresponding sockets. The shell interior is porcelaneous, and each valve has two large, subequal and ovate adductor muscles scars. There is a pallial line without a sinus. They have a very reduced foot. They are collected for food and the shells are sometimes used as lime material or for shell craft.

***Chama coralloides*** (Reeve, 1846)

**Common name(s):** Not designated

**Distribution**: Western Central Atlantic: Baja California to Panama

**Habitat**: Attached to rocks at low water

**Description:** Color is the best diagnostic feature of this species. Interior is dark purple. Exteriorly, it is grayish-white. Shell is covered with short, irregular spines. Common length is 3.5 cm and height is 4.5 cm.

**Biology:** No information

**Fisheries/Aquaculture:** Not reported

**Countries where eaten**: It is edible in areas of its occurrence.

*Chama macerophylla* (Gmelin, 1791)

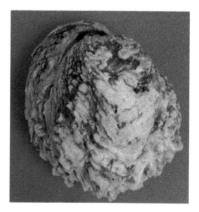

**Common name(s):** Leafy jewel box clam, Bermuda rock cockle

**Distribution:** Atlantic coast of North America, ranging from North Carolina to the West Indies

**Habitat:** Gulf; attaches to platforms and other hard surfaces; coastal waters ranging from the mean low tide to a depth of 20 m

**Description:** Shell of this species is circular, inflated and is variable depending on the shape of whatever it attaches to. Inside margin is finely crenulate (scalloped). Radiating projections are seen from shell. Leaf-like, projections may be long or short and longer projections may have radial grooves. Projections have no regular pattern and are very irregular. Muscle scars are large. Color of exterior shell is usually yellow to orange, sometimes pinkish or purplish. Inside shell is white, except around edges. Maximum shell length is 7.5 cm and dia. is 10 cm.

## Biology

**Feeding:** This sessile species gains its sustenance by filter feeding particles from the water column.

**Reproduction:** Members of this species are gonochoric, and fertilization occurs externally. The resulting larva is a trochophore, and when it develops it becomes a veliger. Both larval stages are planktonic and settle to hard substrata to become sessile adults.

**Fisheries/Aquaculture:** It is commercially harvested.

**Countries where eaten:** Venezuela

**Remarks:** It costs fishermen and corporations a great deal of money each year to prevent and remove this species from ships hulls and docks because of its fouling nature.


### 3.9.12   OTHER SALTWATER VENERID CLAMS

#### 3.9.12.1   BIVALVIA: CYRENIDAE

Bivalves of this family are found in fresh or brackish water and in the mangroves. These clams are free-living with no byssus. The symmetrical shells are somewhat subtriangular, with two or three cardinal teeth and several laterals. The surface is usually marked with concentric growth lines. The foot is large and strong, while the siphon is short. Many species are collected for food.

*Villorita cyprinoides* (Gray, 1825)

**Common name(s):** Black clam

**Distribution:** Western Indian Ocean: India

**Habitat:** Freshwater; brackish water; fine sand, clay and silt just below the surface of soft bottom sediments; depth range 2–3 m

**Description:** Shell of this species is large and thick. Anterior margin is short; regularly curved above and is almost straight in the middle. Surface is with thick concentric ridges. Umbones are prominent, near the anterior side and recurved. A large, thick external ligament is seen posteriorly. Maximum size is 2 cm.

**Biology:** No information

**Fisheries/Aquaculture:** It is harvested extensively from the wild

**Countries where eaten:** It is a very important shell fish. Thousands of people in Kerala (India) depend on this species for their livelihood and basic protein supply.

### 3.9.12.2   BIVALVIA: GLUCONOMIDAE

Bivalves of this family are usually small and found under rocks or coral rubble. They have very reduced and flexible shells which are mostly enclosed within the well-developed mantle. The hinge is irregular, with tiny cardinal teeth and somewhat obscure lateral teeth.

*Glauconome virens* (Linnaeus, 1767)

**Common name(s):** Sea-green mussel, greenish glauconomya

**Distribution:** Indo-West Pacific: from Thailand and the Philippines to northern Australia.

**Habitat:** Embedded in muddy bottoms of brackish water areas, in estuaries and in mangroves

**Description:** Shell of this species is moderately thin, equivalve, elongate-ovate in outline, and narrowly gaping. Posteriorly, it is strongly inequilateral and its anterior part is relatively short and widely rounded. Posterior part is elongate and narrowly pointed. Umbones are low. Outer surface is with irregular concentric lines and grooves and is finely wrinkled. There is no lunule or escutcheon. Periostracum is conspicuous and is often corroded on umbones. Ligament is external. Hinge is with three cardinal teeth in each valve. Lateral teeth are wanting. Posterior right and median left cardinals are strongly bifid. There are two adductor muscle scars; anterior is somewhat narrow and the posterior is rounded. Pallial line is with a deep and narrow sinus, extending forward almost to the level of umbones. Pallial sinus is meeting ventrally the pallial line at its posterior extremity. Internal margins are smooth. *Color*: outside of shell is white or cream, under a greenish periostracum. Inner side is porcelaneous white. Maximum shell length is 7 cm and common length is 5 cm.

**Biology:** It is embedded and is also found in nipa palm swamps. It normally stays in an upright position, with the siphons pointing slightly above the surface of the bottom, forming in it a wedge-shaped hole by their whirling movements.

**Fisheries/Aquaculture:** It is locally exploited in the Philippines for food and is very commonly marketed together with solenids of the genus *Pharella* in Cebu.

**Countries where eaten:** Malaysia and Philippines

## 3.9.12.3   *BIVALVIA: MESODESMATIDAE (WEDGE CLAMS)*

Shell of the members of this family is equivalve, inequilateral and is subtrigonal to wedge-shaped. Umbones are opisthogyrate. Internal ligament is in a deep pit of hinge plate. There are one or two cardinal teeth and lateral teeth. Two adductor muscle scars are seen. Pallial line is with a short sinus.

*Atactodea striata* (Gmelin, 1791) (=*Atactodea glabrata*)

**Common name(s):** Striate beach clam, wedge clam, surf clam

**Distribution:** Indo-Pacific: from East Africa, including Madagascar and the Red Sea, to eastern Polynesia; north to Japan and south to central Queensland

**Habitat:** Intertidal areas; lives as a shallow burrower in littoral and sublittoral sands mainly littoral

**Description:** Shell of this species is small, solid, equivalve and is slightly inequilateral. Outline is very variable, oval, or subtrigonal. Sculpture is of weak, dense concentric lines. Hinge is strong and is contracted with short lateral teeth. Cardinals of left valve are not bifid. Ligament is internal on a narrow and deep resilium. Pallial sinus is very short. Margin is smooth. Color of exterior shells is dirty white. Maximum shell length is 4 cm.

**Biology:** No information

**Fisheries/Aquaculture:** It is commercially exploited in some areas by coastal inhabitants.

**Countries where eaten**: It is an important food source in Fiji

*Mesodesma donacium* Lamarck, 1818

**Common name(s):** Macha clam

**Distribution:** Southern Pacific: Peru and Chile

**Habitat:** Sandy beaches; mainly in the breakers; depth range 0–50 m

**Description:** Its valves are triangular and elongated. Posterior is truncated and short. Each valve has both. In its dorsal portion as above, a protuberance called umbo. Exterior of shell is polychromatic and interior has a porcelain-pearly color. Intervalvar union is by a hinge. Closure of shells is secured by a pair of divergent side teeth. Maximum length is 9 cm.

**Biology**

**Behavior:** This species lives in the swash zone of exposed high-energy intermediate and dissipative sandy beaches. It usually burrows to a depth of about 10 cm but can reach 25 cm if disturbed. Adults are mainly restricted to the surf zone while most of the juveniles to the swash zone.

**Food and feeding:** It consumes phytoplankton and zooplankton, larvae of molluscs, crustaceans and polychaetes, and it feeds on filtered organic particles.

**Reproduction:** Fertilization is external and its larvae are planktonic between 30 and 45 days. The female puts the ovules and a substance through the water which determines the ejaculation of sperm. The larval stages trochophores and velígers are pelagic. Minimum spawning occurs when the shell reaches a size of 4.7–5.7 cm. Sexual maturity of this species is at an age of 2–6 years.

**Predators:** Striped eagle, *Astobatus peruvianus*, and seabirds like the seagull, *Larus dominicanus*.

**Fisheries/Aquaculture:** It is commercially exploited for human consumption. It is the most valuable commercial species in USA, Europe, Japan, Chile, and Peru and is extensively cultivated in Canada.

**Countries where eaten**: Peru to Chile. The dried product is a source of protein, vitamins, and minerals.

*Paphies australis* (Gmelin, 1791)

**Common name(s):** Pipi wedge clam

**Distribution:** Southwest Pacific: New Zealand

**Habitat:** Intertidal; common near the mouths of estuaries; coarser sediments; depth range 0–9 m

**Description:** This species has a solid white to light brown/orange and elongated symmetrical shell with the apex at the middle. It is covered by a thin yellow periostracum. Maximum length is 8 cm and height is 5 cm.

## Biology

**Ecology:** Pipi are often most abundant near subtidal channels where they burrow to about 3 cm below the sediment surface. Juveniles are normally found in fine sandy habitats while adults prefer coarser sediment and fast currents.

**Locomotion:** By releasing a thread of mucus, which makes them more buoyant, these animals are able to float in the water column and move to new locations. Where they find good living conditions, their density may exceed more than 1000 individuals per square meter.

**Feeding:** It is a suspension feeder with short siphons and is living near the sediment surface.

**Reproduction:** Maturity is reached at the size of around 4 cm.

**Fisheries/Aquaculture:** Pipi are easily collected for food.

**Countries where eaten:** Pipi are an important food source for birds and humans in areas of its occurrence.

**Indicator species:** This species prefers mainly a maximum sediment mud content of 5% and are very sensitive to high turbidity. They are also sensitive to zinc contamination. Where sediment becomes muddier (>5% mud) and/or more polluted (particularly with zinc) the abundance of these animals is likely to be reduced. For this reason, this species serves as a good indicator species with which to assess changes in the input of sediment and pollutants into our estuaries.

**Remarks:** This species is a commonly eaten species of bivalve that was found to contain TTX at levels up to 0.80 mg/kg (McNabb et al., 2014).

*Paphies subtriangulata* (Gray, 1828)

**Common name(s):** Northern tuatua

**Distribution:** Southwest Pacific: New Zealand

**Habitat:** Buried in fine clean sand on ocean beaches; from the low intertidal to the shallow subtidal (down to about 4 m depth).

**Description:** Outer surface of the shell of this species is stained a reddish brown from iron-rich sand. Maximum dimensions ($H \times W \times D$): 1.5 × 9.6 × 5.6 cm.

**Biology**

**Locomotion:** This clam burrows beneath the sand and does so very quickly, making it a challenge to dig for at times. It also squirts water when it is threatened.

**Feeding:** The tuatua is a suspension feeder with short siphons.

**Reproduction:** These animals have separate sexes (1:1 sex ratio) and reproduce by broadcast spawning, synchronously releasing eggs and sperm into the water column for external fertilization. Planktonic larval development takes about 2–3 weeks. Larval settlement occurs high in the intertidal, but spat and juveniles are highly mobile, moving around with the tidal flow before reburying themselves rapidly. Juveniles migrate down the beach to occupy the lower intertidal and shallow subtidal as they grow larger. Growth appears to be rapid reaching 4–7 cm shell length in about 3 years. Maximum age of this species is about 5 or more years.

**Fisheries/Aquaculture:** Tuatua support an extensive recreational fisheries and commercial fisheries.

**Countries where eaten:** Areas of its occurrence. Soft parts of the animal are an edible delicacy

*Paphies ventricosa* (Gmelin, 1790)

**Common name(s):** Long tongue, toheroa clam

**Distribution:** Southwest Pacific: New Zealand

**Habitat:** Wide fine-sand beaches where there are extensive sand dunes enclosing freshwater, which percolates to the sea, there promoting the growth of diatoms and other plankton.

**Description:** It is a very large shellfish with a solid white, elongated shell with the apex at the middle. Outer surface is stained with a reddish brown color from iron-rich sand. Maximum length is 12 cm, height 8 cm, and thickness 4 cm.

**Biology:** No information

**Fisheries/Aquaculture:** Not reported

**Countries where eaten**: Areas of its occurrence. It is an extremely popular seafood, often made into a greenish soup, for which it has an international reputation.

### 3.9.12.4 BIVALVIA: PETRICOLIDAE

The symmetrical shell of this family is usually ovate, thin, and white. The hinge has two or three cardinal teeth, but there is no lateral tooth. They have a deep pallial sinus, long siphons, and a narrow foot. The umbones are clearly protruding, with concentric growth lines expanding from the dorsal end to the anterior end. The animal is usually free-living but is often found boring into clay, soft rocks, or corals.

*Petricola lithophaga* (Retzius, 1788)

**Common name(s):** Boring petricola, rock-eating petricola

**Distribution:** Mediterranean Sea: Greece; Black Sea

**Habitat**: It is an endolithic species (endolith is an organism that lives inside rock, coral, animal shells); common on many hard substrates

**Description:** Shape of this shell is forcibly very regular since it lives in niches that dig itself in hard substrates. Valves have an oval shape and inequilateral and is elongated at rear. Surface of shell has several radial ridges with corrugations and irregular accretion striae. Color of exterior shell is creamy white. It is quite common to find whole rocks perforated by Petricola, or collect old valves of thick enough dead oysters. Common size is 2.5 cm.

**Biology**

**Epizoan communities on marine turtles:** A female *Caretta caretta*, the loggerhead turtle had an oyster, *Ostrea edulis* on its carapace. On the oyster was a *Petricola lithophaga* together with annelid and sipunculid worms and algae (Frazier et al., 1985).

**Fisheries/Aquaculture:** No information

**Countries where eaten:** It is eaten in Montenegro (Mediterranean) coast

*Petricola fabagella* (Lamarck, 1818)

**Common name(s):** Not designated

**Distribution:** Western Pacific: China, Japan, and Taiwan

**Habitat:** Intertidal in rocks

**Description:** Shell of this species is thin, inequilateral, and equivalve. Outline of shell is subovate. Umbones are located in front of mid-line. There is no lunule or escutcheon. Ligament is external. Radial sculpture is of riblets anteriorly and coarser and scaly ribs posteriorly. Pallial sinus is large and not confluent with pallial line. Internal margin is smooth. Outside and inside of shell is white. Maximum length is 14 mm.

**Biology:** No information

**Fisheries/Aquaculture:** Not reported

**Countries where eaten:** Alexandria, Egypt

### 3.9.12.5 BIVALVIA: PSAMMOBIIDAE

Shell of this family is ovate to subelliptical or trapezoidal and is somewhat gaping. Ligament is external, on projecting nymphs. There are two small cardinal teeth in either valve and lateral teeth are absent. Two adductor muscle scars are seen. Pallial sinus is deep. Cruciform muscle scars are often obscure.

***Gari kazusensis*** (Yokoyama, 1922) (=*Gari californica*)

**Common name(s):** California sunset clam

**Distribution:** Northeast Pacific: Canada

**Habitat:** Buries in gravel about 15–20 cm deep along the intertidal zone; depth range 0–170 m

**Description:** Shell of this species is ovate-subquadrate to ovate elongate, moderately large, thin, eqivalve, and slightly flattened. Outer surface is rather smooth, with fine commarginal growth lines. Posterodorsal and anterodorsal margins are broadly rounded but the former one is sometimes nearly straight. Ventral margin is also nearly straight. Beaks are small, low and are placed nearly midpoint of dorsal margin. Right valve is with anterior cardinal and narrow posterior cardinal teeth. Left valve is with narrow, weak anterior cardinal and narrow posterior cardinal teeth. Pallial sinus is moderately deep, rounded, and broad. Adductor scars are ovate; the anterior one is more narrow and posterior is broader. Color of exterior shell is white. Periostracum is brownish and light in young specimens. Maximum length is 15 cm.

**Biology:** No information

**Fisheries/Aquaculture:** Not reported

**Countries where eaten:** It is edible in areas of its occurrence.

*Gari solida* (Gray, 1828)

**Common name(s):** Pacific clam

**Distribution:** Southeast Pacific: Peru to Chile

**Habitat:** Intertidal to subtidal to a depth of 20 m

**Description:** Shell of this species is thick, elongated oval and truncated toward the rear end shell. Front end is shorter and angled. Externally, the shell has fine concentric grooves which are thickened toward the rear end. Ligament is elongated and is located behind the umbones. Adductors are large and muscular impression shows that the former is more oval than later. Hinge has two cardinal teeth in each valve. Edges are contiguous, which serve for insertion of the ligament and are outstanding. Internally, the color of the shell is white with cream-colored spots. Approximate length of the shell is up to 10 cm.

**Biology**: It is an infaunal burrower in sandy substrates. It feeds on plankton and organic detritus.

**Fisheries/Aquaculture:** It is commercially harvested in Peru

**Countries where eaten**: Chile

*Gari togata* (Deshayes, 1855)

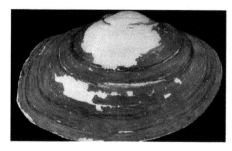

**Common name(s):** Courtesan sunset clam

**Distribution:** Western Pacific: from Indonesia to Western Polynesia; north to China and south to northern New South Wales and New Caledonia

**Habitat:** Deep burrower in thick muddy substrates; depth range 0–10 m.

**Description:** Shell of this species is elongate, moderately inflated, variable in shape, subelliptical and ovate in outline. Posterior margin is rounded to slightly truncate. Ventral margin is broadly rounded. Umbones are decidedly anterior to midline of valves. Outer surface is dull, with numerous, irregular, fine to coarse, concentric growth marks. Periostracum is fibrous. Pallial sinus is not sloping forward, extending just in front to level with umbones. Ventral limb of pallial sinus is only partly confluent with pallial line, generally meeting it a rather short distance to posteroventral end. *Color*: outside of shell is whitish. Interior is whitish. *Size*: Maximum shell length is 10.5 cm.

**Biology:** This species is commonly forming mixed populations with the greenish glauconmya (*Glauconome virens*), a species with considerable external similarity, but it is absent from reduced salinity areas of estuaries.

**Fisheries/Aquaculture:** No information

**Countries where eaten**: It is edible in areas of its occurrence

*Nuttallia nuttallii* (Conrad, 1837) (=*Sanguinolaria nuttalli*)

**Common name (s):** Mahogany Clam, California mahogany clam

**Distribution:** East Pacific.

**Habitat:** Deep-burrowing bivalve

**Description**: Shell of this species is subovate, thin and is much compressed. Posterior margin is obliquely truncate. Extremity is angular and basal margin is regularly arcuate. Beaks are small; distant from the anterior margin; slightly prominent and acute. Ligament is short and very prominent. Nymphs are very prominent. Exterior shell color is whitish, with purple zones and rays. Epidermis is polished and is horn colored, with paler spots and rays. Cardinal teeth are prominent, slender and fragile. Maximum shell length is 15 cm.

**Biology:** This clam is a nonselective suspension feeder that keeps its siphons slightly below the surface of the sand or muddy bays that it occupies.

**Fisheries/Aquaculture:** No information

**Countries where eaten**: Central California to Baja California

*Nuttallia obscurata* (Pohlo, 1972)

**Common name(s):** Purple mahogany clam, purple varnish clam, dark mahogany clam, varnish clam, savory clam

**Distribution:** Northeast Pacific: Gulf of Alaska, Canada, Baja California Sur

**Habitat:** High-to-mid intertidal; brackish water; sand or sand/gravel, often in areas of freshwater seepage; depth range 0–20 cm

**Description:** It is a clam with thin, oval shells that are relatively flat. Outside is covered with a shiny brown periostracum, which is usually worn off near the umbo and hinge. Hinge ligament is prominent and external. Inside of the shell is purple (sometimes it may be nearly white). Living clam has two long, separate siphons, which is reflected by the deep pallial sinus on the interior. Maximum length is 5.5 cm.

**Biology:** The varnish clam is more freshwater tolerant than other species.

**Fisheries/Aquaculture:** This species is commercially harvested and cultivated.

**Countries where eaten**: It is eaten in Korea, Japan, British Columbia, and Washington, USA; marketed in Washington state as the "savory clam."

### 3.9.12.6   BIVALVIA: SEMELIDAE

Shell of this family is rather compressed and is often slightly inequivalve, with a rightward flexure posteriorly. Internal ligament is in a small pit of hinge plate. Hinge is with two cardinal teeth and lateral teeth. Two adductor muscle scars are present. Pallial line is with a deep sinus. Cruciform muscles are leaving small paired scars near pallial line.

*Abra alba* (Wood, 1802)

**Common name(s):** White furrow shell

**Distribution:** North eastern Atlantic Ocean and the Mediterranean Sea

**Habitat:** Occasionally on the lower shore but most abundant in shallow, offshore waters (to about 60 m); buried in soft sediments

**Description:** Shell of this species is thin and brittle and is broadly oval. Umbones are posterior to midline. Sculpture is of very fine concentric lines and growth stages are clear. Chondrophore is elliptical, posteriorly directed. There are two small, peg-like cardinal teeth anterior to chondrophore in right valve and a single one in left valve. Single, thin anterior, and posterior lateral teeth are present in right valve. These teeth are present but less prominent in left valve. Adductor scars and pallial line are clear. Cruciform muscle scars are visible with hand lens. Pallial sinus is deep and is rather irregular. Color of the exterior shell is white and glossy. Inner surfaces are white. Maximum shell length is 2.5 cm.

### Biology

**Food and feeding:** It feeds by means of a pair of separate long, extensible siphons, elongations of the mantle. It is mostly a deposit feeder but is also able to feed on suspended particles.

**Reproduction:** A large number of very small eggs are produced by this species from May to August. The larvae may y have a lengthy period as part of the zooplankton before settling out. Juveniles grow fast during the summer but the ones that settle in autumn seem to delay their further development until the following spring. The life-span of this species is about 2 years.

**Associated species:** *Tellina fabula*, *Phaxas pellucidus*, and *Lagis koreni*

**Fisheries/Aquaculture:** No information

**Countries where eaten:** It is edible in areas of its occurrence.

*Semele decisa* (Conrad, 1837)

**Common name(s):** Clipped semele

**Distribution:** East Pacific; from southern California down to the Baja California peninsula

**Habitat:** Lives primarily in intertidal zones buried shallowly in the sand

**Description:** Valves of this species are nearly round in outline, firm and fairly heavy, little arched, and different on the two sides at the siphonate end. Ligament is fairly large and external. Pallial sinus is large and rounded. Exteriorly the shell is roughened, giving a granular effect. It is faintly tinged with pink and shows more or less of a brown periostracum. Within, the valves are tinged with a faint purple particularly near the margins, making in all a very handsome shell. Maximum shell length is 9 cm.

**Biology:** No information

**Fisheries/Aquaculture:** It is commercially harvested. Due to overfishing and habitat destruction, this species has become rare.

**Countries where eaten:** It is listed as one of the Pacific coast edible species. However, it is not at present abundant enough to be of economic importance.


### 3.9.12.7   BIVALVIA: SOLECURTIDAE

Shell of this family is elongate, equivalve, and inequilateral. Gaping is anteriorly and posteriorly. Ligament is external. Hinge line is with peg-like cardinal teeth which are no more than two in each valve. Lateral teeth are present or absent. Pallial line is with sinus. Cruciform muscles are present and scars are indistinct.

*Azorinus chamasolen* (da Costa, 1778)

**Common name(s):** Antique razor clam

**Distribution:** Norway and of Sweden; from the south and west of the British Isles south to the Azores; Canary Islands and Angola; and Mediterranean

**Habitat:** Lives in bottoms of thick, solid, black mud, muddy sand, or muddy gravel; offshore to considerable depths; depth range 5–400 m

**Description:** Shell of this species is solid, equivalve and is almost equilateral. Beaks are just in front of the midline and are directed inward. Shell is cylindrical in outline with anterior and posterior extremities rounded and gaping widely. Ligament is a dark brown arched band extending one-third the way to the posterior margin. Sculpture is of fine concentric lines and ridges. Growth stages are not clear. Right valve is with two cardinal teeth. Left valve is with one similar cardinal. Cruciform muscle scars are obscure. Pallial sinus is deep, extending to a point below the beaks. Margin is smooth. Color of exterior shell is white or fawn. Periostracum is thick, wrinkled, and dark brown or greenish yellow. Inside of shell is white. Maximum shell length is 6 cm.

**Biology:** No information

**Fisheries/Aquaculture:** Not reported

**Countries where eaten**: Adriatic, Mediterranean, and Spain

*Solecurtus divaricatus* (Lischke, 1869)

**Common name(s):** Divaricate short razor, manifold pattern hair razor

**Distribution:** Western Pacific: from the Malay Peninsula to the Philippines; north to Japan and south to northern New South Wales

**Habitat:** Intertidal zone and shallow sandy seabed; depth range 0–20 m

**Description**: Shell of this species is rectangular and its exterior color is white to ivory yellow. It has narrow rectangular body, a long fleshy foot and basally fused siphons. Maximum shell length is 7.5 cm.

**Biology**: No information

**Fisheries/Aquaculture:** It is collected locally for food in Southeast Asian countries where the species is common.

**Countries where eaten**: Yugoslavia

***Solecurtus strigilatus*** (Linnaeus, 1758)

**Common name(s):** Rosy razor clam, rasp tagelus, scrapper sloecurtus, rasp shorts razor

**Distribution**: Eastern Central Atlantic and the Mediterranean: Congo and Greece

**Habitat:** Subtidal species

**Description:** Shell of this species is subrectangular and is elongated. Clear cut oblique striations run through the valves. Hinge is formed by two cardinal teeth on each valve. Inside very broad palleale breast is seen. This is a very striking and characteristic shell, for its intense pink color with the two typical radial white bands. Common shell length is 8 cm.

**Biology:** It is a jet-propelled burrowing bivalve.

**Fisheries/Aquaculture:** No information

**Countries where eaten:** Adriatic, Mediterranean, Atlantic Spain, Portugal, Norway to Senegal; and Congo

***Tagelus californianus*** (Conrad, 1837)

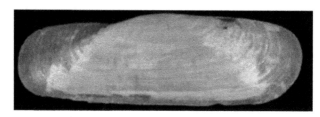

**Common name(s):** California Jackknife clam, California tagelus

**Distribution**: Eastern Pacific: USA and Mexico

**Habitat:** Brackish water; low intertidal zone; muddy sand or mud in back bays, sloughs and estuaries; flats and channels of protected bays; occupies a permanent burrow which is 10–50 cm deep; mud-flat species

**Description:** This species has yellowish-white shell with rust colored stains and is with a dark periostracum. Posterior dorsal margin is slightly sinuous, not sloping downward from beaks. Pallial sinus is shorter not extending past the beaks. Common length is 12 cm and height is 3 cm.

**Biology**

**Nutrition and locomotion:** It feeds on suspended material. If disturbed, this clam pulls in its siphons and by means of its powerful foot rapidly retreats to the bottom of its burrow. If still followed, it will commence to dig into the softer underlying mud or sand.

**Fisheries/Aquaculture:** No information

**Countries where eaten:** This clam is said to have a fair flavor and it may be considered a potentially valuable species. Its use as food, however, was not observed in any of the localities visited.

*Tagelus dombeii* (Lamarck, 1818)

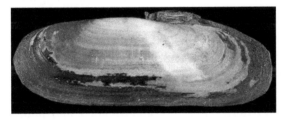

**Common name(s):** Hard razor clam

**Distribution:** Eastern Pacific: Panama to Chile

**Habitat**: Infaunal species; tide pools, sandy, pebbly, muddy, and clay substrates from the intertidal area to a depth of 37 m

**Description:** Edges of the shell are arranged almost parallel with each other and have rounded ends. This species presents a bulky foot and a pair of long siphons which are used for feeding. Maximum shell length is 10 cm.

## Biology

**Feeding:** It is mainly a filter feeder, particularly a suspension feeder. It can switch to surface deposit-feeding based on food availability.

**Life cycle:** Embryos develop into free-swimming trocophore larvae, succeeded by the bivalve veliger, resembling a miniature clam.

**Fisheries/Aquaculture:** It is commercially harvested.

**Countries where eaten:** It is an edible species, which is extracted manually on the Chilean coast at low tide. Fresh are used in the preparation of soups and pies or smoked.

*Tagelus plebeius* (Lightfoot, 1786)

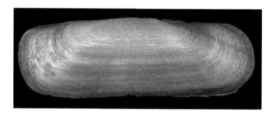

**Common name(s):** Stout tagelus

**Distribution:** Western Atlantic: from Cape Cod to Argentinian Patagonia and Gulf of Mexico, San Matias Gulf, Argentina

**Habitat:** Brackish water; intertidal to shallow subtidal, in muddy sand or mud; stable sediments with silts and clays and are covered with film of benthic microalgae; estuaries; depth range 0–24 m

**Description:** Shell of this species is light, elongate, semicylindrical, and inflated. Posterior margins are rounded and anterior margin is straight but oblique. Weak radial ridge are present posteriorly. Surface is smooth except for fine concentric lines. Umbones are slightly removed from center of shell in posterior direction and are indistinct. *Color*: periostracum is olive green to brownish yellow. Maximum shell length is 12 cm.

**Biology:** It is a suspension feeder and builds permanent passageways from 50 to 70 cm deep.

**Fisheries/Aquaculture:** It is commercially harvested and is a potential species for aquaculture.

**Countries where eaten**: This species is an economically important species as a source of food in many countries such as USA, Argentina, West Indies, Surinam, Brazil, Chile, China, and Spain.

**Other uses:** It is occasionally used as a bait.

## KEYWORDS

- **profile of individual species**
- **biology**
- **fisheries**
- **aquaculture**

# CHAPTER 4

# AQUACULTURE VALUES OF EDIBLE MARINE BIVALVE MOLLUSCS

## CONTENTS

## ABSTRACT

The present status, and species of bivalve aquaculture, culture methodologies, aquaculture production in different countries, and the positive and negative effects of bivalve aquaculture have been dealt within this chapter.

## 4.1 STATUS OF WORLD BIVALVE AQUACULTURE PRODUCTION

Edible marine bivalve molluscs have been a major component of world aquaculture, especially during the last 50 years. While in 1999 over 8.8 million tons of cultured bivalves were landed, it was 12.8 million tons in 2010 and 13.2 million metric tons in 2012. Based on their production in 2012, they accounted for 87% of the total molluscan aquaculture production. The number of bivalve species being cultured worldwide is also increasing annually and more and more developing countries are involved in this venture. Molluscs are presently cultured in 76 countries worldwide (FAO, 2014). The top 10 countries for aquaculture production are China (81.4%), followed by Vietnam (2.6%), Korea Republic (2.5%), Japan (2.3%), Chile (1.7%), Spain (1.4%), Thailand (1.4%), USA (1.1%), France (1.1%), and Italy (0.7%).

## 4.2 COMMERCIAL EDIBLE MARINE BIVALVE AQUACULTURE IN WORLD COUNTRIES

### 4.2.1 MUSSELS

Edible marine mussel species under intensive/commercial aquaculture.

| Species | Countries of production |
| --- | --- |
| *Aulacomya ater* | Peru, Chile |
| *Choromytilus chorus* | Chile |
| *Mytilus coruscus* | Korea, China |
| *Mytilus edulis* | Netherland, Spain, France, UK |
| *Mytilus galloprovincialis* | Italy, Greece, France |
| *Perna canaliculus* | New Zealand |
| *Perna perna* | Venezuela |
| *Perna viridis* | India, China, Taiwan, SE Asia |

**Source:** Berthou et al. (http://www.eolss.net/sample-chapters/c10/E5-05-02-05.pdf).

## 4.2.2   OYSTERS

Edible marine oyster species under intensive/commercial aquaculture.

| Species | Countries of production |
|---|---|
| *Alectryonella plicatula* | China |
| *Crassostrea columbiensis* | Mexico, Panama |
| *Crassostrea gigas* | China, Japan, Taiwan, USA, France |
| *Crassostrea iredalei* | Philippines, Malaysia |
| *Crassostrea rhizophorae* | Cuba, Venezuela |
| *Crassostrea virginica* | USA, Canada, Mexico |
| *Ostrea puelchana* | Chile, New Zealand |
| *Ostrea conchaphila* | USA |
| *Ostrea edulis* | France, Spain, UK, Turkey, Greece |
| *Saccostrea cuccullata* | India, SE Asia, Australia, W. Africa, Philippines, New Zealand |

**Source:** Berthou et al. (http://www.eolss.net/sample-chapters/c10/E5-05-02-05.pdf).

## 4.2.3   SCALLOPS

Edible marine scallop species under intensive/commercial aquaculture.

| Species | Countries of production |
|---|---|
| *Aequipecten opercularis* | UK, France |
| *Argopecten irradians* | China, USA |
| *Argopecten purpuratus* | Peru, Chile |
| *Chlamys farreri* | China, Japan |
| *Patinopecten yessoensis* | China, Japan, Russia |
| *Pecten maximus* | France, UK, Ireland |

**Source:** Berthou et al. (http://www.eolss.net/sample-chapters/c10/E5-05-02-05.pdf).

## 4.2.4   PEARL OYSTERS

Edible marine pearl oyster species under intensive/commercial aquaculture.

| Species | Countries of production |
|---|---|
| *Pinctada margaritifera* | Indo-West Pacific |
| *Pinctada maxima* | Southeast Asia |
| *Pinctada radiate* | Japan, China, India |

**Source:** Berthou et al. (http://www.eolss.net/sample-chapters/c10/E5-05-02-05.pdf).

## 4.2.5 CLAMS

Edible marine clam species under intensive/commercial aquaculture.

| Species | Countries of production |
|---|---|
| *Anadara granosa* | China, Malaysia, Thailand |
| *Anadara tuberculosa* | Korea |
| *Scapharca globosa* | Venezuela |
| *Scapharca inaequivalvis* | Japan, Philippines |
| *Scapharca subcrenata* | Indonesia |
| *Placuna placenta* | Japan, Taiwan, Philippines |
| *Cerastoderma edule* | Philippines, Bangladesh |
| *Hippopus hippopus* | Portugal |
| *Tridacna derasa* | Japan |
| *Tridacna gigas* | Australia, Philippines, Papua-New Guinea |
| *Mactra veneriformis* | Palau, Indonesia |
| *Pseudocardium sybillae* | Japan |
| *Mesodesma donacium* | Canada |
| *Mercenaria mercenaria* | Italy, Turkey, Spain, France |
| *Meretrix lusoria* | USA, Canada |
| *Protothaca thaca* | India, Korea, China, Taiwan |
| *Protothaca staminea* | Japan |
| *Ruditapes decussates* | Thailand |
| *Ruditapes philippinarum* | Chile |
| *Venerupis pullastra* | China, Korea, Japan, USA |
| *Panopea abrupta* | Canada |

**Source:** Berthou et al. (http://www.eolss.net/sample-chapters/c10/E5-05-02-05.pdf).

## 4.3   HABITATS FOR MARINE BIVALVE AQUACULTURE

Successful aquaculture of edible bivalve species requires suitable substrates for growout in open sea environments. During growout, oysters are cultured in a variety of systems, which vary from on-bottom to off-bottom to suspended or floating. Floating or suspended cages are used for scallop growout. Mussel growout culture usually occurs on ropes in floating rafts or off-bottom poles without cages.

## 4.4   BIVALVE AQUACULTURE STAGES AND METHODOLOGIES

Bivalve aquaculture includes three stages, namely, production of seed, juvenile nursery culture, and growout of subadults to harvest size.

### 4.4.1   SEED PRODUCTION

The seed production of bivalve molluscs may be either through hatchery or from natural resources. In hatchery production, seed are produced under controlled conditions. The practice of collecting seed from the wild is performed by providing suitable substrates for metamorphosing larvae, such as rock piles, bamboo poles, and shell strings. These methods have been found to be suitable for oyster, mussel, and clam aquaculture. Placing shell substrate, or cultch, during the natural spawning season is still a viable method of collecting oyster seed.

### 4.4.2   NURSERY

The nursery stage is an intermediate stage between the hatchery and growout and is performed to increase the seed size in a protected environment prior to out-planting. Based on the habitat requirements of the aquaculture species, nursery methods may include land-based flow-through systems (up- or down-wellers, raceways, floating or submerged nets, or trays in intertidal ponds) with enhanced microalgae culture, floating or bottom nets, or trays in open waters. Management practices during the nursery stage include cleaning the culture systems, measuring water quality, and monitoring for disease.

## 4.4.3   GROWOUT

After the nursery stage, larger seed or juveniles are out-planted and grown to marketable size. The first step of this growout stage is site selection. Growout may take place in open waters, offshore, in coastal ponds, or in intertidal areas. All growout operations involve suitable culture environmental conditions, such as wave and tidal height, water depth, substrate, salinity, temperature, phytoplankton biomass, etc. Growout culture methods are species-specific and they include on-bottom, off-bottom poles or racks, floating rafts or longline systems, and cages or nets of various designs. The selection of growout culture gear must consider the biological characteristics of the cultured species, and the physical, chemical, and biological characteristics of the culture environment (Yang et al., http://edis.ifas.ufl.edu/pdffiles/FA/FA19100.pdf).

## 4.5   CULTURE PRACTICES OF EDIBLE, MARINE BIVALVE MOLLUSCS

### 4.5.1   MUSSEL CULTURE METHODS IN DIFFERENT COUNTRIES

| Country | Species cultured | Methods of culture |
| --- | --- | --- |
| China | Sea mussels | Longline |
| Spain | *Mytilus galloprovincialis* | Raft |
| New Zealand | *Perna canaliculus* | Longline |
| Italy | *Mytilus galloprovincialis* | Longline, hanging park |
| Italy | *Mytilus galloprovincialis* | Longline, hanging park |
| France | *Mytilis edulis* | Bouchot, longline |
|  | *Mytilius galloprovincialis* | Bouchot, longline |
| Thailand | *Perna viridis* | Bamboo pole |
| Chile | *Mytilus chilensis* | Raft, longline |
| Greece | *Mytilus galloprovincialis* | Raft, longline |
| Canada | *Mytilus edulis* | Raft, longline |
| Ireland | *Mytilus edulis* | On-bottom, raft, longline |
| Korea | *Mytilus crassistesta* | Longline |
| Philippines | *Perna viridis* | Bamboo pole |

**Source:** Gosling (2003).

## 4.5.2   GREEN MUSSEL (PERNA VIRIDIS) FARMING

### 4.5.2.1   SITE SELECTION AND ENVIRONMENTAL CONDITIONS

The proper selection of culture sites is important for green mussel culture. The site for mussel cultivation should be well-protected or sheltered coves and bays rather than open unprotected areas. Sites affected by strong wind and big waves must be avoided. Culture methods vary with different water depths. Bottom culture can be practiced in areas where the mean tide level is less than 1.5 m. For off-bottom culture, methods such as raft and long-line usually need a minimum water column height during low water spring tide. The hanging ropes with mussel seeds of these culture methods should be at least 1 m above the sea floor during extreme low water spring tides. The favorable water depth for mussel cultivation is 2 m or more.

**Turbidity**: A site having a Secchi-disk reading of less than 25 cm should be considered unsuitable for mussel culture.

**Salinity**: Green mussel is reported to tolerate a high range of salinity. The green mussel shows a good growth performance in estuarine habitats with salinities ranging from 18 to 33 ppt and temperature from 1 to 32°C.

**Food organisms:** As filter feeders, green mussels mainly feed on a wide range of phytoplankton species, small zooplankton, and other suspended fine organic materials. High primary productivity areas lead to high productivity and biomass of mussels.

### 4.5.2.2   CULTURE METHODS

Cultivation of mussels is of two categories; they are on-bottom cultivation and off-bottom cultivation.

**On-bottom culture:** On-bottom culture or seabed culture is largely practiced in Europe especially in the Netherlands, Germany, Ireland, and the United Kingdom. Bottom culture is based on transferring wild mussels to a sheltered culture plot where the density is reduced to improve growth and fattening. In Philippines, this method is used in shallow areas from 0.6 m at low tide and 3.6 m at high tide. The mussel seeds are collected from the bay using bamboo poles and after 1 or 2 months, the mussels are removed from the bamboo poles and laid at the bottom of estuary.

**Off-bottom culture:** The culture methods under this category are practiced in intertidal zones and/or mussels are grown above the seabed.

The three principle methods of off-bottom culture include pole, raft, and longline. The farming methods of this type of culture are fixed suspended cultivation, floating suspended cultivation, and deep-water cultivation.

i) **Fixed suspended cultivation:** The culture methods under this subcategory include rack culture, tray culture, wig-wam culture, rope-web culture, and pole cultivation. All these methods are practiced in the Philippines except pole culture, which is practiced in France. These methods of cultivation require a fixed platform or structures for settlement and growth of the mussels. Further, the cultivation occurs in soft and muddy sea beds with water depths of 2–3 m. The collected spats grow to marketable size, 5–10 cm in 6–10 months. The pole cultivation or "Bouchot" culture method is the most significant culture practiced under fixed suspended cultivation.

ii) **Floating suspended cultivation:** It is done in deeper coastal waters of high primary productivity. The two main methods of this category are the raft-culture method and the long-line culture method. The raft-culture method is popular in Spain and in many other countries, notably Australia, China, Chile, Canada, USA, India, Ireland, Malaysia, New Zealand, Scotland, and Venezuela. With this method of culture, the mussels are grown attached to suspended ropes, which are tied to a raft.

### 4.5.3  OYSTER CULTURE METHODS

Oyster culture consists of gathering their seeds and growing them to marketable size. The oyster culture methods are on-bottom and off-bottom.

#### 4.5.3.1  ON-BOTTOM CULTURE

This is growing oysters directly on the bottom subtidally or intertidally. This method requires a stable, nonshifting bottom within the correct tidal range. This method consists of planting the seedling directly on the bottom where they are left to grow to the marketable size.

## 4.5.3.2  OFF-BOTTOM CULTURE

Where bottom conditions are not suited (due to softness and unsuitable wave action and tidal level), oysters are held in suspension or off-bottom in several ways. Off-bottom culture is divided into three methods: raft, rack, and stake.

i)  **Raft culture:** Oysters are suspended from floating structures such as raft. They may be held in tray or stringed. The rafts can be of any shape or material and styrofoam, oil drums, or polyethylene floats are used the float the raft.

ii)  **Raft-tray culture:** To grow single, well-shaped oysters for a particular market such as the half-shell trade, this method is used. A tray may be made of wire or plastic mesh with a wooden frame, or entirely made of bamboo. Single oysters are laid on trays and allowed to grow until marketable size.

iii)  **Rack culture:** In this method, racks of wood, bamboo, or metal, imbedded in the ground either subtidally or intertidally, are used to hold vertically or horizontally oysters which are on trays or strings or sticks. The rack-string is very productive, as experienced in the Philippines.

iv)  **Stake culture:** It is suitable for shallow lagoons which are too shallow or too soft-bottomed for other culture methods. The stakes hold oysters vertically off-bottom.

## 4.5.4  SCALLOP-CULTURE METHODS

### 4.5.4.1  SUSPENDED CULTURE

Culturing scallops in hanging devices is a common practice in Asia. The primary apparatus used in this method are lantern nets with mesh sizes of 15 and 21 mm. Lantern nets commonly have 5, 10, or 15 tiers. They are generally suspended from a floating or submerged longline that is anchored at both ends and held at or below the surface by floats placed along the backbone. Lantern nets can also be suspended from floating rafts. Bay scallops can also be grown in pocket nets, a rectangular sheet of mesh hung vertically in the water column with pockets of mesh stitched on one side.

## 4.5.4.2   CAGE CULTURE

Floating or bottom cages are used to grow bay scallops to market size.

i)   **Floating Cages:** The floating scallop cage consists of a wooden frame with plastic mesh covering the sides. The recently developed scallop consists of a plastic mesh bag with two flotation "logs" arranged along the sides of the bag. The flotation logs range from foam swim "noodles" to rigid plastic cylinders.

ii)  **Bottom Cages:** Bottom cages for scallop culture are generally constructed from plastic coated wire and provide a rigid frame with a series of shelves holding semirigid polyethylene bags. Cages should be designed to optimize the number of scallops held, allow adequate water movement through the bags, and elevate the bags off the bottom to prevent sediments from smothering the scallops.

## 4.5.5   CLAM-FARMING METHODS

Calm species such as native littleneck clams (*Protothaca staminea*), Manila clams (*Venerupis japonica*), and geoduck clams (*Panope abrupta*) are normally cultivated commercially.

## 4.5.5.1   BEACH CULTURE

This is the simplest method of growing clams and is employed on a natural beach without any protective netting or bags. When adult clams are harvested, the substrate is loosened and the crop is thinned to improve the growth of the remaining clams. Clams planted at lower densities (30–60 clams per square foot) may have better growth and survival than clams planted at higher densities.

## 4.5.5.2   IN-GROUND BAG CULTURE

On beaches where moon snails are serious predators, clam seed are enclosed in mesh bags which are partially buried in the substrate. This method is known as in-ground bag culture and is more suitable for native littleneck or Manila clams at lower tidal levels. In-ground culture, the bags are made of heavy plastic mesh. This size mesh facilitates water flow through the bags while excluding predators. The bags are placed in rows, with individual bags spaced approximately 0.7 m apart. An aisle should exist between the rows to facilitate harvesting. Seed should be large enough (about 8 mm) so that it will not fall through the holes in the mesh. An optimal planting density ranges between 500 and 700 clams per bag. The seeded bags are placed in shallow depressions in the beach sediment and secured with metal rods or rebar pins.

## 4.6 MAJOR MARINE BIVALVE PRODUCING COUNTRIES AND THEIR PRODUCTION

China is by far the leading producer of bivalve molluscs, with 10 million tons in 2010, representing 80% of the global bivalve mollusc aquaculture production. Other major bivalve producers in 2010 were Japan (0.8 million tons), the USA (0.7 million tons), the Republic of Korea (0.4 million tons), Thailand (0.3 million tons), France (0.2 million tons), and Spain (0.2 million tons).

## 4.7 THE QUANTITY AND PERCENTAGE OF WORLDWIDE MARINE BIVALVE AQUACULTURE PRODUCTION

| Species groups | Quantity (million tons) | | |
|---|---|---|---|
| | 1992 | 2010 | 2012 |
| Clams, cockles, and ark shells | 2.8 (31%) | 4.8 (38%) | 5.0 (38%) |
| Oysters | 3.7 (42%) | 4.5 (35%) | 4.7 (36%) |
| Mussels | 1.5 (17%) | 1.8 (14%) | 1.8 (14%) |
| Scallops | 1.0 (11%) | 1.7 (13%) | 1.7 (13%) |

**Source:** Gosling (2003), FAO (2012, 2014).

## 4.8 MAJOR FARMED EDIBLE MARINE BIVALVE SPECIES (AS ON 2011)

| Species | Production (million tons) | Largest producer | % |
|---|---|---|---|
| *Crassostrea gigas* | 4.5 | China | 83 |
| *Ruditapes philippinarum* | 3.7 | China | 98 |
| *Patinopecten yessoensis* | 1.4 | China | 92 |
| *Mytilus chilensis* | 0.3 | Chile | 100 |
| *Sinonovacula constricta* | 0.7 | China | 100 |
| *Argopecten purpuratus* | 0.06 | Peru | 83 |
| *Anadara granosa* | 0.4 | China | 66 |
| *Mytilus galloprovincialis* | 1.0 | China | 69 |
| *Mytilus edulis* | 0.2 | France | 35 |
| *Perna canalicula* | 0.2 | New Zealand | 45 |
| *Meretrix* spp. | 0.1 | Vietnam | 50 |
| *Mercenaria mercenaria* | 0.03 | USA | 100 |
| *Crassostrea virginica* | 0.07 | USA | 95 |
| *Ruditapes decussatus* | 0.004 | Portugal | 56 |
| *Saccostrea commercialis* | 0.006 | Australia | 100 |
| *Mytilus coruscus* | 0.07 | Rep. Korea | 100 |
| *Pteria penguin* | 0.05 | Indonesia | 100 |
| *Panopea generosa* | 0.0006 | USA | 100 |
| *Scapharca broughtonii* | 0.002 | Rep. Korea | 100 |
| *Atrina* spp. | 0.03 | China | 100 |
| *Perna perna* | 0.02 | Brazil | 100 |

**Source:** Shinn et al. (2015).

## 4.9 AQUACULTURE PRODUCTION OF TOP COUNTRIES AND THEIR MAJOR FARMED SPECIES (AS ON 2012)

**China**—10.82 million tons.

| Species/groups farmed |
|---|
| *Crassostrea* spp. |
| *Ruditapes philippinarum* |
| Pectinidae |
| Mytildae (mussels) |
| *Sinonovacula* spp. |
| *Anadara granosa* |
| *Atrina* spp. |

**Source:** Yang et al. (http://edis.ifas.ufl.edu/pdffiles/FA/FA19100.pdf).

## Vietnam—0.4 million tons.

| Groups farmed |
| --- |
| Clams and oysters |

Source: Yang et al. (http://edis.ifas.ufl.edu/pdffiles/FA/FA19100.pdf).

## Korea Republic—0.36 million tons.

| Species farmed |
| --- |
| *Crassostrea gigas* |
| *Mytilus coruscus* |
| *Ruditapes philippinarum* |
| *Anadara granosa* |
| *Scapharca broughtonii* |
| *Patinopecten yessoensis* |
| *Cyclina sinensis* |

Source: Yang et al. (http://edis.ifas.ufl.edu/pdffiles/FA/FA19100.pdf).

## Japan—0.34 million tons.

| Species farmed |
| --- |
| *Patinopecten yessoensis* |
| *Crassostrea gigas* |

Source: Yang et al. (http://edis.ifas.ufl.edu/pdffiles/FA/FA19100.pdf).

## Chile—0.25 million tons.

| Species farmed |
| --- |
| *Mytilus chilensis* |
| *Argopecten purpuratus* |
| *Aulacomya ater* |
| *Choromytilus chorus* |
| *Ostrea chilensis* |
| *Crassostrea gigas* |

Source: Yang et al. (http://edis.ifas.ufl.edu/pdffiles/FA/FA19100.pdf).

## Spain—0.21 million tons.

| Species/groups farmed |
| --- |
| Mytilidae |
| *Ruditapes philippinarum* |
| *Ostrea edulis* |
| *Crassostrea gigas* |
| *Cerastoderma edule* |
| *Venerupis pullastra* |
| *Ruditapes decussatus* |
| *Venus verrucosa* |
| *Chlamys varia* |
| *Venerupis rhomboides* |
| *Pecten maximus* |
| *Ensis ensis* |
| *Donax* spp. |
| *Crassostrea* spp. |
| *Chamelea gallina* |
| *Aequipecten opercularis* |

**Source:** Yang et al. (http://edis.ifas.ufl.edu/pdffiles/FA/FA19100.pdf).

## Thailand—0.20 million tons.

| Species/groups farmed |
| --- |
| *Perna viridis* |
| *Anadara granosa* |
| *Crassostrea* spp. |
| *Ostrea edulis* |
| *Modiolus* spp. |
| *Ostrea* spp. |

**Source:** Yang et al. (http://edis.ifas.ufl.edu/pdffiles/FA/FA19100.pdf).

## USA—0.17 million tons.

| Species/groups farmed |
| --- |
| *Crassostrea virginica* |
| *Crassostrea gigas* |
| *Mercenaria mercenaria* |
| *Ruditapes philippinarum* |
| *Mytilus edulis* |
| *Crassostrea* spp. |
| *Mya arenaria* |
| *Panopea abrupta* |
| *Protothaca staminea* |
| *Saxidomus giganteus* |
| *Cardiidae* spp. |
| *Ostrea conchaphila* |
| *Pectinidae* |
| *Tresus nuttallii* |

**Source:** Yang et al. (http://edis.ifas.ufl.edu/pdffiles/FA/FA19100.pdf).

## France—0.16 million tons.

| Species/groups farmed |
| --- |
| *Crassostrea gigas* |
| *Mytilus edulis* |
| *Mytilus galloprovincialis* |
| *Cerastoderma edule* |
| *Ostrea edulis* |
| *Ruditapes philippinarum* |
| *Ostrea edulis* |
| *Pecten maximus* |
| *Ruditapes decussatus* |

**Source:** Yang et al. (http://edis.ifas.ufl.edu/pdffiles/FA/FA19100.pdf).

## Italy—0.11 million tons.

| Species/groups farmed |
| --- |
| *Mytilus galloprovincialis* |
| *Ruditapes philippinarum* |
| *Ruditapes decussatus* |
| *Crassostrea* spp. |

**Source:** Yang et al. (http://edis.ifas.ufl.edu/pdffiles/FA/FA19100.pdf).

## 4.10   BIVALVE AQUACULTURE AND ENVIRONMENT

**Positive effects:** Edible bivalves (both in wild and culture) have been reported to play significant roles in the food chain of the marine environment. Filtration or grazing of bivalves has been shown to control phytoplankton growth by removing them from the water. When not sufficiently grazed, phytoplankton populations can bloom excessively, which often leads to the deterioration of water quality. The various gears associated with marine bivalve aquaculture act as refugia for a variety of marine organisms, including the juvenile stages of various species of commercially valuable finfish.

**Negative effects:** Shellfish in suspended culture enhance fish and crab populations on the bottom and that fouling organisms on the mussel longlines can enhance populations of grazing and predatory fish. While introduced species for commercial aquaculture have been economically successful and have posed no environmental problems. There are great concerns that the movement of cultured species (broodstock or seed) may facilitate the movement of disease-causing organisms and exotic species which pose potential dangers for both cultured and wild stocks (Rice, 2008).

## KEYWORDS

- culturable species of bivalves
- culture practices
- hatchery production
- aquaculture yield
- bivalve aquaculture and environment

# CHAPTER 5

# NUTRITIONAL VALUES OF EDIBLE MARINE BIVALVE MOLLUSCS

## CONTENTS

## ABSTRACT

The nutritional values of edible marine bivalve molluscs, namely, such as carbohydrate, protein, lipid, ash, minerals, and vitamins have been dealt within this chapter.

*Abra alba*

| Proximate composition (%) | | | | |
|---|---|---|---|---|
| Protein (kcal/g dw) | Lipid (kcal/g dw) | CHO (kcal/g dw) | Ash (kcal/g dw) | Calorific value (kcal/g dw) |
| 53.4–61.9 | 3.3–6.8 | 7.0–17.2 | 13.5–29.6 | 3.9–4.5 |

**Source:** Singh et al. (2012).

*Amusium pleuronectes*

| Nutrition Information (per 100 g) | |
|---|---|
| Protein | 11.6 g |
| Fat | 0.9 mg |
| Sodium | 163 mg |
| Cholesterol | 102 mg |
| Alpha-linoleic acid | 21 mg |
| Docosahexaenoic acid | 116 mg |
| Eicosapentaenoic acid | 166 mg |
| Saturated fat | 32% of total fat |
| Monounsaturated fat | 15% of total fat |
| Polyunsaturated fat | 53% of total fat |
| Energy | 53 cal |

**Source:** Anon. (2014).

*Anadara kagoshimensis*

| Proximate composition (%) | | | | |
|---|---|---|---|---|
| Moisture | Crude protein | CHO | Lipid | Ash |
| 83.8 | 8.8 | – | 0.4 | 4.0 |

**Source:** Mohammad and Yusuf (2016).

## *Anadara ovalis*

| Nutritional composition (per 100 g) | |
|---|---|
| Calories | 35 |
| Total fat (g) | 0.5 |
| Protein (g) | 7 |
| Carbohydrates (g) | 0 |
| Cholesterol (mg) | 35 |
| Sodium (mg) | 740 |

**Source:** Sturmer et al. (ndb).

## *Arca noae*

| Proximate composition (% of dry weight) | |
|---|---|
| Protein | 54.4–62.1 |
| Carbohydrate | 4.1–8.1 |
| Lipid | 3.5–8.6 |

| Fatty acid composition (% of total fatty acid) | |
|---|---|
| SFA | 32.9 |
| MUFA | 20.7 |
| PUFA | 44.1 |

*SFA*, saturated; *MUFA*, monounsaturated; *PUFA*, polyunsaturated fatty acids.

**Source:** Radić et al. (2014).

## *Argopecten gibbus*

| Protein | Fat | Glycogen |
|---|---|---|
| 15.8 | 0.6 | 1.5 |

**Source:** Anon. (1989).

## *Argopecten irradians (=Aequipecen irradians)*

| Protein | Fat | Glycogen |
|---|---|---|
| 15.7 | 0.5 | 1.5 |

**Source:** Anon. (1989).

## *Atrina zelandica*

| | Proximate composition (%) | | | |
|---|---|---|---|---|
| **Protein** | **Lipids** | **Moisture** | **Ash** | **CHO** |
| 14.8 | 0.5 | 81.0 | 1.3 | 2.4 |

**Source:** Anon. (ndt).

## *Callista florida*

| | Proximate composition (%) | | | |
|---|---|---|---|---|
| **Moisture** | **Crude protein** | **CHO** | **Lipid** | **Ash** |
| 79.8 | 9.5 | 6.7 | 1.1 | 2.9 |

**Source:** Mohammad and Yusuf (2016).

## *Cerastoderma edule (=Cardium edule)*

| Proximate composition (%) | | |
|---|---|---|
| Protein | CHO | Fat |
| 18.5 | 0.8 | 4.2 |
| Minerals (mg/kg) | | |
| Fe | Cu | Co |
| 225 | 53 | 24 |

**Source:** Abdel-Salam (2013).

| Amino acids (g/100 g) (FW) | |
|---|---|
| His* | 1.6 |
| Thr* | 4.5 |
| Arg | 5.6 |
| Tyr | 1.1 |
| Val* | 4.5 |
| Phe* | 4.4 |
| Ile* | 4.5 |
| Lys* | 11.2 |
| Leu* | 9.0 |
| Met* | 2.2 |
| Trp* | 1.2 |

*FW*, fresh weight. *Essential amino acids.

**Source:** Anon. (ndt).

## *Cerastoderma glaucum*

| Proximate composition (%) | | | | |
|---|---|---|---|---|
| Moisture | Crude protein | CHO | Lipid | Ash |
| 86.9 | 5.9 | 3.2 | 0.5 | 3.5 |

**Source:** Mohammad and Yusuf (2016).

## *Clinocardium nuttallii (=Cardium corbis)*

| Proximate composition (%) | | | |
|---|---|---|---|
| Moisture | Protein | Fat | Ash |
| 82.4 | 11.8 | 1.0 | 1.6 |

**Source:** Krzynowek and Murphy (1987).

## *Codakia orbicularis*

| Proximate composition (%) | | | | |
|---|---|---|---|---|
| Moisture | Fat | Ash | PUFA (%TFA) | Cholesterol (mg%) |
| 74.8 | 1.1 | 2.9 | 19.5 | 20.6 |

*TFA*, Total fatty acids.

**Source:** Krzynowek and Murphy (1987).

## *Corbicula japonica*

| Proximate composition (%) | | |
|---|---|---|
| Protein | Fat | Glycogen |
| 11.1 | 1.6 | 4.8 |

**Source:** Anon. (1989).

| Fatty acids (% of total fatty acids) | |
|---|---|
| 13:0 | 0.1 |
| 14:0 | 1.9 |
| 15:0 | 0.3 |
| 16:0 | 16.9 |
| 17:0 | 0.6 |
| 16:2$(n-4)$ | 0.2 |
| 18:1$(n-11)$ | 0.3 |

| Fatty acids (% of total fatty acids) | |
|---|---|
| $18:1(n-9)$ | 2.7 |
| $18:1(n-7)$ | 2.7 |
| $18:2(n-6)$ | 1.5 |
| $18:2(n-3)$ | 0.2 |
| $20:1(n-9)$ | 1.1 |
| $20:1(n-7)$ | 0.4 |
| $18:4(n-3)$ | 1.3 |
| $20:2(n-6)$ | 1.0 |
| $20:4(n-6)$ | 3.3 |
| $20:5(n-3)$ | 7.8 |
| $24:1(n-9)$ | 0.4 |
| $22:5(n-3)$ | 3.4 |
| $22:6(n-3)$ | 13.5 |

**Source:** Teshima et al. (1990).

### *Crassostrea cuttackensis (=Crassostrea gryphoides)*

| Extraction | Protein (µg/ml/mg) | Carbohydrate (µg/ml/mg) | Lipid (µg/ml/mg) |
|---|---|---|---|
| MeOH | 270 | 27.6 | 0.2 |
| PBS | 370 | 15 | 0.07 |
| AEH-MeOH | 744 | 6.5 | 0.03 |
| AEH-Aq | 126 | 16.7 | 0.02 |

*MeOH*, methanol; *PBS*, phosphate buffer saline; *AEH*, extracts with protease inhibitors and acid enzyme hydrolysate.

**Source:** Sharma et al. (2009).

### *Crassostrea gigas*

| Nutritional facts (per 85 g) | |
|---|---|
| Protein | 16 g |
| CHO | 8 g |
| Total fat | 4 g |
| Cholesterol | 85 mg |
| Sodium | 180 mg |
| Vitamin A | 8% |

| Nutritional facts (per 85 g) | |
| --- | --- |
| Vitamin C | 20% |
| Calcium | 2% |
| Iron | 45% |
| Selenium | 190% |
| Calories | 140 |

**Source:** Anon. (ndc).

| Total lipids and classes of lipids | | | |
| --- | --- | --- | --- |
| Total lipids (% dw) | Natural lipids (% TL) | Glycolipids (% TL) | Phospholipids (% TL) |
| 7.1 | 40 | 9 | 50.4 |

**Source:** Dagorn et al. (2016).

| Fatty acid compositions (%TFA) | |
| --- | --- |
| 14:0 | 1.3 |
| 16:0 | 19.1 |
| 16:1$(n-7)$ | 1.9 |
| 17:0 | 1.9 |
| 17:1 | 4.1 |
| 18:0 | 3.8 |
| 18:1$(n-7)$ | 3.9 |
| 18:1$(n-9)$ | 2.1 |
| 18:4$(n-3)$ | 2.3 |
| 20:1$(n-11)$ | 2.9 |
| 20:1$(n-9)$ | 2.1 |
| 20:1$(n-7)$ | 2.1 |
| 20:4$(n-6)$ | 4.3 |
| 20:5$(n-3)$ | 19.5 |
| 22:5$(n-3)$ | 2.1 |
| 22:6$(n-3)$ | 19.3 |

*TFA*, total fatty acids.

**Source:** Pogoda et al. (2013).

## *Crassostrea madrasensis*

| Nutritional facts | |
|---|---|
| Moisture | 82.6% |
| Protein | 9.4% |
| Fat | 3.25% |
| Carbohydrate | 3.2% |
| Ash | 1.0% |
| Total amino acid | 99.33 g/100 g crude protein EAA |
| Valine | 0.17 g/100 g crude protein |
| Threonine | 3.62 g/100 g crude protein |

*EAA*, essential amino acids.

**Source:** Asha et al. (2014).

| Proximate composition (% dw) | |
|---|---|
| Protein | 59.1 |
| Carbohydrate | 18.3 |
| Lipid | 14.0 |
| Ash | 8.6 |

**Source:** Salaskar and Nayak (2011).

## *Crassostrea tulipa (=Crassostrea gasar)*

| Proximate composition (%) | |
|---|---|
| Moisture | 72.1 |
| Crude protein | 21.0 |
| Ash | 5.7 |
| Fat | 0.3 |

**Source:** Davies and Jamabo (2016).

| Amino acids (mg/100 g) | |
|---|---|
| **Nonessential** | |
| Aspartic acid | 3190 |
| Glutamic acid | 3730 |
| Serine | 1160 |
| Glycine | 1730 |

| Amino acids (mg/100 g) | |
|---|---|
| Arginine | 3740 |
| Alanine | 1460 |
| Tyrosine | 980 |
| Proline | 1160 |
| **Essential** | |
| Phenylalanine | 1130 |
| Valine | 1430 |
| Histidine | 1080 |
| Threonine | 1420 |
| Lysine | 1935 |
| Leucine | 1780 |
| Isoleucine | 1390 |
| Methionine | 720 |

*NEE*, nonessential amino acids; *EAA*, essential amino acids.

**Source:** Silva et al. (nd).

| Minerals (mg/g) | | | | | | |
|---|---|---|---|---|---|---|
| **Zn** | **Fe** | **Cu** | **Mn** | **K** | **Mg** | **Ca** |
| 520 | 826 | 10 | 36 | 290 | 470 | 4920 |

**Source:** Silva et al. (nd).

## *Crassostrea virginica*

| Nutritional facts (per 85 g) | |
|---|---|
| Protein | 6 g |
| Total CHO | 6 g |
| Total fat | 2 g |
| Cholesterol | 30 mg |
| Sodium | 140 mg |
| Vitamin A | 2% |
| Vitamin C | 8% |
| Calcium | 4% |
| Iron | 35% |
| Selenium | 90% |

**Source:** Anon. (ndc).

| Proximate mean composition (% NW) | |
|---|---|
| Protein | 5.8 |
| Carbohydrate | 2.7 |
| Fat | 1.1 |
| Ash | 1.9 |

*NW*, net weight.

*Source*: Anon. (ndo).

| Metals (mg/kg dw) | |
|---|---|
| Fe | 136 |
| Cu | 65 |
| Zn | 1420 |
| Mn | 4 |

**Source:** Anon. (ndo).

## *Donax cuneatus*

| Proximate composition (%) | | |
|---|---|---|
| Protein | Lipid | CHO |
| 57–68 | 5–7 | 11–26 |

**Source:** Singh et al. (2012).

| Amino acids | |
|---|---|
| Essential amino acids (% of amino acids) | |
| Phenylalanine | 6.5 |
| Lysine | 7.1 |
| Histidine | 2.4 |
| Methionine | 9.5 |
| Arginine | 1.3 |
| Leucine | 6.6 |
| Threonine | 6.0 |
| Isolucine | 0.2 |
| Valine | 1.9 |
| Tryptophan | 0.7 |

| Amino acids | |
|---|---|
| Nonessential amino acids (% of amino acids) | |
| Glycine | 5.7 |
| Serine | 7.6 |
| Glutamic acid | 4.0 |
| Cystine | 1.7 |
| Glutamate | 5.2 |
| Alanine | 0.7 |
| Proline | 0.7 |
| Aspartate | 3.6 |
| Tyrosine | 6.5 |
| Aspertic acid | 2.6 |

| Fatty acids | |
|---|---|
| Saturated fatty acids (% of fatty acids) | |
| Palmitic acid | 20.1 |
| Margaric acid | 6.2 |
| Stearic acid | 9.6 |
| Monounsaturated fatty acids (% of fatty acids) | |
| Oleic acid | 14.2 |
| Polyunsaturated fatty acids (% of fatty acids) | |
| Linolenic acid | 10.7 |
| Alpha-linolenic acid | 16.3 |
| Stearidonic or moroctic acid | 7.8 |

**Source:** Gopalsamy et al. (2014).

*Donax incarnatus*

| Proximate composition (%) | |
|---|---|
| Protein | 23.5 |
| Carbohydrate | 10.2 |
| Lipid | 1.3 |

| Amino acids | | | |
|---|---|---|---|
| Essential and nonessential amino acids (% of amino acids) | | | |
| EAA | | NEAA | |
| Phenylalanine | 2.3 | Glycine | 2.1 |
| Lysine | 13.3 | Serine | 0.1 |
| Histidine | 8.9 | Glutamic acid | 1.4 |
| Methionine | 9.0 | Cystine | 0.6 |
| Arginine | 0.1 | Glutamate | 9.0 |
| Leucine | 6.6 | Alanine | 6.0 |
| Threonine | 1.4 | Proline | 3.3 |
| Isolucine | 5.6 | Aspartate | 4.4 |
| Valine | 7.3 | Tyrosine | 3.3 |
| Tryptophan | 3.6 | Aspertic acid | 5.2 |

*EAA*, essential amino acids.

| Fatty acids | |
|---|---|
| Saturated fatty acids (% of fatty acids) | |
| Palmitic acid | 28.1 |
| Margaric acid | 5.7 |
| Stearic acid | 10.3 |
| Monounsaturated fatty acids (% of fatty acids) | |
| Oleic acid | 12.3 |
| Polyunsaturated fatty acids (% of fatty acids) | |
| Linolenic acid | 18.7 |
| Alpha-linolenic acid | 11.2 |
| Stearidonic or moroctic acid | 7.8 |

| Vitamins (mg/g) | |
|---|---|
| Retinol (A) | 105.6 |
| Calciferol (D) | 14.2 |
| Tocopherol (E) | 1.3 |
| Vitamin (K) | 0.6 |
| Pyridoxin (B6) | 0.3 |
| Cobalamin (B12) | 2.1 |
| Vitamin (C) | 23.8 |

| Minerals (mg/g) | |
| --- | --- |
| Macrominerals | |
| Calcium | 315.2 |
| Sodium | 91.7 |
| Potassium | 20.4 |
| Copper | 1.5 |
| Magnesium | 60.5 |
| Trace minerals | |
| Iron | 1.4 |
| Zinc | 0.3 |

**Source:** Periyasamy et al. (2014).

## *Donax scortum*

| Proximate composition (%) | | | | |
| --- | --- | --- | --- | --- |
| Protein (kcal/g dw) | Lipid (kcal/g dw) | CHO (kcal/g dw) | Ash (kcal/g dw) | Calorific value (kcal/g dw) |
| 59.9–66.2 | 4.1–9.9 | 15.0–26.9 | 6.2–9.7 | 4.9–5.3 |

**Source:** Singh et al. (2012).

## *Donax trunculus*

| Proximate composition (%) | | | | |
| --- | --- | --- | --- | --- |
| Protein (kcal/g dw) | Lipid (kcal/g dw) | CHO (kcal/g dw) | Ash (kcal/g dw) | Calorific value (kcal/g dw) |
| 52.9–64.1 | 2.9–6.9 | 6.5–14.7 | 18.7–30.0 | 3.9–4.5 |

**Source:** Singh et al. (2012).

| Nutritional facts (g/100 g edible portion) | |
| --- | --- |
| Protein | 9.5 |
| Lipid | 1.0 |
| Minerals (mm/100 g) | |
| Iron | 6.3 |
| Zinc | 1.2 |
| Magnesium | 48.5 |

**Source:** FISHEAT (nd).

*Ensis siliqua*

| Amino acid composition (% dry weight) | |
|---|---|
| Essential amino acids | |
| Threonine | 2.5 |
| Valine | 2.2 |
| Methionine | 0.9 |
| Isoleucine | 1.9 |
| Leucine | 3.5 |
| Phenylalanine | 1.9 |
| Histidine | 1.0 |
| Lysine | 3.6 |
| Arginine | 3.7 |
| Nonessential amino acids | |
| Aspartic acid | 4.8 |
| Tyrosine | 1.7 |
| Serine | 2.4 |
| Glutamic acid | 7.1 |
| Glycine | 5.2 |
| Alanine | 3.7 |
| Proline | 1.8 |
| Fatty acid composition (mg/g dw) | |
| Saturated | 33.0 |
| Monosaturated | 17.3 |
| Polyunaaturated | 37.6 |
| Total | 87.9 |

**Source:** Baptista et al. (2014).

*Gafrarium divaricatum*

| Proximate composition (%) | |
|---|---|
| Protein | 26.3 |
| Carbohydrate | 11.2 |
| Lipid | 1.3 |
| Ash | 5.6 |
| Moisture | 6.1 |

| Essential amino acids (% of total) | |
|---|---|
| Threonine | 1.4 |
| Arginine | 0.2 |
| Histidine | 9.0 |
| Valine | 7.3 |
| Methionine | 8.9 |
| Isoleucine | 5.5 |
| Phenylanine | 2.3 |
| Leucine | 6.5 |
| Lysine | 14.4 |
| Tryptophan | 3.9 |
| **Nonessential amino acids (% of total)** | |
| Aspartic acid | 5.0 |
| Glutamic acid | 1.5 |
| Asparagine | 3.8 |
| Serine | 0.1 |
| Glycine | 2.2 |
| Alanine | 5.9 |
| Cysteine | 1.0 |
| Tyrosine | 3.5 |
| Proline | 3.2 |

| Fatty acids (% of total) | |
|---|---|
| Palmitic acid | 27.2 |
| Stearic acid | 11.0 |
| Oleic acid | 12.5 |
| Linolenic acid | 18.0 |
| Moroctic acid | 8.1 |

| Vitamins (mg/g) | |
|---|---|
| Vitamin A | 112.3 IU |
| Vitamin D | 14.0 IU |
| Vitamin E | 1.1 |
| Vitamin B6 | 0.3 |
| Vitamin B1 | 22.0* |
| Vitamin C | 24.1 |
| Vitamin K | 0.6 |

*µg/g.

**Source:** Eswar et al. (2016).

## Gafrarium tumidum

| Proximate composition of viscera (%) | |
|---|---|
| Protein | 24.8 |
| Carbohydrate | 13.5 |
| Fat | 7.3 |

| Essential amino acids of viscera (g amino acid/100 g protein) | |
|---|---|
| Phenylalanine | 1.2 |
| Threonine | 0.9 |
| Valine | 0.5 |
| Histidine | 0.7 |
| Isoleucine | 1.1 |
| Methionine | 1.0 |
| Leucine | 0.9 |
| Lysine | 0.4 |
| Proline | 0.3 |
| Tryptophan | 1.0 |
| **Nonessential amino acids (g amino acid/100 g protein)** | |
| Alanine | 1.1 |
| Arginine | 1.1 |
| Asparagine | 0.5 |
| Aspartic acid | 0.7 |
| Cystine | 1.0 |
| Glutamic acid | 1.1 |
| Glutamine | 0.6 |
| Glycine | 1.1 |
| Serine | 0.3 |
| Tyrosine | 0.9 |

| Fatty acids of viscera (% of total) | |
|---|---|
| SFA | |
| Palmitic | 0.7 |
| Margaric | 0.4 |
| Stearic | 0.8 |
| MUFA | |
| Oleic | 0.9 |
| PUFA | |
| Linoleic | 1.4 |
| α-Linolenic | 1.1 |
| Morotic | 0.2 |

*MUFA*, monounsaturated fatty acids; *PUFA*, polyunsaturated fatty acids; *SFA*, saturated fatty acid.

**Source:** Babu et al. (2012).

## *Geloina erosa*

| Proximate composition (μg/ml/mg) | | | |
|---|---|---|---|
| Extraction | Protein | Carbohydrate | Lipid |
| MeOH | 420 | 24.5 | 0.6 |
| PBS | 640 | 2.3 | 0.1 |
| AEH-MeOH | 528 | 7.1 | 0.2 |
| AEH-Aq | 154 | 6.5 | 0.0 |

*MeOH*, methanol; *PBS*, phosphate buffer saline; *AEH*, extracts with protease inhibitors and acid enzyme hydrolysate.

**Source:** Sharma et al. (2009).

## *Leukoma staminea*

| Proximate composition (%dw) | |
|---|---|
| Water | 79.4 |
| Protein | 65.5 |
| Fat | 4.8 |
| Carbohydrate | 17.0 |
| Ash | 12.6 |

**Source:** Miller and Boxt (nd).

| Proximate composition (% dw) | |
|---|---|
| Moisture | 79.4 |
| Protein | 65.5 |
| Fat | 4.8 |
| Carbohydrate | 17.0 |
| Ash | 12.6 |

**Source:** Anon. (ndb).

| Proximate composition (%) | | | |
|---|---|---|---|
| Moisture | Protein | Fat | Ash |
| 79.4 | 13.5 | 1.0 | 2.6 |

**Source:** Krzynowek and Murphy (1987).

## Limaria hians

| Proximate composition (%) | | | | |
|---|---|---|---|---|
| Protein (kcal/g dw) | Lipid (kcal/g dw) | CHO (kcal/g dw) | Ash (kcal/g dw) | Calorific value (kcal/g dw) |
| 11.3–18.7 | 4.9–8.2 | 1.7–6.9 | 13.1–27.4 | 3.8–4.7 |

**Source:** Singh et al. (2012).

## Lunarca ovalis (=Anadara ovalis)

| Nutritional facts (per 100 g) | |
|---|---|
| Calories | 35 |
| Protein (g) | 7 |
| Total fat (g) | 0.5 |
| Carbohydrates (g) | 0 |
| Cholesterol (mg) | 35 |
| Sodium (mg) | 740 |

**Source:** Leslie et al. (nd).

## Macomangulus tenuis

| Proximate composition (% FW) | | | | |
|---|---|---|---|---|
| Protein | Lipids | Moisture | Ash | Carbohydrate |
| 7.9–10.9 | 3.9–12.8* | 71.0–83.0 | 5.8–27.7* | 6.4–25.6* |

*FW*, fresh weight. *The percentage on dry matter basis.

**Source:** Anon. (ndt).

## Mactra violacea

| | Proximate composition (%) |
|---|---|
| Protein | 11.9 |
| Fat | 1 |
| Ash | 3.2 |
| Moisture | 80 |

**Source:** Laxmilatha (2009).

## Marcia recens

| | Proximate composition (%) | |
|---|---|---|
| **Protein** | **Lipid** | **CHO** |
| 33.1–40.5 | 3.3–10.8 | 1.9–7.7 |

**Source:** Singh et al. (2012).

## Mercenaria mercenaria

| | | Proximate composition (%) | | |
|---|---|---|---|---|
| **Moisture** | **Protein** | **Fat** | **Ash** | **Na (mg%)** |
| 81.7 | 9.70 | 1.2 | 1.5 | 55.8 |

**Source:** Krzynowek and Murphy (1987).

| Nutritional composition (per 100 g) | |
|---|---|
| Calories | 70 |
| Calories from fat | 9 |
| Protein (g) | 16 |
| Carbohydrates (g) | 0 |
| Total fat (g) | 0.9 |
| Saturated fat (g) | 0 |
| Cholesterol (mg) | 40 |
| Sodium (mg) | 57 |

**Source:** Sturmer et al. (nda).

*Meretrix casta*

| Proximate composition (%) | | | | |
|---|---|---|---|---|
| Protein (kcal/g dw) | Lipid (kcal/g dw) | CHO (kcal/g dw) | Ash (kcal/g dw) | Calorific value (kcal/g dw) |
| 27.2–41.1 | 5.3–19.1 | 2.9–14.0 | 1.5–11.2 | 3.8–5.1 |

**Source:** Singh et al. (2012).

| Proximate composition (%) | | | |
|---|---|---|---|
| Protein | Lipids | Carbohydrates | Moisture |
| 11.8 | 2.0 | 6.2 | 81.1 |

| Composition of trace elements (µg/g) | | | | | | | | |
|---|---|---|---|---|---|---|---|---|
| Co | Cr | Cu | Fe | Mg | Mn | Ni | Cd | Zn |
| 7.5 | 26.8 | 24.0 | 430 | 1.1 | 35.1 | 2.6 | 1.4 | 70.8 |

**Source:** Marichamy et al. (2011).

| Proximate composition (%) | |
|---|---|
| Protein | 41.0 |
| CHO | 5.6 |
| Lipid | 4.2 |
| Ash | 2.8 |

| Amino acids (%) | |
|---|---|
| EAA | |
| Phenylalanine | 8.5 |
| Lysine | 9.1 |
| Histidine | 5.5 |
| Methionine | 6.1 |
| Arginine | 3.6 |
| Leucine | 7.2 |
| Threonine | 4.1 |
| Isolucine | 2.5 |
| Valine | 1.5 |
| Tryptophan | 0.4 |

*EAA*, essential amino acids.

| NEAA | |
|---|---|
| Glycine | 8.4 |
| Serine | 7.2 |
| Glutamic acid | 6.2 |
| Cysteine | 5.4 |
| Glutamate | 4.5 |
| Alanine | 4.2 |
| Proline | 3.4 |
| Aspartate | 2.3 |
| Tyrosin | 1.3 |
| Asparagine | 111 |

*NEAA*, Nonessential amino acids.

| Fatty acids | |
|---|---|
| SFA | |
| Palmitic acid | 15.4 |
| Margaric acid | 12.4 |
| Stearic acid | 36.2 |
| MUFA | |
| Oleic acid | 14.6 |
| PUFA | |
| Linolenic acid | 3.4 |
| Alpha-linolenic acid | 12.7 |
| Stearidonic or moroctic acid | 2.6 |

*MUFA*, monounsaturated fatty acids; *PUFA*, polyunsaturated fatty acids; *SFA*, saturated fatty acid.

| Vitamins | |
|---|---|
| Retinol (A) | 14.4 IU |
| Calciferol (D) | 200 IU |
| Tocopherol (E) | 1.2 mg/g |
| Vitamin (K) | 0.6 mg/g |
| Pyridoxin (B6) | 5.9 mg/g |
| Cobalamin (B12) | 0.4 μg/g |
| Vitamin (C) | 19.6 mg/g |

| Macrominerals (mg/g) | |
|---|---|
| Calcium | 345.5 |
| Sodium | 155.6 |
| Potassium | 13.4 |
| Copper | 2.3 |
| Magnesium | 145.6 |
| **Trace minerals** | |
| Iron | 15.6 |
| Zinc | 12.3 |

**Source:** Srilatha et al. (2013).

| Proximate composition (µg/ml/mg) | | | |
|---|---|---|---|
| **Extraction** | **Protein** | **Carbohydrate** | **Lipid** |
| MeOH | 190 | 5.8 | 0.2 |
| PBS | 960 | 24 | 0.1 |
| AEH-MeOH | 356 | 7.2 | 0.0 |
| AEH-Aq | 194 | 32.8 | 0.0 |

*MeOH*, methanol; *PBS*, phosphate buffer saline; *AEH*, extracts with protease inhibitors and acid enzyme hydrolysate.

**Source:** Sharma et al. (2009).

## *Meretrix lusoria*

| Proximate composition | |
|---|---|
| Moisture | 76.2–84.2% |
| Protein | 9.1–12.8% |
| Carbohydrate | 0.3–7.9% |
| Fat | 1.6–6.6% |
| Ash | 1.2–2.6% |

| | |
|---|---|
| Essential amino acids | 167.7–187.6 mg/g |
| Leucine | 30.9–37.0 mg/g) |
| Lysine | 35.2–36.0 mg/g |

| Fatty acids (% of total fatty acids) | |
|---|---|
| Polyunsaturated fatty acids | 46.8–49.2 |
| DHA | 13.3–16.5 |
| EPA | 4.8–7.1 |
| Cholesterol | 0.1–0.2% wet weight |

**Source:** Karnjanapratum et al. (2013).

| Minerals (mg/g) | | | | | | |
|---|---|---|---|---|---|---|
| **Fe** | **Zn** | **Ca** | **Mg** | **Na** | **K** | **P** |
| 42.6 | 1.2 | 92.4 | 56.7 | 1.3 | 0.3 | 0.7 |

**Source:** Ademolu et al. (2015).

## Meretrix meretrix

| Proximate composition (%) | | | | |
|---|---|---|---|---|
| **Moisture** | **Ash** | **Fat** | **Protein** | **Carbohydrate** |
| 80.0 | 1.4 | 0.1 | 9.4 | 9.0 |

**Source:** Abdullah et al. (2016).

## Modiolus moduloides

| Nutritional facts (%) | |
|---|---|
| Moisture | 78.7 |
| Crude protein | 71.5 |
| Crude carbohydrate | 9.3 |
| Crude fat | 6.6 |
| Ash | 12.5 |
| Crude fiber | 0.1 |
| Calcium | 0.4 |
| Phosphorus | 1.2 |

**Source:** Buranon (1980).

## *Mya arenaria*

| Nutritional facts (per 100 g) | |
|---|---|
| Calories | 74 |
| Total fat | 0.9 g |
| Saturated fat | 0.2 g |
| Cholesterol | 34 mg |
| Sodium | 56 mg |
| Protein | 12.7 g |
| Omega 3 | 0.2 g |

**Source:** Anon. (ndp).

| Amino acids (g/100 g) (FW) | |
|---|---|
| Asp | 9.3 |
| Glu | 13.0 |
| Ser | 4.3 |
| His* | 1.8 |
| Gly | 6.9 |
| Thr* | 4.4 |
| Arg | 7.0 |
| Ala | 6.8 |
| Tyr | 3.2 |
| Val* | 4.2 |
| Phe* | 3.3 |
| Ile* | 4.5 |
| Lys* | 7.7 |
| Leu* | 6.5 |
| Pro | 3.4 |
| Met* | 2.4 |
| Trp* | 1.2 |

*EAA*, essential amino acids.

**Source:** Anon. (ndt).

## *Mytilus edulis*

| Proximate composition (g/100 g FW) | |
|---|---|
| Total carbohydrate | 2.6 |
| Total protein | 5.6 |
| Total fat | 1.0 |

**Source:** Sirbu et al. (2011).

## *Mytilus edulis*

| Free amino acids (mg/100 g) | |
|---|---|
| Alanine | 340 |
| Arginine | 416 |
| Aspartic acid | 200 |
| Glutamic acid | 317 |
| Glycine | 399 |
| Histidine | 12 |
| Isoleucine | 25 |
| Leucine | 15 |
| Lysine | 39 |
| Methionine | 10 |
| Phenylalanine | 10 |
| Proline | 29 |
| Threonine | 40 |
| Tyrosine | 13 |
| Valine | 14 |

**Source:** Anon. (ndo).

## *Mytilus galloprovincialis*

| Nutritional facts (g/100 g FW) | |
|---|---|
| Total carbohydrate | 1.9 |
| Total protein | 9.0 |
| Total fat | 1.8 |
| Mineral | 0.8 |
| Water (%) | 85.9 |
| Dry matter (%) | 13.2 |

**Source:** Sirbu et al. (2011).

| Amino acids (g/100 g) | |
|---|---|
| Aspartic acid | 1.4 |
| Threonine* | 0.7 |
| Serine | 0.8 |
| Glutamic acid | 1.8 |
| Proline | 1.5 |
| Glycine | 0.8 |
| Alanine | 0.7 |
| Cystine | 0.3 |
| Valine* | 0.7 |
| Methionine* | 0.3 |
| Isoleucine* | 0.6 |
| Leucine* | 1.0 |
| Tyrosine | 0.7 |
| Phenylalanine* | 0.8 |
| Histidine* | 0.4 |
| Lysine* | 1.1 |
| Arginine* | 0.9 |

*EAA*, essential amino acids.

**Source:** Fengör et al. (2008).

| Fatty acid composition (% of total fatty acids) | |
|---|---|
| C14:0 | 2.4 |
| C15:0 | 0.5 |
| C16:0 | 17.7 |
| C17:0 | 1.1 |
| C18:0 | 3.9 |
| ΣSFA | 25.7 |
| C14:1 | 1.2 |
| C15:1 | 0.5 |
| C17:1 | 1.4 |
| C16:2n4 | 1.9 |
| C16:4n1 | 1.1 |
| C18:3n3 | 1.9 |
| C18:3n6 | 2.2 |

| Fatty acid composition (% of total fatty acids) | |
| --- | --- |
| C18:4$n$3 | 2.7 |
| C20:2$n$6 | 4.5 |
| C20:3$n$3 | 1.5 |
| C20:3$n$6 | 1.2 |
| C20:4$n$6 | 0.8 |
| C20:5$n$3 | 12.8 |
| C22:4$n$6 | 3.8 |
| C22:5$n$6 | 2.2 |
| C22:5$n$3 | 1.0 |
| C22:6$n$3 | 15.3 |

**Source:** Freites et al. (2002).

## *Noetia ponderosa*

| Nutritional composition (per 100 g) | |
| --- | --- |
| Calories | 50 |
| Total fat (g) | 1 |
| Protein (g) | 11 |
| Carbohydrates (g) | 10 |
| Cholesterol (mg) | 55 |
| Sodium (mg) | 480 |

**Source:** Leslie et al. (nd).

## *Ostrea edulis*

| Nutritional facts (% WW) | |
| --- | --- |
| Moisture | 78.7 |
| Protein | 10.0 |
| Carbohydrate | 7.8 |
| Glycogen | 6.8 |
| Fat | 2.2 |
| Ash | 1.3 |

**Source:** Anon. (ndo).

| Proximate composition(% FW) | | | | |
|---|---|---|---|---|
| Protein | Lipids | Moisture | Ash | Carbohydrate |
| 12.9 | 3.0 | 79.4 | 1.6 | 3.1 |

Source: Anon. (ndt).

| Free amino acids (mg/l00 g) | |
|---|---|
| Alanine | 646.0 |
| Arginine | 66.6 |
| Aspartic acid | 26.1 |
| Glutamic acid | 264.0 |
| Glycine | 248.0 |
| Histidine | 22.9 |
| Isoleucine | 19.2 |
| Leucine | 12.9 |
| Lysine | 22.0 |
| Methionine | 8.4 |
| Phenylalanine | 8.5 |
| Proline | 166.0 |
| Threonine | 9.7 |
| Tyrosine | 10.3 |
| Valine | 10.8 |

Source: Anon. (ndo).

| Fatty acid compositions (% TFA) | |
|---|---|
| 14:0 | 2.0 |
| 16:0 | 18.9 |
| 16:1$(n-7)$ | 1.5 |
| 17:0 | 2.3 |
| 17:1 | 2.5 |
| 18:0 | 5.6 |
| 18:1$(n-7)$ | 1.7 |
| 18:1$(n-9)$ | 3.0 |
| 18:4$(n-3)$ | 1.6 |
| 20:1$(n-11)$ | 1.4 |
| 20:1$(n-9)$ | 1.4 |

| Fatty acid compositions (% TFA) | |
|---|---|
| 20:1($n-7$) | 3.9 |
| 20:4($n-6$) | 3.9 |
| 20:5($n-3$) | 12.5 |
| 22:5($n-3$) | 1.6 |
| 22:6($n-3$) | 23.7 |

**Source:** Pogoda et al. (2013).

## *Ostrea lurida*

| Proximate composition (% DW) | |
|---|---|
| Moisture | 81.0 |
| Protein | 50.5 |
| Carbohydrate | 28.4 |
| Fat | 13.2 |
| Ash | 7.9 |

## *Panopea abrupta*

| Nutritional composition (per 100 g) | |
|---|---|
| Protein (g) | 14.4 |
| Iron (mg) | 12.5 |
| Fat (g) | 1.3 |
| Ca (mg) | 36.0 |
| Calories | 80 |

## *Panopea generosa*

| Proximate composition (%) | | | |
|---|---|---|---|
| Moisture | Protein | Fat | Ash |
| 78.8 | 15.0 | 3.2 | 1.8 |

**Source:** Anon. (1989).

## *Paratapes undulatus*

| Proximate composition (%) | | | | |
|---|---|---|---|---|
| Moisture | Crude protein | CHO | Lipid | Ash |
| 88.3 | 10.3 | – | 0.5 | 2.4 |

**Source:** Mohammad and Yusuf (2016).

## Pecten maximus

| Proximate composition (%) | | |
| --- | --- | --- |
| Protein | Fat | Glycogen |
| 16.7 | 0.5 | 3.7 |

**Source:** Anon. (1989).

## Pecten novaezelandiae

| Proximate composition(% FW) | | | | |
| --- | --- | --- | --- | --- |
| Protein | Lipids | Moisture | Ash | CHO |
| 14.3–15.4 | 0.7–1.3 | 78.7–82.4 | 1.6–1.9 | 2.7 |

## Perna canaliculus

| Proximate composition (% FW) | | | | |
| --- | --- | --- | --- | --- |
| Protein | Lipids | Moisture | Ash | CHO |
| 11.9–13.9 | 2.1–2.2 | 78.2–80.9 | 1.7–2.2 | 3.4 |

## Perna canaliculus

| Amino acids (g/100 g protein) | |
| --- | --- |
| Asp | 1.3 |
| Glu | 1.7 |
| Ser | 0.7 |
| His* | 0.2 |
| Gly | 1.1 |
| Thr* | 0.5 |
| Arg | 1.0 |
| Ala | 0.6 |
| Tyr | 0.4 |
| Val* | 0.5 |
| Phe* | 0.4 |
| Ile* | 0.5 |
| Lys* | 1.0 |
| Leu* | 0.8 |
| Pro | 0.5 |
| Cys | 0.1 |
| Met* | 0.3 |

*EAA, essential amino acids.

## Perna canaliculus

| Fatty acids | |
|---|---|
| Polyunsaturated fatty acids | 45–46% |
| $n-3$ PUFA | 39–41% |
| DHA | 19% |
| EPA | 15% |
| Palmitic acid | 5% |
| Cholesterol | 31% (of total sterols) |

**Source:** Murphy et al. (2002).

## Perna viridis

| Nutritional facts (%) | |
|---|---|
| Moisture | 81.5 |
| Crude protein | 63.9 |
| Crude carbohydrate | 15.1 |
| Crude fat | 9.5 |
| Ash | 11.5 |
| Crude fiber | 0.1 |
| Calcium | 0.31 |
| Phosphorus | 0.89 |

**Source:** Buranon (1980).

| Proximate composition (%) | | | | |
|---|---|---|---|---|
| Protein (kcal/g dw) | Lipid (kcal/g dw) | CHO (kcal/g dw) | Ash (kcal/g dw) | Calorific value (kcal/g dw) |
| 57.8–66.5 | 4.6–9.8 | 14.7–29.3 | 4.8–9.0 | 4.9–5.4 |

**Source:** Singh et al. (2012).

| Biochemical composition (µg/ml/mg) | | | |
|---|---|---|---|
| Extraction | Protein | Carbohydrate | Lipid |
| MeOH | 280 | 24.7 | 0.1 |
| PBS | 830 | 32.1 | 0.0 |
| AEH-MeOH | 440 | 2.0 | 0.1 |
| AEH-Aq | 170 | 1.1 | 0.0 |

*MeOH*, methanol; *PBS*, phosphate buffer saline; *AEH*, extracts with protease inhibitors and acid enzyme hydrolysate.

**Source:** Sharma et al. (2009).

| Amino acids % (W/W) | |
| --- | --- |
| Phenyl alanine | 1.0 |
| Glycine | 0.8 |
| Lysine | 1.0 |
| Serine | 1.4 |
| Histidine | 1.2 |
| Glutamic acid | 1.9 |
| Methionine | 2.2 |
| Cystine | 1.1 |
| Arginine | 1.2 |
| Glutamine | 1.0 |
| Leucine | 1.2 |
| Alanine | 0.5 |
| Threonine | 1.3 |
| Proline | 0.9 |
| Isoleucine | 1.1 |
| Asparagine | 1.0 |
| Valine | 1.1 |
| Tyrosine | 1.1 |
| Tryptophan | 1.2 |
| Aspartic acid | 1.2 |

| Vitamins (mg/100 g) | |
| --- | --- |
| Vitamin A | 14.5 |
| Vitamin B6 | 3.5 |
| Vitamin B12 | 1.8 |
| Vitamin B1 | 1.0 |
| Vitamin B2 | 1.3 |
| Vitamin B3 | 2.7 |
| Vitamin C | 13.6 |

| Minerals (mg/100 g) | |
|---|---|
| Sodium | 156.7 |
| Calcium | 235.9 |
| Potassium | 156.7 |
| Magnesium | 24.6 |
| Iron | 5.1 |
| Zinc | 2.9 |
| Copper | 1.3 |
| Iodine | 112.7 |

**Source:** Saritha (2015).

| Fatty acid (as wt%) | |
|---|---|
| Saturated fatty acids | 30.3 |
| Monounsaturated fatty acids | 27.3 |
| **Polyunsaturated fatty acids (% TFA)** | |
| 20;5$n$3 | 12.8 |
| 22;6$n$3 | 9.9 |
| PUFA | 35.0 |

**Source:** Chakraborty et al. (2011).

## *Pholas dactylus*

| Proximate composition (%) | | | | |
|---|---|---|---|---|
| Moisture | Ash | Fat | Protein | CHO |
| 83.8 | 1.2 | 0.1 | 11.4 | 3.6 |

**Source:** Abdullah et al. (2016).

## *Pinctada imbricata radiata*

| Proximate composition (%) | |
|---|---|
| Protein | 13.9 |
| Fat | 1.0 |
| Moisture | 82.0 |
| Ash | 2.7 |

**Source:** Al-Sayed (2010).

## *Placopecten magellanicus*

| Proximate composition (%) | | |
|---|---|---|
| Protein | Fat | Glycogen |
| 17.6 | 0.5 | 1.8 |

Source: Anon. (1989).

## Politapes aureus

| Proximate composition (%) | | | | |
|---|---|---|---|---|
| Moisture | Crude protein | CHO | Lipid | Ash |
| 79.0 | 11.4 | 5.5 | 0.9 | 3.2 |

Source: Mohammad and Yusuf (2016).

## *Protapes gallus*

| Proximate composition | |
|---|---|
| Protein | 15.6–86.8% |
| Carbohydrate | 4.3–72.7% |
| Lipid | 2.3–5.1% |
| Calories | 4.6–6.5 kcal/g |

Source: Nagvenkar and Jagtap (2013).

## *Ruditapes decussatus (=Tapes decussatus)*

| Proximate composition (%) | | |
|---|---|---|
| Protein | Fat | Ash |
| 10.8 | 1.1 | 1.7 |

Source: Anon. (2004).

## *Ruditapes decussatus*

| Proximate composition (%) | |
|---|---|
| Moisture | 84.0 |
| Protein | 9.6 |
| Carbohydrate | 1.1 |
| Lipid | 1.0 |
| Ash | 3.3 |

Source: Dincer (2006).

| Amino acids (g/100 g) (FW) | |
|---|---|
| His* | 1.6 |
| Thr* | 3.2 |
| Arg | 4.0 |
| Val* | 3.2 |
| Phe* | 3.2 |
| Ile* | 5.4 |
| Lys* | 8.0 |
| Leu* | 8.2 |
| Met* | 2.3 |
| Trp* | 1.1 |

*EAA, essential amino acids.

| Lipid class content (% of total lipids) | |
|---|---|
| Phospholipids | 60.89 |
| Triacylglycerols | 23.49 |
| Free fatty acids | 7.6 |
| Sterols | 8.2 |
| **Fatty acid composition (% of total fatty acids)** | |
| $\Sigma$Saturated | 29.6 |
| $\Sigma$Monoenoic | 31.2 |
| $\Sigma$Polyenoic | 39.2 |
| $\Sigma n - 6$ | 7.0 |
| $\Sigma n - 9$ | 18.0 |
| $\Sigma n - 11$ | 1.1 |
| $\Sigma n - 3$ PUFA | 22.4 |

**Source:** Fernández-Reiriz et al. (1999).

## Ruditapes phillippinarum

| Proximate composition (%) | |
|---|---|
| Moisture | 85.9 |
| Protein | 8.7 |
| Carbohydrate | 1.1 |
| Lipid | 0.8 |
| Ash | 3.1 |

**Source:** Dincer (2006).

| Proximate composition (%) | | | | |
|---|---|---|---|---|
| Moisture | Protein | Fat | Ash | Carbohydrate |
| 84.9 | 9.2 | 1.7 | 1.2 | 1.8 |

Source: Krzynowek and Murphy (1987).

| Fatty acids (% of total fatty acids) | |
|---|---|
| 13:0 | 0.5 |
| 14:0 | 1.4 |
| 15:0 | 0.4 |
| 16:0 | 18.4 |
| 17:0 | 0.2 |
| 16:2($n - 4$) | 0.5 |
| 18:1($n - 11$) | 0.4 |
| 18:1($n - 9$) | 2.1 |
| 18:1($n - 7$) | 2.7 |
| 18:2($n - 6$) | 1.3 |
| 18:2($n - 3$) | 0.4 |
| 20:1($n - 9$) | 1.3 |
| 20:1($n - 7$) | 2.2 |
| 18:4($n - 3$) | 1.0 |
| 20:2($n - 6$) | 1.3 |
| 20:4($n - 6$) | 2.1 |
| 20:5($n - 3$) | 9.9 |
| 24:1($n - 9$) | 0.9 |
| 22:5($n - 3$) | 2.2 |
| 22:6($n - 3$) | 14.4 |

Source: Teshima et al. (1990).

## Saccostrea cucullata

| Proximate composition (%) | | | | |
|---|---|---|---|---|
| Protein (kcal/g dw) | Lipid (kcal/g dw) | CHO (kcal/g dw) | Ash (kcal/g dw) | Calorific value (kcal/g dw) |
| 42.2–55.1 | 11.8–22.9 | 9.1–20.8 | 12.2–26.2 | 4.4–5.4 |

Source: Singh et al. (2012).

## *Saxidomus nuttalli*

| Proximate composition (%) | | | |
|---|---|---|---|
| Moisture | Protein | Fat | Ash |
| 80.0 | 13.0 | 1.23 | 1.7 |

**Source:** Krzynowek and Murphy (1987).

## *Solen stirctus*

| Nutritional facts (% DW) | |
|---|---|
| Protein | 66.0 |
| Fat | 9.3 |
| Ash | 13.8 |
| Amino acids (total) | 58.3 |
| Essential amino acids | 35.6% (of total amino acids) |

## *Spisula sachalinensis*

| Proximate composition (%) | |
|---|---|
| Moisture | 77.3 |
| Protein | 14.1 |
| Lipid | 1.0 |
| Ash | 1.4 |

| Amino acids (mg/100 g) | |
|---|---|
| Alanine | 1121 |
| Arginine | 79 |
| Aspartic acid | 4 |
| Glutamic acid | 81 |
| Glycine | 677 |
| Histidine | 28 |
| Isoleucine | 7 |
| Leucine | 9 |
| Lysine | 9 |
| Methionine | 15 |
| Phenylalanine | 8 |
| Proline | 31 |
| Serine | 21 |
| Taurine | 1245 |
| Threonine | 14 |
| Tyrosine | 36 |
| Valine | 8 |

| Fatty acids (mg/100 g tissue) of total lipid | |
|---|---|
| 14:0 | 14.8 |
| 16:0 | 95.5 |
| 16:1$(n-7)$ | 27.3 |
| 18:0 | 27.7 |
| 18:1$(n-5)$ | 0.1 |
| 18:1$(n-7)$ | 9.9 |
| 18:1$(n-9)$ | 16.5 |
| 18:2$(n-6)$ | 1.2 |
| 18:3$(n-3)$ | 0.8 |
| 18:4$(n-3)$ | 4.2 |
| 20:1$(n-11)$ | 24.5 |
| 20:1$(n-9)$ | 8.8 |
| 20:1$(n-7)$ | 24.3 |
| 20:2$(n-6)$ | 0.2 |
| 20:4$(n-6)$ | 8.0 |
| 20:4$(n-3)$ | 4.1 |
| 20:5$(n-3)$ | 130.2 |
| 22:5$(n-3)$ | 14.4 |
| 22:6$(n-3)$ | 50.5 |

**Source:** Sasaki and Ohta (1999).

*Spisula solidissima*

| Proximate composition | | | | |
|---|---|---|---|---|
| Moisture (%) | Fat (%) | Ash (%) | PUFA (% of total fatty acids) | Cholesterol (mg%) |
| 73.4 | 1.4 | 2.8 | 41.7 | 63.4 |

**Source:** Krzynowek and Murphy (1987).

| Nutritional facts (per 100 g) | |
|---|---|
| Calories | 74 |
| Fat calories | 8 |
| Total fat | 0.9 g |
| Saturated fat | 0.2 g |
| Cholesterol | 34 mg |
| Sodium | 56 mg |
| Protein | 12.7 g |
| Omega 3 | 0.2 g |

## *Striostrea prismatica*

| Proximate composition (%) | |
| --- | --- |
| Proteins | 49.3 |
| Carbohydrates | 19.3 |
| Lipids | 14.1 |
| Ash | 17.4 |

**Source:** Paez-Osuna et al. (1993).

## *Tegillarca granosa*

| Proximate composition (%) | | | | |
| --- | --- | --- | --- | --- |
| Moisture | Crude protein | CHO | Lipid | Ash |
| 81.3 | 11.7 | – | 1.1 | 2.4 |

**Source:** Mohammad and Yusuf (2016).

## *Tellina angulata*

| Proximate composition (%) | | | | |
| --- | --- | --- | --- | --- |
| Protein (kcal/g dw) | Lipid (kcal/g dw) | CHO (kcal/g dw) | Ash (kcal/g dw) | Calorific value (kcal/g dw) |
| 39.0–65.7 | 8.9–19.8 | 10.0–16.2 | – | – |

**Source:** Singh et al. (2012).

## *Villorita cyprinoides*

| Proximate composition (µg/ml/mg) | | | |
| --- | --- | --- | --- |
| Extraction | Protein | Carbohydrate | Lipid |
| MeOH | 246 | 34.28 | 0.15 |
| PBS | 415 | 19.58 | 0.05 |
| AEH-MeOH | 720 | 22.08 | 0.06 |
| AEH-Aq | 176 | 5.25 | 0.02 |

*MeOH*, methanol; *PBS*, phosphate buffer saline; *AEH*, extracts with protease inhibitors and acid enzyme hydrolysate.

**Source:** Sharma et al. (2009).

*Zygochlamys patagonica*

| Nutritional facts (per 84 g) | |
|---|---|
| Proximate composition | |
| Protein | 20 g |
| Carbohydrate | 0 g |
| Dietary fiber | 0 g |
| Fat | 1 g |
| Cholesterol | 45 mg |
| Energy | |
| Calories | 100 kcal |
| Calories from fat | 10 kcal |
| Minerals | |
| Calcium | 100 mg |
| Iron | 2.7 mg |
| Sodium | 230 mg |
| Vitamins | |
| Vitamin A | 100 IU |
| Vitamin C | 0 mg |

## KEYWORDS

- **protein**
- **lipid**
- **fat**
- **ash**
- **calorific value**

# CHAPTER 6

# PHARMACEUTICAL VALUES OF EDIBLE MARINE BIVALVE MOLLUSCS

## CONTENTS

## ABSTRACT

The pharmaceutical compounds produced by the edible species of marine bivalve molluscs with anticancer, antiulcer, antioxidant, anticoagulant, antibacterial, antifungal, anti-HSV, and antiviral values are dealt within this chapter.

### *Anadara antiquata*

**Artificial bones:** The bioceramic powder prepared from the clam shell (*Anadara antiquata*) has been reported to contain high-quality hydroxyapatite and has the potential to proceed as implant or artificial bone that has physical and mechanical property approaches like natural human bones (Gunawarman et al., 2015).

### *Anadara broughtonii*

**Antitumor:** A new in vitro antitumor polypeptide, coded as J2-C3, has been isolated from this species. The IC50 values of J2-C3 were found to be 65.57, 93.33, and 122.95 µg/ml against A549, HT-29, and HepG2 cell lines, respectively. Therefore, J2-C3 might be developed as a potential antitumor agent (Xu et al., 2013).

**Antimicrobial and hemolytic:** The heterotrophic bacteria associated with this species were found to possess antimicrobial activity. They were the Gram-positive isolates belonging to the genera *Bacillus*, *Paenibacillus*, and *Saccharothrix* and two Gram-negative strains related to *Pseudomonas* and *Sphingomonas*. Substances with hemolytic and surface activities were isolated from strain *Bacillus pumilus* An 112 (Romanenkoa et al., 2008).

**Antiulcer:** The shell of this species has been reported to reduce gastric ulcer pain, which contains calcium carbonate can neutralize stomach acid.

### *Anadara kagochimensis*

**Antitumor:** P2, a polypeptide fraction, has been isolated from this species. It inhibited the proliferation of seven tumor cell lines, especially in HeLa and HT-29 cell lines. The IC50 values were 11.43 µg/ml for HeLa and 13.00 µg/ml for HT-29 treated by P2 for 48 h. P2 had little cytotoxicity on normal liver cells (L-02). The results demonstrated that P2 might be a potential antitumor agent with high efficiency in dose-dependent and time-dependent manners and low toxicity (Hu et al., 2012).

**Antioxidant:** A new antitumor and antioxidant peptide (H3) has been isolated from this species (Chen et al., 2013).

**Hypoglycemia and hypolid effects:** The hydrolysate isolated from this species has shown hypoglycemia and hypolid effects (Changgui, 1996).

### Anadara satowi

**Immunological:** The heterogeneity (dimeric and tetrameric hemoglobins) of hemoglobin from Arca was demonstrated. Hb I has a dimer structure and Hb II has a tetramer structure. These hemoglobins have immunological properties (Ohnoki et al., 1973).

### Anomalocardia flexuosa

**Anticoagulant:** Heparin with high anticoagulant activity has been isolated from this species. Three significant quantitative differences were observed for this heparin when compared with the ones from mammalian origin, namely, a higher degree of binding with antithrombin III (45%), higher molecular weight (27–43 kDa), and higher anticoagulant activity (320 IU/mg) (Dietrich et al., 1985).

### Argopecten ventricosus

**Antibacterial:** The antibacterial activity (against *Vibrio alginolyticus*) of the hemolymph of this species has been reported (Luna-González et al., 2007).

### Azumapecten farreri

**Agglutination:** Agglutination plays a crucial role in eliminating potential pathogens in marine invertebrates. Multiple types of lectins exist in this species. The mannan-binding lectins present in the hemolymph of this species has shown the highest agglutination activity (Tan et al., 2013).

**UVB-induced apoptosis:** The polypeptide from this species has been reported to avert UVB-induced apoptosis in thymocytes by modulating c-fos and c-jun expression, cytochrome *c* release, and the consequent activation of caspase-3, which were essential components of the UV-induced cell apoptotic pathway. The results suggested that PCF is a promising protective substance against UV radiation (Chen et al., 2007).

### Callista chione

**Anticoagulant and antithrombin:** The anticoagulant properties were measured as APTT (97 ± 12.1 IU/mg) and anti-Xa activity (antithrombin activity) (52 ± 7.4 IU/mg) (Luppi et al., 2005).

## Chionista fluctifraga

**Anticancer:** The protein extracts of this species showed antiproliferative effect against HeLa and MDA-MB-231 cell lines (García-Morales et al., 2016).

## Conus spp.

**Muscle relaxant**: The venom from *Conus* can be used as a muscle relaxant during heart operations (Ramesh et al., 2012).

## Corbicula japonica

**Antihepatitis:** Fired shell powder of this species has been traditionally used in Japan as a folk medicine to improve liver disorder (Sasaki et al., 2011).

## Crassostrea gigas

**Anticancer:** The hexane extracts of this species with palmitic acid, margaric acid, and stearic acid exhibited the highest anticancer activity. These lipids effectively inhibit in vitro human prostate tumor growth by inducing apoptosis of cancer cells (Kim et al., 2013)

**Antioxidant activity:** The purified peptide of this species was found to significantly scavenge cellular radicals and protective effect on DNA damage caused by hydroxyl radicals generated (Qian et al., 2008).

**Anti-HSV-1:** The hemolymph of this species has shown anti-HSV-1 activity (Green et al., 2015).

**Antitumor:** The hydrolysates isolated from this species produced strong immunostimulating effects in mice, which might result in its antitumor activity. The antitumor and immunostimulating effects of this species reveal its potential for tumor therapy and as a dietary supplement with immunostimulatory activity (Wang et al., 2010).

**Dietary supplement**: The Pacific oyster *Crassostrea gigas* powder containing natural taurine is used as a dietary supplement for treating liver disorders, arthritis, and skin problems.

## Crassostrea madrasensis

**Antibacterial:** Extracts of *Crassostrea madrasensis* exhibits effective antibacterial activity against *Pseudomonas aeruginosa*, *Staphylococcus aureus*, *Escherichia coli*, and *Salmonella typhi* water and methanol extracts showed highest activity against *Proteus mirabilis* (10 mm) (Annamalai et al., 2007).

### *Crassostrea rhizophorae*

**Antiviral:** The extracts of this species have shown antiviral activity (Dang et al., 2015).

### *Crassostrea rivularis*

**Functional food:** The content of EPA in this is higher than that of DHA. Therefore, this species is suitable for developing the functional food of prevention of cardiovascular disease (Cheng et al., 2009).

### *Crenomytilus grayanus*

**Antiviral:** The extracts of this species have shown antiviral activity (Dang et al., 2015).

**Antifungal:** A lectin with antifungal activity has been isolated from this species (Chikalovets et al., 2015).

### *Cyclina sinensis*

**Antioxidant activity and hepatoprotective**: The polysaccharides isolated from this species had potent antioxidant and hepatoprotective activities (Jiang et al., 2013).

**Lung infection:** Powder of this species is used to clear heat in the lungs and resolve phlegm (Wu et al., 2014).

### *Donax faba*

**Anticoagulant:** Glycosaminoglycans (GAGs) isolated from this species have shown more anticoagulant activity, and it is suggested that the GAG from *Donax faba* could be an alternative source of heparin (Periyasamy et al., 2013).

**Antibacterial:** The ethanol extract of this species showed a higher degree of inhibition to the human and fish pathogens. Maximum zone of inhibition was exhibited against *Salmonella paratyphi* and *Shigella* sp. (7 mm) by the ethanol extract (Giftson and Patterson, 2014).

**Antioxidant:** *D. faba* exhibited the highest antioxidant activities in iron-chelating assay with EC50 (half maximal effective concentration) of 0.76 mg/ml.

### *Donax scortum*

**Antioxidant:** The chitosan of the shell of this species could serve as a potential source for obtaining natural antioxidant. The chelating ability and reducing power of the sulfated chitosan of the shell of this species was

43.24 at 10 mg/ml and 0.578 at 0.75 mg/ml. IC50 values of scavenging abilities on superoxide anion and hydroxyl radicals were 0.398 and 0.527 mg/ml, whereas those of chelating ability and reducing power were 5.01 and 0.728 mg/ml, respectively (Subhapradha et al., 2013).

### *Gafrarium divaricatum*

**Antimicrobial:** Crude methanolic extract of this species showed consistent antimicrobial activity against *Cryptococcus neoformans*, *Aspergillus flavus*, and *Aspergillus niger* with minimum inhibitory concentration (MIC) of 0.1, 0.3, and 0.3 mg/ml, respectively. The compounds isolated from this extract with therapeutic properties include (1) octadecane, 2,2,4,15,17,17-hexamethyl-7,12-bis(3,5,5-trimethylhexyl) for anticancer properties; (2) L-(+)-ascorbic acid 2,6-dihexadecanoate for antitumor and antibacterial properties; (3) 2-Cyclopentene-1-tridecanoic acid for antibacterial activity; and (4) 17-(1,5-Dimethylhexyl)-10,13-dimethyl-2,3,4,7,8,9,10,11,12,13,14,15,16,17-tetradecahydro-1*H*-cyclopenta[*a*] phenanthren-3-ol as drug for mineral disorders and antibacterial properties (Babar et al., 2016).

### *Galatea paradoxa*

**Antimicrobial and antioxidant:** The crude peptide extracted from this species showed antimicrobial activity against eight strains of bacteria (*E. coli, S. aureus, Bacillus subtilis, S. typhi, Enterococcus feacalis, Klebseilla pneumoniae, Streptococcus pneumoniae,* and *P. aeruginosa*) and one strain of fungi (*Candida albicans*). The MICs of the extracts determined were 17 mg/ml of *Galatea paradoxa* against the entire spectrum of microorganisms tested except for *C. albicans* which was 20 mg/ml. Antioxidant activity using the 2,2-diphenyl-1-picrylhydrazyl (DPPH) assay showed scavenging ability on the DPPH radical was 56.77% at 0.39 mg/ml for this species (Borquaye et al., 2015). The zone of inhibition (mm) of extracts of this species against *E. coli, S. aureus, S. pneumonia, P. aeruginosa,* and *B. subtilis* was 20 mm. For *S. typhi, E. feacalis,* and *K. pneumonia,* the values were found to be 15.3, 16.0, and 14.7 mm, respectively. However, the zone of inhibition of this extract for *C. albicans* was nil.

### *Geloina erosa*

This species has shown antiviral, antioxidant, and hepatoprotective activities (Nair et al., 2015).

## *Geloina expansa*

**Antibacterial:** The ethanolic crude extract of this species showed 24-h activity on *E. coli* with inhibition zone range of 27.65–32.57 mm (Argente and Ilano, 2015).

## *Lithophaga teres*

**Antioxidant:** It exhibited the highest antioxidant activities in DPPH and iron-chelating assays with EC50 of 1.25 and 5.23 mg/ml, respectively.

## *Mactra quadrangularis*

**Antioxidant:** The protein polypeptides isolated from this species have shown antioxidant activity.

**Antithrombotic:** The GAG isolated from this showed antithrombotic effect. The antiplatelet aggregation experiments using big-eared rabbits manifested that GAG can markedly decrease the maximum aggregation percentage of rabbits blood platelet (Cui et al., 2014).

**Antihyperglycemic:** The crude polysaccharide fraction of this species exhibited glycemia inhibition activity, and 300 mg/kg dose has the optimal effect among all the studied doses. It is concluded that the crude polysaccharide fraction of this species can be explored as a novel health product that possesses potential as an antihyperglycemic agent (Wang et al., 2011).

**ACE:** Angiotensin I-converting enzyme (ACE)-inhibitory peptides have been isolated from this species. The cysteine thiol group of the most potent peptide VVCVPW of this species may play a key role in the binding of this peptide to the ACE active site (Liu et al., 2014).

## *Marcia opima*

**Anticoagulant:** The heparin isolated from this species showed metachromatic shift. Further, the GAGs of this species exhibited prominent anticoagulant activity (Pandian and Thirugnanasambandan, 2008).

## *Mercenaria campechiensis*

**Anticancer:** The mercenene isolated from this species degenerated HeLa cells completely in 72 h (Schmeer and Beery, 1965).

## *Mercenaria mercenaria*

**Anticoagulant:** Heparin with antithrombin activity was isolated from this species. It exhibited a molecular weight of approximately 18,000 and contained approximately 2.5 sulfate groups per disaccharide (Jordan et al., 1986).

**Anticancer:** The biologically active principle, namely, mercenene isolated from this species has shown antitumor activity with human HeLa cancer cells (Schmeer and Beery, 1966).

### Meretrix casta

**Antioxidant:** The protein hydrolysate isolated from this species showed high antioxidant activity (Nazeer et al., 2013).

### Meretrix lusoria

**Anticancer:** The ethyl acetate extract of this species showed anticancer activity and the compounds showing strong apoptosis-inducing activity were identified as epidioxysterols (Pan et al., 2006). The epidioxysterols have also been reported to induce apoptosis in HL-60 cells showing its potential for treating liver disease and hepatitis also (Kim, 2012).

### Meretri meretrix

**Lung infection:** Powder of this species is used to clear heat in the lungs and resolve phlegm.

**Immunological:** The water-soluble polysaccharide MMPX-B2 isolated from this species was found to stimulate the murine macrophages to release various cytokines (Li et al., 2016).

**Antitumor:** Novel antitumor protein from the coelomic fluid of *Meretrix meretrix* exhibited significant cytotoxicity to several cancer cell types, including human hepatoma BEL-7402, human breast cancer MCF-7, and human colon cancer HCT116 cells (Ning et al., 2009).

### Minnivola pyxidata

The extracts of this species have shown antibacterial, analgesic activity, anti-inflammatory, and antihistamine activities.

### Modiolus modiolus

**Antibacterial:** The antibacterial activity (against *V. alginolyticus*) of the hemolymph of this species has been reported (Luna-González et al., 2007).

### Mya arenaria

**Antiviral:** This species showed inhibition of viral infection (LT-1) at intracellular level. Further, inhibition of tumors in hamsters by adenovirus 12 has also been reported (Dang et al., 2015).

## *Mytilus chilensis*

**Antibacterial:** The antibacterial activity (against *V. alginolyticus*) of the hemolymph of this species has been reported (Luna-González et al., 2007).

## *Mytilus edulis*

**Antibacterial:** The antibacterial activity (against *V. alginolyticus*) of the hemolymph of this species has been reported (Luna-González et al., 2007).

**Antithrombin activity:** Adhesive protein present in byssus is used as adhesive for surgical applications (Ninan et al., 2007).

**Anticoagulant:** Edible part of this species showed this activity (Jung and Kim, 2009).

**Antifungal:** Blood of this species has shown this activity.

**Antihypertensive:** Fermented sauce of this species has shown this activity.

**Antioxidant:** Mussel-derived radical-scavenging peptide is present in fermented sauce of this species. An antioxidative peptide with a molecular mass of 620 Da has been purified from this species (Jung et al., 2005).

**Antimicrobial:** The following compounds have shown antimicrobial activity.

- **Mytilin A**—present in hemocytes
- **Mytilin B**—present in hemocytes
- **Mytilus defensin A**—present in blood
- **Mytilus defensin B**—present in blood.

## *Mytilus galloprovincialis*

**Antibacterial:** The antibacterial activity (against *V. alginolyticus*) of the hemolymph of this species has been reported (Luna-González et al., 2007)

**Antimicrobial:** Antimicrobial compounds such as MGD-1,2, mytilin A–D and G1, myticin B, C are present in the extract of this species.

**Antiviral:** The compound mytilin has been reported to inhibit viral transcription (Dang et al., 2015).

**Hepatoprotective and anti-inflammatory**: This species has been reported to possess hepatoprotective and anti-inflammatory compounds (Nair et al., 2015).

**Cytotoxic:** The plasma of this species contains cytotoxic activity against both vertebrate (erythrocytes and mouse tumor) and protozoan cells (Hubert et al., 1996).

*Mytilus unguiculatus*

**Antimicrobial:** An antibacterial peptide, mytichitin-A has been isolated from this species and this was found to involve in the host immune response against bacterial infection and might contribute to the clearance of invading bacteria (Jing-jing et al., 2014).

*Nodipecten nodosus*

**Anticoagulant and antithrombotic:** A HS-like GAG, with anticoagulant and antithrombotic properties, has been isolated from this species. In vivo assays demonstrated that at the dose of 1 mg/kg, this mollusc HS inhibited thrombus growth in photochemically injured arteries (Gomes et al., 2010).

*Ostrea edulis*

**Antibacterial:** This species has shown antibacterial activity against both Gram-negative and Gram-positive bacteria. When tested against the marine pathogenic *V. alginolyticus*, hemocyte lysates of this species were more active than cell-free plasma (Hubert et al., 1996).

*Paratapes undulatus*

**Hepatoprotective:** The bioactive compounds, namely, hydrolysates isolated from this species have shown potential hepatoprotective roles through antioxidant activities (Nair et al., 2015).

**EPA:** The content of EPA in this species is higher than that of DHA. Therefore, this species is suitable for developing the functional food of prevention for cardiovascular disease (Cheng et al., 2009).

*Perna canaliculus*

**Pain relief for osteoarthritis:** Most conventional treatments for osteoar-thritis, such as nonsteroidal anti-inflammatory drugs and simple analge-sics, have shown side effects. On the other hand, PCSO-524™, a nonpolar lipid extract from this species, is rich in omega-3 fatty acids and has been shown to reduce inflammation in both animal studies and patient trials.

The PCSO-524™ patients were found to show a statistically significant improvement compared with patients who took fish oil. There was an 89% decrease in their pain symptoms and 91% reported an improved quality of life. These results suggest that PCSO-524™ might offer a potential alternative complementary therapy with no side effects for osteoarthritis patients (Zawadzki et al., 2014; Brien et al., 2008). Freeze-dried extracts of this species have been promoted extensively as a treatment for rheumatoid arthritis for some years (Caughey et al., 1983).

**Anti-inflammatory:** Lipids of this species have been reported to possess anti-inflammatory activity in vitro and in vivo. The anti-inflammatory activity has been reported to reside in either the free fatty acid fraction with fatty acids containing four, five, and six double bonds, or sterols, or a polysaccharide fraction (Murphy et al., 2002).

**Antithrombin:** Pernin (a protein composed of 497 amino acids) present in the cell-free hemolymph of this species has shown antithrombin activity (Scotti et al., 2001).

*Perna viridis*

**Antibacterial:** The antibacterial activity of the extracts of this species was tested against human pathogens *Acinetobacter baumannii*, *P. aeruginosa*, and *E. coli* K1. The bactericidal activity (%) of this species is given below:

|        | *A. baumannii* | *E. coli* K1 | *P. aeruginosa* |
|--------|----------------|--------------|-----------------|
| Gills  | 82.4           | 47.4         | 99.4            |
| Gut    | 94.4           | 34.2         | 99.6            |
| Gonad  | 92.2           | 21.2         | 95.6            |

The above results indicated that *P. aeruginosa* was maximally inhibited by the methanol extracts of *Perna viridis* (Kiran et al., 2014).

**Antibacterial:** The 10:10 (methanol:ethanol) fractionated extracts of this species has shown highest activity against *P. mirabilis* (8 mm) and 14:6, 4:16, and 2:18 fractions showed prominent activity against *P. aeruginosa*, *E. coli*, and *K. pneumoniae* (Annamalai et al., 2007).

**Antibacterial and antioxidant:** The antibacterial activity of the extract of this species showed maximum zone of inhibition (15 mm) against *Vibrio* *cholerae* and minimum activity (5 mm) was observed in *K. pneumoniae.* The antioxidant activity of crude protein tissue extract from the *P. viridis* were measured in different system of assay such as DPPH assay (76.9%), total reducing power (27.8 μg), total antioxidant activity (174 μg), hydrogen peroxide scavenging (88.12%), and nitrous oxide scavenging activity (62.5%) at 100 μg/ml (Madhu et al., 2014).

**Antioxidant and cytotoxic:** The ethyl acetate extract of this species was found to possess both antioxidant and cytotoxic activities, which can be attributed to the presence of secondary metabolites like phenolics and alkaloids. The extract showed good scavenging activity against DPPH radical (IC50 0.616 mg/ml). It also proved significantly cytotoxic toward brine shrimp, Dalton's lymphoma ascites, and Ehrlich ascites carcinoma cells (Sreejamole and Radhakrishnan, 2013).

**Antiviral:** Extracts, prepared from this species, have shown the inhibition of HIV virus replication. Maximum difference in the expected infectious dose (EID50) value was observed in the extract prepared from this species for in vitro studies conducted with influenza virus type-A (2.50 lg) and virus type-B (3.00 lg) (Chatterji et al., 2002).

**Antimalarial activity:** Extracts, prepared from this species, showed the inhibition of replication of *Plasmodium falsiparum* (Malhotra et al., 2003).

**Anticancer:** The extract of this species was found to inhibit the formation of endothelial cell capillary tube in a concentration-dependent manner in vitro. At 100 μg/ml concentration, this extract showed a significant inhibitory effect on the formation of intercellular junctions and tube. Substances or molecules, having anti-angiogenic properties, are useful in cancer therapy. Anti-angiogenic substances can be used to inhibit the growth of tumor mass by preventing neo-vascularization of an early developing tumor. Therefore, such substances are of importance, not only to prevent the development of tumor but also to get rid of the subsequent metastasis (Mirshahi et al., 2009).

**Treatment of osteoporosis:** Extracts, prepared from this species, showed an inhibition on the formation of osteoclast, and on the basis of this finding, it could therefore be used to control osteoporosis (Rao et al., 2003).

### *Pholas orientalis*

**Antifungal:** Cold methanol and PBS (phosphate buffer saline) extracts of body meat and cold methanol extract of the mantle of this species have shown antifungal activities at 1000 mg/ml dose. Of the three fungal strains tested, *Trichoderma* sp., a fungus that caused skin dermatitis in humans, exhibited a zone of inhibition. So, there is potential for the extract of this species as an antifungal agent (Seraspe and Abaracoso, 2014).

### *Pinctada imbricata*

**DHA:** The content of DHA in this species is higher than that of EPA. Therefore, this species is suitable for developing the functional food of healthy brain and prevention of Alzheimer's (Cheng et al., 2009).

### *Placuna placenta*

**Youth and vitality:** It helps in the preparation of "Mouktik Bhasma" which helps in gaining of youth and vitality.

### *Polymesoda erosa*

**Antiviral:** For in vivo studies, maximum difference in EID50 values was observed in the extracts of this species (4.00 lg) with influenza virus type-A (Chatterji et al., 2002).

### *Protapes gallus*

**Antioxidant:** The scavenging potential of 2,2-diphenyl-2-picryl-hydrazyl radical and reducing action increased in a dose-dependent manner. Inhibition of lipid peroxidation shown by methanol extract and its significant correlation with OH-scavenging activity implies its potential to protect the cell damage against reactive oxygen species (Pawar et al., 2013).

**Antihypertensive, anti-inflammatory, and antidiabetic:** This species may serve as potential source of bioactive leads for use as functional food supplements to deter deleterious free-radical-induced diseases, such as inflammation, diabetes and hypertension (Joy et al., 2016).

### *Pteria avicular*

**Antibacterial:** Highest activity of the crude acetone extract of this species was exhibited against *K. pneumoniae* (5 mm) and *Staphylococcus epidermidis* (5 mm). Similarly, highest against *S. paratyphi* B (5 mm) was with the chloroform extract. For this species, 100% acetonated fraction of the extract can be considered as potent antimicrobial compound against these pathogens (Gnanambal et al., nd).

## Pteria penguin

**Antibacterial:** Highest activity was exhibited against *K. pneumoniae* (6 mm) and *S. epidermidis* (5.5 mm) by the crude acetone extract of this species and against *S. paratyphi* B (5.5 mm) by the chloroform extract. For this species, 100% acetonated fraction of the extract can be considered as potent antimicrobial compound against these human pathogens (Mohanraj et al., 2012).

## Ruditapes philippinarum

**Anticancer:** A novel anticancer peptide has been purified from the hydrolyzates of this species. This peptide effectively induced apoptosis on prostate, breast, and lung cancer cells (Kim et al., 2013).

**Anti-inflammatory:** The water extract of this species containing the active compound taurine has shown anti-inflammatory effects (Cheong et al., 2015).

**Anticoagulant activity:** Heparin (at a concentration of 2.1 mg/g dry animal) with high anticoagulant activity has been isolated from this species. This clam heparin was found to have the ATIII binding site mainly identical to that of human and porcine intestinal mucosal heparins and bovine intestinal mucosal heparin (Luppi et al., 2005).

## Saccostrea cucculata

**Antioxidant:** It exhibited the highest antioxidant activities in DPPH and Son of Sevenless (SOS) assays with EC50 of 1.15 and 0.756 mg/ml, respectively.

**Antioxidant and anticancer:** A total of seven antioxidant peptides have been isolated from this species. Among these peptides (SCAP 1–7), three peptides (SCAP 1, 3, and 7) showed the highest scavenging ability on DPPH radicals. The hydrolysate of this species was tested for cell cytotoxicity on HT-29 (human colon carcinoma) cell lines, and it was found that there was a significant cytotoxic effect on these cell lines. It is suggested that the anticancer and antioxidative hydrolysate from this species may be useful ingredients in food and nutraceutical applications (Umayaparvathi et al., 2014).

## Sinonovacula constricta

**Immunomodulator:** A water-soluble polysaccharide fraction (SCP-1) was isolated from this species. It was found that SCP-1 could significantly increase the viability of macrophages, enhance the capability of macrophage phagocytosis, increase the activity of acid phosphatase, and promote the production of nitric oxide, mouse tumor necrosis factor-$\alpha$, mouse interferon-$\gamma$, and mouse interleukin-1$\beta$. The results suggest that SCP-1 possesses potent immunomodulating effect and may be explored as a potential biological response modifier (Yuan et al., 2015).

**ACE-inhibitor:** A novel ACE-inhibitory peptide (VQY) with an IC50 of 9.8 µM was identified from this species. The inhibitory kinetics investigation by Lineweaver–Burk plots demonstrated that the peptide acts as a competitive ACE inhibitor (Li et al., 2016).

**Health supplement:** A ceramide aminoethylphosphonate has been isolated from this species. On comparison with the synthetic compound, the aqueous hydrolysis product was found to behave like 2-aminoethylphosphonic acid. The results suggest that a person may ingest and absorb 2-aminoethylphosphonic acid, through the food chain, on the consumption of this edible shellfish (Tamari and Kandatsu, 1986).

### *Spondylus varius*

**Liver protection:** This species was found to suppress the carbon tetrachloride-induced increase in serum aspartate and alanine aminotransferase activities in mice. It is suggested that this species exerts a protective effect against liver injury (Koyama et al., 2006).

### *Sunetta scripta*

**Antimicrobial:** Sunettin, a novel antimicrobial peptide, has been isolated from this species (Sathyan et al., 2012).

### *Tapes philippinarum*

**Anticoagulant:** Sulfated polysaccharides with anticoagulant activities have been isolated from this species.

### *Tegillarca granosa*

**Inducement of apoptosis**: Haishengsu isolated from the extract of this species has been found to inhibit the proliferation of human hepatocellular carcinoma BEL-7402 cells in a dose- and time-dependent manner (Chen et al., 2015).

**Orthopedic applications**: The hydroxyapatite formed from the shells of this species could find potential orthopedic and biomedical applications (Khiri et al., 2016).

| Bioactive compound | Activity |
|---|---|
| 3-Phenyl-2,3-dihydrobenzo[*b*]-furan-2-ol—furan compound | Antimicrobial |
| *n*-Hexadecanoic acid | Antioxidant, hyporcholesterolemic, and hemolytic |
| 4,7,10,13,16,19-Docosahexaenoic acid | Antidiabetic, anti-asthma, anticancer, and antiheart disease |
| Phenol, 2,4-bis (1-phenylethyl) | Analgesic, anesthetic, antioxidant, antiseptic, antibacterial, antiviral, diuretic, cancer preventive and vasodilator |
| 1,2-Benzenedicaryboxylic acid | Antimicrobial |

**Source:** Ramasamy and Balasubramanian (2012).

### *Tivela mactroides*

**Anticoagulant:** Sulfated polysaccharides with anticoagulant activities have been isolated from this species. Heparin with high anticoagulant activity was isolated from this species. A large portion of the polysaccharide chain of this species showed high affinity to immobilized antithrombin (Pejler et al., 1987).

### *Tridacna* sp.

**Anti-atherosclerotic:** The methanolic crude extracts of *Tridacna* have shown anti-atherosclerotic activities. The effective concentrations were found to be 1.56 µg/ml for diethyl ether fraction and 25.00 µg/ml for butanol fraction (Sarizan, 2013).

### *Villorita cyprinoides*

**Antiulcer:** The extract (100 and 200 mg/kg) showed significant ($p < 0.001$) reduction in gastric volume, free acidity, and ulcer index as compared to control. These results may further suggest that *Villorita cyprinoides* extract was found to possess anti-ulcerogenic as well as ulcer-healing properties, which might be due to its antisecretory activity by its antioxidant mechanism (Ajithkumar et al., 2012).

**Antiviral:** For in vivo studies, maximum difference in EID50 values was observed in the extracts of this species (4.00 lg) with influenza virus type-A (Chatterji et al., 2002).

## KEYWORDS

- bioceramic powder
- hydroxyapatite
- antitumor polypeptide
- heterotrophic bacteria
- peptide

# CHAPTER 7

# POTENTIAL DISEASES AND PARASITES IN EDIBLE MARINE BIVALVE MOLLUSCS

## CONTENTS

## ABSTRACT

The disease-causing agents, such as microbes, parasitic protozoans, helminths, copepods, and crabs infecting the edible marine bivalve molluscs, have been dealt within this chapter.

The major disease-causing agents of commercially important and edible marine bivalve molluscs are bacteria, viruses, fungi, protozoans, helminths, and parasitic crustaceans. Most bacterial diseases of these bivalves are mainly caused by a range of *Vibrio* species. The helminth parasites of bivalves are turbellarians, trematodes, cestodes, and nematodes. Among trematodes, their larval stages are more important. Among parasitic crustaceans, the parasitic copepods and pinnotherid crabs are worth-mentioning. The above disease-causing organisms can cause mortalities in natural and captive populations. Oysters have suffered frequent and extensive mass mortalities due to epizootic disease, namely, "dermocystidium disease," caused by a fungus, in the Gulf of Mexico, and "Delaware Bay disease," caused by a protozoan, in the Middle Atlantic States. Parasitic fungi have also been implicated in "shell disease" of European oysters and a fatal disease of bivalve larvae in hatcheries. Several species of haplosporidan protozoans cause serious mortalities of oysters and mussels. Larger animal parasites, such as larval trematodes and parasitic copepods, have been reported to affect reproduction, growth, and survival of edible bivalve molluscs.

## 7.1 POTENTIAL DISEASES IN EDIBLE MARINE BIVALVE MOLLUSCS

### 7.1.1 MICROBIAL DISEASES

#### 7.1.1.1 BACTERIAL DISEASES

##### 7.1.1.1.1 Brown Ring Disease

**Host(s)**: *Rudtapes philippinarum, Ruditapes decussatus, Tapes rhomboïdes, Venerupis aurea, Venerupis corrugata, Dosinia exoleta, Pecten maximus*, and *Chlamys varia*

**Causative agent**: *Vibrio tapetis*

**Symptoms/Impact**: *V. tapetis* penetrates into the extrapallial space through disruption of the periostracal lamina. From the extrapallial space, the bacteria eventually penetrate through the mantle epithelium and into the soft tissues where bacterial proliferation can cause severe damage and subsequent death. Infected animals may also have a significant decrease in glycogen suggesting that mass mortalities could be exacerbated by the degeneration of metabolic activity (http://www.dfo-mpo.gc.ca/science/aah-saa/diseases-maladies/brdcc-eng.html). A characteristic symptom, namely, a brown deposit may appear on the inner surface of the valves.

### 7.1.1.1.2 Foot Disease

**Host(s)**: *Crassostrea cuttackensis*

**Causative agent**: *Myotomus ostrearum*

**Symptoms/Impact**: The disease is localized in the shell under the attachment of the adductor muscle, where it causes roughening and blistering of the shell and degeneration of adjacent muscle tissue.

### 7.1.1.1.3 Nocardiosis

**Host(s)**: *Crassostrea gigas*

**Causative agent**: *Nocardia crassostrea*

**Symptoms/Impact**: It is characterized by few to no external signs of infection in naturally or experimentally infected oysters. Diseased animals show some mantle lesions with focal areas of discoloration in light infections and with large internal nodules in most tissues in heavy infections (Friedman et al., 1998).

### 7.1.1.1.4 Vibriosis

**Host(s)**: *Crassostrea gigas*, *Ostrea edulis*, *Perna canaliculus*, *Pecten maximus*, and *Venerupis decussatus*

**Causative agent**: *Vibrio splendidus*

**Symptoms/Impact**: Summer mortality outbreaks (Marie-Agnès et al., 2015).

### 7.1.1.1.5 Vibriosis

**Host(s)**: *Venerupis rhomboids*

**Causative agent**: *Vibrio tasmaniensis*

**Symptoms/Impact**: It induces mortality (Marie-Agnès et al., 2015).

### 7.1.1.1.6 Necrosis

**Host(s)**: *Crassostrea gigas*

**Causative agent**: *Achromobacter* sp.

**Symptoms/Impact**: Large-scale mortalities (Moribund oysters had diffuse cell infiltration, massive increase in bacterial numbers, and tissue necrosis) (Sindermann and Rosenfield, 1967).

### 7.1.1.1.7 Rickettsiosis

**Host(s)**: *Anomalocardia brasiliana, Crassostrea rhizophora, Donax trunculus, Iphigenia brasiliana, Mesodesma mactroides, Mytella bicolor, Mytilus edulis*, and *Pteria penguin*

**Causative agent**: Rickettsia-like organisms.

**Symptoms/Impact**: Digestive epithelia; gills, mantle, gonad, digestive gland, and foot muscle are damaged and mortality occurs (Zeidan et al., 2012).

### 7.1.1.1.8 Rickettsiosia

**Host(s)**: *Mytilus trossulus, Mytilus galloprovincialis, Mytilus californianus*, and *Mytilus edulis*, and a wide variety of other marine bivalves including oysters, clams, cockles, and scallops.

**Causative agent**: Rickettsia-like Chlamydia-like organisms

**Symptoms/Impact**: Microcolonies appear in the epithelial cells of the gills and digestive gland. Infections are usually of light intensity and are not associated with disease, and usually with no hemocytic response. However, large colonies can cause host cell hypertrophy with displacement and compression of the host cell nucleus against the basal membrane.

## 7.1.1.2   VIRAL DISEASES

### 7.1.1.2.1   Gill Disease, Gill Necrosis Virus Disease

**Host(s)**: *Crassostrea angulata*, *Crassostrea gigas*, and *Ostrea edulis*

**Causative agent**: Icosahedral DNA virus (also protistan, *Thanatostrea polymorpha*)

**Symptoms/Impact**: Extensive gill erosion leads to high mortalities. Initial clinical signs include yellow spots on the gills progress to brown discoloration with associated necrosis and degeneration. Yellow or green pustules may also occur on the mantle or adductor muscle. Oysters with nonproliferating infections, or in a state of recovery, may show gill lesions but no tissue necrosis. Gill disease is considered as an important factor in the elimination of *C. angulata* from the culture areas on the Atlantic coast of France (Renault, 1996).

### 7.1.1.2.2   Icosahedrical Virus-Like Disease

**Host(s)**: *R. decussatus* and *Venerupis corrugata*

**Causative agent**: Icosahedrical-virus-like organism

**Symptoms/Impact**: Impact on host is not known but mortality may occur.

## 7.1.1.3   FUNGAL DISEASES

### 7.1.1.3.1   Fungal Disease

**Host(s)**: *Crassostrea virginica*, *Ostrea frons*, *Ostrea stentina*, *O. edulis*, *O. lurida*, *Crassostrea rhizophorae*, *C. commercialis*, and *C. angulata*

**Causative agent**: *Dermocystidium marinum*

**Symptoms/Impact**: Infections are initiated in the gill, mantle, or gut epithelia or in the connective tissue near the basement membrane resulting in progressive degeneration among juxtaposed host cells. Adductor and smooth muscle are also invaded and extensive multiplication of the pathogen often occurs therein (Perkins, 1976).

### 7.1.1.3.2   Fungal Disease

**Host(s)**: *Mercenaria mercenaria*

**Causative agent**: Olpidium-like, chytrid (phycomycete) fungus (quahaug parasite X-QPX)

**Symptoms/Impact**: Mass mortalities (>90%) of 1- and 2-year-old hatchery-held quahaugs occurred. QPX infects the gills, and connective tissue of the digestive gland, gonad, and mantle (Anon., 2014).

### 7.1.1.3.3   Larval Mycosis

**Host(s)**: *Crassostrea virginica* and *Mercenaria mercenaria*

**Causative agent**: *Sirolpidium zoophthorum*

**Symptoms/Impact**: Veliger and postmetamorphic larvae up to 0.4 mm in length are particularly susceptible and can suffer over 90% mortality within 48 h (Anon., 2014).

### 7.1.1.3.4   Hinge Disease (Shell Disease)

**Host(s)**: *Ostrea edulis*, *Saccostrea cucullata*, and *Crassostrea angulata*

**Causative agent**: *Ostracoblabe implexa*

**Symptoms/Impact**: Severe shell damage occurs. Infections first appear as raised white spots, with translucent centers, which coalesce into conchiolin warts on the inner shell surface. In *O. edulis*, surface irritation results in heavier conchiolin deposition (Anon., 2014).

### 7.1.1.3.5 Digestive-Tract Impaction

**Host(s)**: *Crassostrea gigas* larvae

**Causative agent**: *Dermocystidium*-like fungus

**Symptoms/Impact**: It is affecting oyster-hatchery production. *C. gigas* larvae failed to complete metamorphosis and over 90% mortality occurred. The infection was characterized by erosion of the mantle and velar epithelia and occlusion of the stomach by thick-walled spheres. The intestine was also dilated or ruptured (Anon., 2014).

### 7.1.1.4 PROTOZOAN DISEASES

### 7.1.1.4.1 Perkinsiosis (Dermo Disease)

**Host(s)**: *Crassostrea virginica*, *C. rhizophorae*, *C. corteziensis*, and *C. gigas*

**Causative agent**: *Perkinsus marinus*

**Symptoms/Impact**: Early infections occur in the digestive tract, gills, palps, and mantle. Advanced infections are associated with diffuse systemic parasitism, pronounced hemocytosis, and tissue lysis. Further, infections reduce feeding, growth, and reproduction followed by mortality (Kern, 2011).

### 7.1.1.4.2 Perkinsiosis (Dermo Disease)

**Host(s)**: *Anadara trapezia*, *Austrovenus stutchburyi*, *R. decussatus*, *Ruditapes philippinarum*, *Tridacna maxima*, *Tridacna crocea*, and *Pitar rostrata*; oysters *Crassostrea gigas*, *C. ariakensis*, and *C. sikamea*; and pearl oysters *Pinctada margaritifera* and *Pinctada imbricata*

**Causative agent**: *Perkinsus olseni*

**Symptoms/Impact**: It infects the connective tissue of all organs. Infection with *P. olseni* can be fatal depending on host and environmental conditions. Death may occur 1 or 2 years after infection (Anon., 2009).

### 7.1.1.4.3  MSX Disease

**Host(s)**: *Crassostrea virginica* and *Crassostrea gigas*

**Causative agent**: *Haplosporidium nelsoni*

**Symptoms/Impact**: Mortalities may occur in severe conditions. At death, the oysters are grossly emaciated and the storage cells appear shrunken and disrupted.

### 7.1.1.4.4  Shell Disease

**Host(s)**: *Crassostrea virginica*

**Causative agent**: Not known

**Symptoms/Impact**: Affected oysters show mantle retraction and deposition of an anomalous conchiolin layer on the inner surface of the shell. Mortalities may occur (Renault et al., 2002).

### 7.1.1.4.5  Bonamiosis

**Host(s)**: *Ostrea chilensis, Crassostrea viriginica*, and *Crassostrea ariakensis*

**Causative agent**: *Bonamia exitiosa*

**Symptoms/Impact**: These tiny parasites infect and proliferate in oyster hemocytes or blood cells. Transmission occurs directly from oyster to oyster and infections produce inflammation. In advanced cases, it can be highly disruptive of host tissue structure and function. Mortality may occur in susceptible hosts like *Crassostrea ariakensis*.

### 7.1.1.4.6  Bonamiosis

**Host(s)**: *Ostrea chilensis, Ostrea edulis, Ostrea puelchana, Ostrea angasi, O. lurida, Crassostrea angulate*, and *Crassostrea ariakensis*

**Causative agent**: *Bonamia ostreae*

**Symptoms/Impact**: Among the host species, *Ostrea edulis* is the only known natural susceptible species and infection intensity increases concurrently to mortality with age and/or size of the oysters. Infection of wild and cultured flat oysters is often lethal, and death usually occurs concurrently with the highest intensity infection level.

### 7.1.1.4.7 Aber Disease (Digestive Gland Disease)

**Host(s)**: *Crassostrea glomerata, Ostrea edulis, Ostrea denselamellosa, Ostrea chilensis, Ostrea angasi, Ostrea puelchana, Mytilus edulis, Mytilus galloprovincialis, Mytilus edulis, Mytilus galloprovincialis, Cerastoderma edule, Venerupis corrugata, Polititapes rhomboides, R. decussatus, Ruditapes philippinarum, Ensis minor, Ensis siliqua, Solen marginatus, Chamelea gallina*, and *Argopecten gibbus*

**Causative agent**: *Marteilia refringens*

**Symptoms/Impact**: Infection causes a poor condition index with glycogen loss (emaciation), discoloration of the digestive gland, cessation of growth, tissue necrosis, and mortalities (Renault, 1996).

### 7.1.1.4.8 Marteiliosis

**Host(s)**: *Saccostrea glomerata*

**Causative agent**: *Marteilioides branchialis*

**Symptoms/Impact**: Focal patch of discoloration and swelling occurs on the gill lamellae. Severe gill lesions can cause significant economic losses.

### 7.1.1.4.9 Marteiliosis

**Host(s)**: *Crassostrea gigas*

**Causative agent**: *Marteilioides chungmuensis*

**Symptoms/Impact**: The parasite invades the oyster through the epithelial tissue of the labial palp, replicates in the connective tissue, and then moves to the gonad, producing spores inside the oocytes. The reproductive output is reduced due to this invasion (Tun et al., 2008).

### 7.1.1.4.10 Mikrocytosis (Denman Island Disease)

**Host(s)**: *Crassostrea gigas*, *Crassostrea virginica*, *Ostrea edulis*, and *Ostrea lurida*

**Causative agent**: *Mikrocytos mackini*

**Symptoms/Impact**: Formation of green or yellow-green focal pustules throughout the body but are most frequently observed on the surface of the body, labial palps, and occasionally within the adductor muscle. Often, brown scars occur on the white nacre layer of the shell adjacent to pustules on the mantle surface (Bower and Meyer, 2004).

### 7.1.1.4.11 Mikrocytosis

**Host(s)**: *Crassostrea gigas*

**Causative agent**: *Microcylos roughleyi*

**Symptoms/Impact**: It causes systemic infection of hemocytes leading to gills, connective tissues, gonads and digestive tract (Sani and Vijayan, nd).

### 7.1.1.4.12 SSO (Seaside Organism Disease)

**Host(s)**: *Crassostrea virginica* and *Crassostrea gigas*

**Causative agent**: *Haplosporidium costale*

**Symptoms/Impact**: It inhibits the growth of the infected oysters and can cause a sharp (4–6 weeks) seasonal mortality in May and June in Virginia and Maryland which coincides with the sporulation of *H. costale* in *C. virginica* (Anon., ndh). This parasite together with *H. nelsoni* has decimated large populations of Virginia oysters on the Atlantic coast of the USA and now extends northward into Atlantic Canada.

### 7.1.2 HELMINTH DISEASES

#### 7.1.2.1 TREMATODES

There are several trematode families infecting edible, marine bivalve molluscs as intermediate hosts, most notable are the family Bucephalidae and the genus *Bucephalus*. The trematode sporocyst infestations may result in poor growth and debilitation in later stages, weakness in valve closure, reduced byssal thread production (in mussels), infertility, chalky shell deposits, and bivalve mortality. Metacercariae of tremadoes cause shell gaping and deformities, chalky shell deposits, behavioral changes, reduced tolerance to stress, and host mortality in heavy infestations (https://www. adfg.alaska.gov/static/species/disease/pdfs/bivalvediseases/trematode_ metacercariae_and_sporocysts.pdf). Certain adverse effects of trematode sporocysts and metacercariae on the edible bivalve species are given below:

Bucephalidae (sporocysts) damages the mantle, gills, digestive gland, and foot of host species such as *Mytella guyanensis*, *Anomalocardia brasiliana*, and *Iphigenia brasiliana* (Boehs et al., 2010).

*Bucephalus haimeanus* (larvae) affects the growth of European and American oysters.

*Bucephalus cuculus*: Sporocysts of this species occur in the gonad and digestive gland of the European oysters and sterilize the host, *Ostrea chilensis*.

Larval trematodes (Bucephalidae) develop blackish discoloration of oyster mantle and viscera in certain species of oysters.

*Gymnophalloicles tokensis:* The metacercariae of this species affects the mantle and gills of certain species of oysters and thereby their growth is halted and reproduction is inhibited.

## 7.1.2.2   CESTODES

Commercially important host species: *Crassostrea virginica*, *Crassostrea gigas*, *Crassostrea madrasensis*, *Saccostrea glomerata*, *Striostrea mytiloides*, and *Pinctada* sp. as well as other bivalves including scallops (e.g., *Argopecten irradians*), clams (e.g., *Mercenaria mercenaria*).

Impact on the host species: Oysters and other bivalves serve as primary intermediate hosts which become infected by ingesting eggs containing oncospheres released from the intestinal tract of the definitive host.

In some locations, the prevalence of infection with metacestodes (a term used for any larval form between the egg and adult cestode and also referred to as larval cestodes) can reach 100% with several hundred metacestodes per oyster. Heavy infections may cause physiological stress and may affect growth and reproduction because of the poor condition (transparent and watery consistence) of the tissues. However, some oysters with heavy parasite burdens (up to 125 metacestodes per *C. virginica*) may not exhibit any detrimental effects (http://www.dfo-mpo.gc.ca/science/aah-saa/diseases-maladies/cestodparoy-eng.html). Cestode parasitism of certain commercially important edible marine bivalve molluscs is given below:

*Tylocephalum* sp.: The metacestode of this unidentified species of trematode damages the connective tissue of the digestive gland and gonad of *Mytella guyanensis*, *Anomalocardia brasiliana*, and *Iphigenia brasiliana* (Boehs et al., 2010). The metacestode of this unidentified species damages the connective tissue of the digestive gland and mantle of the pearl oyster, *Pinctada maxima* and *Pteria penguin* (Hine and Thorne, 2000). The larval coracidium of this species has been reported to damage the stomach and gills of the American and Pacific oysters.

### 7.1.2.3   NEMATODES

The nematodes are the uncommon parasites of marine bivalves and the two species, namely, *Echinocephalus sinensis* and *Sulcascaris sulcata* are worth mentioning. While the former is in oysters the latter is in scallops and clams. The larval stages of both these species are found in bivalves which are the intermediate hosts and the adult stages are seen in elasmobranchs and turtles.

#### 7.1.2.3.1   Nematode Parasitism

**Host(s)**: *Crassostrea gigas*

**Causative agent**: *Echinocephalus sinensis*

**Symptoms/Impact**: Coiled larvae of this species inhabit the gonoduct lumen of *C. gigas* and cause tissue damage. Mature and young adults of *Echinocephalus sinensis* are found in the intestine of eagle ray, *Aetabatus flagellum* (Ko, 2011; Gosling, 2003).

#### 7.1.2.3.2   Nematode Parasitism

**Host(s)**: *Crassostrea gigas* and *Crassostrea virginica*

**Causative agent**: *Echinocephalus crassostreai*

**Symptoms/Impact**: The second- and third-stage larvae of *E. crassostreai* occur primarily encysted in the gonad of *C. gigas* with minimal pathology.

### 7.1.2.3.3  Nematode Parasitism

**Host(s)**: *Spisula solidissima, Ylistrum balloti,* and *Chlamys* spp.

**Causative agent**: *Sulcascaris sulcata*

**Symptoms/Impact**: In *Spisula solidissima,* it inhabits all tissues and gives a brown color to its meat due to the presence of another haplosporiudian parasite, *Urosporidium spisuli* infecting the worms (Gosling, 2003).

### 7.1.3  PARASITIC COPEPODS

**Copepod parasitism**: These are obligate endoparasites inhabiting alimentary tract. Important species of parasitic copepods include *Mytilicola intestinalis, Mytilicola orientalis,* and *Mytilicola porrecta. M. intestinalis* infects mainly European mussels, clams, cockles, and flat oysters; *M. orientalis* inhabits European flat, Pacific and native oysters; and *M. porrecta* inhabits ribbed and hooked mussels. Poor growth and mortality in mussel to infestation of *M. intestinalis* have been reported. Infestation in gastrointestinal tract may cause metaplasia (columnar epithelium is reduced to cuboida or squamous) of mucosal epithelium (Pogoda et al., 2013; https://www.adfg.alaska.gov/static/species/disease/pdfs/bivalvediseases/gill_and_gut_parasitic_copepods.pdf).

**Host(s)**: *Crassostrea virginica, Crassostrea gigas, Crassostrea angulata, Ostrea edulis, Ostrea tulipa,* and *Mya arenaria.*

**Causative agent**: *Myicola ostreae, Myicola metisiensis, Ostrincola* sp., and *Herrmannella duggani.*

**Symptoms/Impact**: *Myicola ostreae* causes gill lesions in *C. gigas* and *C. angulata* in France; *Myicola metisiensis* has affected *Mya arenaria* eastern Canada; *Ostrincola* sp. has caused severe gill erosion in a remnant introduced population of *C. virginica* in British Columbia; *Herrmannella duggani* has been reported from *O. edulis* which showed reduction in gills and indented gill margins (Humes, 1986; http://www.dfo-mpo.gc.ca/science/aah-saa/diseases-maladies/pcgoy-eng.html).

## 7.1.4 PARASITIC CRABS

The pea crabs are a common parasite of bivalves around the world. These small crabs often live inside oysters and are a type of leptoparasite, meaning they steal food from their hosts. Pea crabs sit on the gills and pick out some of the food the oyster traps before the oyster can consume it. By scurrying around inside oysters, the pea crabs can also damage the gills mechanically. The pea crabs, like most parasites, don't kill their hosts, but they can certainly affect the oysters' overall health (http://blog.wfsu.org/blog-coastal-health/?p=3779). The infestation of these pea crabs frequently causes problems for bivalve aquaculture also through end-consumer complaints and rejections of international consignments. The commercially important pea crabs and their parasitism are given below.

### 7.1.4.1 PEA CRAB PARASITISM

**(i) Host(s)**: *Ostrea edulis*, *Crassostrea virginica*, *Ostrea conchaphila*, and *Crassostrea gigas* and other species of bivalves including mussels, clams, cockles, and scallops.

**Causative agent**: *Pinnotheres pisum*, *Pinnotheres ostreum*, *Pinnixa littoralis*, and *Pinnotheres pholadis*

**Symptoms/Impact**: No evidence of direct pathology has been reported. However, the pea crab parasites rob their hosts' food resulting in a weakened condition of the hosts. Further, the infestation has been reported to develop a characteristic irritating flavor in the parasitized oysters (http://www.dfo-mpo.gc.ca/science/aah-saa/diseases-maladies/pcboy-eng.html).

**(ii) Host(s)**: *Argopecten irradians*

**Causative agent**: *Pinnotheres maculatus*

**Symptoms/Impact**: *P. maculatus* infestation has been reported to lower the reproductive potential of individual scallops, but the low rates of parasitism may only minimally impact the host population (Bologna and Heck, 2000).

**(iii) Host(s)**: *Perna canaliculus*

**Causative agent**: *Nepinnotheres novaezelandiae*

**Symptoms/Impact**: These pea crabs have been found to rapidly colonize farmed mussels and mature quickly to establish a significant breeding population within the mussel farm, with larval output. The latter, in turn, was capable of infecting nearby mussel farms as well as wild populations of bivalves (Trottier and Jeffs, 2012).

**(iv) Host(s)**: *Mytilus edulis*

**Causative agent**: *Pinnotheres pisum*

**Symptoms/Impact**: The presence of this crab causes gill damage and infected mussels show considerably lower tissue weights and slightly greater shell weights (Seed, 1969). Further, this species has been reported to cause nettle rash (reddish itchy weals or swellings) in mussels.

## 7.1.5   OTHER DISEASES

**Disseminated neoplasia:** Disseminated neoplasia, a diffuse tumor of the hemolymph system, is one of the most destructive diseases among bivalve mollusc populations and is characterized by the development of abnormal, rounded blood cells that actively proliferate. This disease has been reported in *Mytilus edulis* collected from an intertidal beach in Connecticut in western Long Island Sound. This disease has caused epizootic mortalities in another mussel species, *M. trossulus* (Galimany and Sunila, 2008). This disease has also been detected in cockle, *Cerastoderma edule*, from various beds in Galicia (NW Spain) (Díaz et al., 2011). A retrovirus could readily produce the chromosomal aberrations associated with disseminated neoplasia in *Mytilus trossulus* (Barber, 2004).

**Hemocytic neoplasia:** This disease otherwise called "blood cell proliferative disorder" has been reported in *Mytilus edulis*, *Mytilus trossulus*, and *Mya arenaria*, and in advanced stages, this disease is fatal in most cases (Elston et al., 1988).

**Gonadal neoplasia**: It consists of small, basophilic, undifferentiated cells that originate as small foci in gonadal follicles where they proliferate and eventually invade surrounding tissues. The most obvious sublethal effect of gonadal neoplasia on bivalves is the alteration of normal gametogenic processes. As the disease progresses, the undifferentiated germ cells proliferate and replace normal gametes, reducing fecundity in affected individuals. Deaths of at least some individuals have been reported in *Mya arenaria* and *Mercenaria* spp. (Barber, 2004).

## 7.2 POTENTIAL PARASITES IN EDIBLE MARINE BIVALVE MOLLUSCS

### 7.2.1 BACTERIAL PARASITES (SINDERMANN AND ROSENFIELD, 1967; HYUN ET AL., 2014; KIM ET AL., 2013)

| Host | Causative agent |
|---|---|
| *Anomalocardia brasiliana* | *Rickettsia-like organisms* |
| *Atrina pectinata* | *Endozoicomonas atrinae* |
| *Chlamys varia* | *Vibrio tapetis* |
| *Crassostrea cuttackensis* | *Myotomus ostrearum* |
| *Crassostrea gigas* | *Nocardia crassostreae* |
| | *Vibrio splendidus* |
| | *Achromobacter* |
| *Crassostrea rhizophora* | Rickettsia-like organisms |
| *Donax trunculus* | Rickettsia-like organisms |
| *Dosinia exoleta* | *Vibrio tapetis* |
| *Fulvia mutica* | *Bizionia fulviae* |
| *Iphigenia brasiliensis* | Rickettsia-like organisms |
| *Mytella bicolor* | Rickettsia-like organisms |
| *Mytilus edulis* | *Vibrios vibriosis* |
| | Rickettsia-like organisms |
| | Chlamydia-like organisms |
| *Ostrea edulis* | *Vibrio splendidus* |
| | *Vibrio kanaloae* |
| *Pecten maximus* | *Vibrio tapetis* |
| | *Vibrio splendidus* |
| *Perna canaliculus* | *Vibrio splendidus* |
| *Perna indica* | *Vibrio parahaemolyticus* |
| *Polititapes rhomboides* | *Vibrio tasmaniensis* |
| | *Vibrio tapetis* |
| *Pteria penguin* | Rickettsia-like organisms |

| Host | Causative agent |
|------|-----------------|
| *Ruditapes decussatus* | *Vibrio tapetis* |
| | *Vibrio splendidus* |
| *Ruditapes philippinarum* | *Vibrio tapetis* |
| *Scapharca broughtonii* | *Aliisedimentitalea scapharcae* |
| *Venerupis aurea* | *Vibrio tapetis* |
| *Venerupis corrugata* | *Vibrio tapetis* |

## 7.2.2 VIRUS (SINDERMANN AND ROSENFIELD, 1967; MAENO ET AL., 2006; BEDFORD ET AL., 1978)

| Host | Causative agent |
|------|-----------------|
| *Atrina pectinata* | Virus-like particles |
| *Crassostrea angulata* | Virus |
| *Crassostrea gigas* (larvae) | Virus |
| *Crassostrea glomerata* | Reovirus type III |
| | Semliki Forest virus |
| *Mytilus edulis* | Picornaviridae-like virus |
| *Ostrea edulis* | Herpes virus like (OHSV-1) |
| *Perna canaliculus* | Herpes virus |
| *Ruditapes decussatus* | Icosahedrical-virus-like organism |
| | Herpes-like virus |
| *Ruditapes philippinarum* | Herpes-like virus |
| *Venerupis corrugata* | Icosahedrical virus-like organism |

## 7.2.3 FUNGI (SINDERMANN AND ROSENFIELD, 1967)

| Host | Causative agent |
|------|-----------------|
| *Crassostrea angulata* | *Dermocystidium marinum* |
| *Crassostrea commercialis* | *D. marinum* |
| *Crassostrea rhizophorae* | *D. marinum* |
| *Crassostrea virginica* | *D. marinum* |
| *Dendostrea frons* | *D. marinum* |

| Host | Causative agent |
| --- | --- |
| *Ostrea edulis* | *D. marinum* |
| | *Ostracoblabe implexa* |
| *O. lurida* | *D. marinum* |
| *O. stentina* | *D. marinum* |
| Oyster and clam larvae | *Sirolpidium zoophthorum* |
| *Ruditapes decussatus* | *S. zoophthorum* |

## 7.2.4  PROTOZOAN PARASITES

(Sindermann and Rosenfield, 1967; Jasim et al., 2011; Moyer et al 1993; Goggin et al., 1987; Dungan et al., 2007; Moss et al., 2007; Renault et al., 2002)

| Host | Causative agent |
| --- | --- |
| *Alectryonella plicatula* | *Perkinsus* sp. |
| *Argopecten gibbus* | *Marteilia* sp. |
| *Argopecten ventricosus* | *Apicomplexa protozoan* |
| *Austrovenus stutchburyi* | *Perkinsus olsenii* |
| *Barbatia coralicola* | *Perkinsus olsenii* |
| *Chama pacificus* | *Perkinsus olsenii* |
| *Chama sulphurea* | *Perkinsus olsenii* |
| *Chlamys islandica* | *Apicomplexa protozoan* |
| *Crassostrea ariakensis* | *Perkinsus marinus* |
| | *Perkinsus olseni* |
| | *Bonamia* spp. |
| *Crassostrea corteziensis* | *Perkinsus marinus* |
| | *Vibrio mexicanus* |
| *Crassostrea gasar* | *Ancistrocoma* sp. |
| *Crassostrea gigas* | *Perkinsus marinus* |
| | *Mikrocytos mackini* |
| *Crassostrea iredalei* | *Nematopsis* sp. |
| *Crassostrea rhizophorae* | *Perkinsus beihaiensis* |
| | *Perkinsus marinus* |
| | *Nematopsis* sp. |
| | *Ancistrocoma* sp. |
| | *Trichodina* sp. |
| | *Sphenophrya* sp. |

| Host | Causative agent |
|------|-----------------|
| *Crassostrea rivularis* | *Bonamia-like protozoan* |
| *Crassostrea tulipa* | *Perkinsus* sp. |
| | *Nematopsis* sp. |
| *Crassostrea virginica* | *Perkinsus marinus* |
| | *Haplosporidium nelson* |
| | *Bonamia* spp. |
| *Leukoma thaca* | *Apicomplexa* |
| *Macoma balthica* | *Perkinsus chesapeaki* |
| *Mya arenaria* | *Perkinsus chesapeaki* |
| *Mytilus californianus* | *Haplosporidium tumifacientis* |
| *Mytilus galloprovincialis* | *Chytridiopsis mytilovum* |
| *Ostrea angasi* | *Bonamia exitiosa* |
| *Ostrea chilensis* | *Bonamia exitiosa* |
| | *Bonamia ostreae* |
| *Ostrea edulis* (eggs) | *Chytridiopsis ovicola* |
| *Pecten maximus* | *Pseudoklossia pectinis* |
| *Pinctada fucata* | *Perkinsus olseni* |
| *Saccostrea palmula* | *Perkinsus marinus* |
| *Spondylus nicobaricus* | *Perkinsus marinus* |
| *Spondylus squamosus* | *Perkinsus marinus* |
| *Tegillarca granosa* | *Nematopsis* sp. |
| *Tridacna crocea* | *Perkinsus marinus* |
| *Tridacna gigas* | *Perkinsus marinus* |
| *Tridacna maxima* | *Perkinsus marinus* |
| *Venerupis philippinarum* | *Perkinsus olseni* |

## 7.2.5  BORING SPONGES

| Host | Causative agent |
|------|-----------------|
| *Mytilus edulis* | *Cliona* spp. |
| *Placopecten magellanicus* | *Cliona vastifica* |
| | *Cliona celata* |
| *Patinopecten yessoensis* | *Cliona vastifica* |
| | *C. celata* |

## 7.2.6 HELMINTHS

### 7.2.6.1 TURBELLARIANS

(Sindermann and Rosenfield, 1967; Queiroga et al., 2015)

| Host | Causative agent |
|---|---|
| *Crassostrea gasar* | *Urastoma* sp. |
| *Crassostrea rhizophorae* | *Urastoma* sp. |
| *Macoma nasuta* | *Graffilla pugetensi* |
| *Mesodesma donacium* | *Paravortex* sp. |
| *Tegillarca granosa* | Unidentified turbellarian |
| *Venerupis corrugata* | *Paravortex* sp. |

### 7.2.6.2 TREMATODES

(Sindermann and Rosenfield, 1967)

| Host | Causative agent |
|---|---|
| *Anomalocardia brasiliana* | Bucephalidae (sporocysts) |
| *Anomalocardia flexuosa* | *Bucephalopsis haimeana* (cercaria) |
| *Aqeuipecten irradians* | Sporocysts and cercariae of trematodes |
| *Cerastoderma edule* | *Gymnophallus* larvae of trematodes |
| *Crassostrea gasar* | *Lobatostoma ringens* |
| | *Stephanostomum* sp. |
| *Crassostrea gigas* | *Proctoeces ostrea* (metacercariae) |
| *Crassostrea virginica* | *Bucephalus haimeanus*(larvae) |
| | *Proctoeces maculates* (metacercariae) |
| *Cyclina sinensis* | *Himasthla alincia* (metacercariae) |
| *Donax gouldii* | *Postmonorchis donacis* (larvae) |
| *Iphigenia brasiliana* | *Bucephalidae* (sporocysts) |
| *Mactra veneriformis* | *Acanthoparyphium marilae* |
| | *Himasthla alincia* (metacercariae) |

| Host | Causative agent |
|------|-----------------|
| *Meretrix petechialis* | *Himasthla alincia* (metacercariae) |
| *Mesodesma donacium* | Monorchiida (metacercariae) |
| *Mya arenaria* | *Hhnasthla quissetensis* |
| *Mytella guyanensi* | *Bucephalidae* (sporocysts) |
| *Mytilus edulis* | *Prosorhynchus* sp. |
| *Perna perna* | *Proctoeces maculatus* |
| *Ostrea chilensis* | *Bucephalus cucullus* (sporocysts) |
| | *Bucephalus haimeanus* |
| | *Bucephalus longicornutus* |
| *Ostrea edulis* | *Bucephalus haimeanus* (larvae) |
| | *Proctoeces maculates* (metacercariae) |
| *Ostrea lutaria* | *Bucephalus cuculus* (sporocysts) |
| *Pinna deltodes* | *Anthobothrium* sp. |
| *Pinctada fucata* | *Mutta margaritiferae* |
| | *Musalia herdmani* |
| | *Aspidogaster margaritifeae* |
| *Pinctada vulgaris* | *Muttua margaritiferae* |
| | *Musalia herdmani* (metacercariae) |
| | *Aspidogaster margaritiferae* |
| *Pinctada martensii* | *Bucephalus margaritae* (sporocysts and cercariae) |
| *Ruditapes philippinarum* | *Cercaria elegans* |
| *Ruditapes philippinarum* | *Himasthla alincia* (metacercariae) |
| | *Proctoeces orientalis* |
| *Saccostrea glomerata* | *Imogine mcgrathi* (ployclad-stylochid flatworm) |
| *Solen grandis* | *Himasthla alincia* (metacercariae) |
| *Tegillarca granosa* | *Trematode sporocysts* |
| *Venerupis corrugata* | Unidentified cercariae and metacercariae of trematodes |

## 7.2.6.3 CESTODES

(Sindermann and Rosenfield, 1967; Cake Jr., 1977; Erazo-Pagador, 2010; Cova et al., 2015)

| Host | Causative agent |
|---|---|
| *Anomalocardia brasiliana* | *Tylocephalum* sp. |
| *Argopecten irradians concentricus* | *Parachristianella* sp. |
| | *Polypocephalus* sp. |
| | *Tylocephalum* sp. |
| | *Rhinobothrium* sp. |
| *Atrina seminuda* | *Nybelinia* sp. |
| | *Rhinebothrium* sp. |
| *Austromacoma constricta* | *Acanthobothrium* sp. |
| *Crassostrea gasar* | *Tylocephalum* sp. |
| *Crassostrea iredalei* | *Tylocephalum* sp. |
| *Crassostrea rhizophorae* | *Tylocephalum* larvae |
| *Crassostrea virginica* | *Tylocephalum* (larval coracidium) |
| *Donax variabilis* | *Nybelinia* sp. |
| | *Acanthobothrium* sp. |
| *Dosinia discus* | *Rhinebothrium* sp. |
| *Iphigenia brasiliana* | *Tylocephalum* sp. |
| *Lunarca ovalis* | *Acanthobothrium* sp. |
| *Macrocallista nimbosa* | *Parachristianella* sp. |
| | *Tylocephalum* sp. |
| | *Acanthobothrium* sp. |
| | *Rhinobothrium* sp. |
| *Mesodesma donacium* | *Rodobothrium mesodesmatidae* |
| *Mytella guyanensi* | *Tylocephalum* sp. |
| *Noetia ponderosa* | *Rhinebothrium* sp. |
| *Ostrea edulis* | *Tylocephalum* (larval coracidium) |
| *Pinctada maxima* | *Tylocephalum* sp. |
| *Pteria penguin* | *Tylocephalum* sp. |
| | Tetrabothriate cestode |
| *Raeta plicatella* | *Acanthobothrium* sp. |
| | *Rhinebothrium* sp. |

| Host | Causative agent |
|---|---|
| *Schizothaerus nuttallii* | *Anabothrium* sp. |
| *Spisula solidissima* | *Parachristianella* sp. |
| | *Tylocephalum* sp. |
| | *Rhinobothrium* sp. |
| *Tagelus plebeius* | *Acanthobothrium* sp. |
| *Venerupis staminea* | *Echeneibothrium* sp. |

## 7.2.6.4 NEMATODES

(Sindermann and Rosenfield, 1967; Machkevsky, 1997; Rudders and Fisher, http://s3.amazonaws.com/nefmc.org/Doc5.VIMS-PDT-Nematode.pdf)

| Host | Causative agent |
|---|---|
| *Argopecten aequisulcatus* | Larvae of *Echinocephalus pseudouncinatus* |
| *Aequipecten maximus* | Larvae of *Sulcascaris sulcata* |
| *Argopecten gibbus* | Larvae of *S. sulcata* |
| *Chlamys bifrons* | Larvae of *Echinocepahlus overstreeti* |
| *Crassostrea gasar* | *Echinocephalus sinensis* |
| *Crassostrea gigas* | *E. sinensis* |
| | Larvae of *Echinocephalus crassostrea* |
| *Marcia opima* | Larvae of *Bucephala* sp. |
| *Margaritifera vulgaris* | Larvae of *E. pseudouncinatus* |
| *Pecten* sp. | Larvae of *S. sulcata* |
| *Pecten albus* | Larvae of *E. overstreeti* |
| *Pinctada fucata* | *Ascaris meleagrinae* |
| | *Cheiracanthus uncinatus* |
| | *Oxyuris* sp. |
| *Pinna* sp. | Larvae of *E. pseudouncinatus* |
| *R. philippinarum* | *P. orientalis* |
| *Ruditapes philippinarum* | *P. orientalis* |
| *Spisula solidissima* | Larvae of *S. sulcata* |
| *Tegillarca granosa* | *P. orientalis* |

## 7.2.7 POLYCHAETES

(Tinoco-Orta and Cáceres-Martínez, 2003)

| Host | Causative agent |
|------|-----------------|
| *Anomalocardia flexuosa* | *Polydora* sp. |
| *Chione fluctifraga* | *Polydora* sp. |
| *Crassostrea rhizophorae* | *Polydora* sp. |
| | *Neanthes succinea* |
| *Mytilus edulis* | *Polydora ciliata* |

## 7.2.8 COPEPOD PARASITES

(Sindermann and Rosenfield, 1967; Caceres-Martinez et al., 2005)

| Host | Causative agent |
|------|-----------------|
| *Argopecten ventricosus* | *Pseudomyicola spinosus* |
| *Chione cancellata* | *pseudomyicola spinosus* |
| *Crassostrea angulata* | *Myticola ostreae* |
| *Crassostrea gigas* | *M. ostreae* |
| | *Mytilicola orientalis* |
| *Crasssostrea virginica* | *Mytilicola intestinalis* |
| *Mactra veneriformis* | *Mytilicola mactrae* |
| *Mercenaria mercenaria* | *Mytilicola porrecta* |
| *Meretrix meretrix* | *Conchyliurus bombasticus* |
| | *Conchyliurus fragilis* |
| | *Lichomolgus similis* |
| | *Ostrincola portonoviensis* |
| *Mesodesma donacium* | *Paranthessius mesodesmatis* |
| *Mytilus crassitesta* | *M. orientalis* |
| *Mytilus edulis* | *M. intestinalis* |
| | *M. orientalis* |
| *Mytilus galloprovincialis* | *M. intestinalis* |
| *Modiolus demissus* | *Mytilicola porrecta* |
| *Mytilus crassistesta* | *M. orientalis* |
| *Myltilus edulis* | *M. orientalis* |
| *Mytilus recurvus* | *Mytilicola porrecta* |
| *Perna canaliculus* | *Lichomolgus uncus* |
| | *P. spinosus* |
| *Ruditapes decussatus* | *M. intestinalis* |
| Isopod parasite | |
| *Mytilus edulis platensis* | *Edotia doellojuradoi* |

## 7.2.9   PARASITIC PINNOTHERID CRABS

(Hornell and Southwell, 1909)

| Host | Causative agent |
|---|---|
| *Anomalocardia flexuosa* | *Holothuriophilus tomentosus* |
| *Atrina rigida* | *Pinnotheres maculatus* (apparently a symbiont) |
| *Chione stutchbury* | *N. novaezelandiae* |
| *Crassostrea belcheri* | *Pinnotheres* sp. |
| *Crassostrea gigas* | *N. novaezelandiae* |
| *Crassostrea virginica* | *Zaops ostreus* |
| *Mytilus edulis* | *N. novaezelandiae* |
| | *Pinnotheres pisum* |
| *Ostrea puelchana* | *Tumidotheres maculatus* |
| *Ostrea vitrifacta* | *Pinnotheres* sp. |
| *Perna canaliculus* | *N. novaezelandiae* |
| *Placuna placenta* | *Pinnoteres placunse* |
| *Saccostrea palmua* | *Juxtafabia muliniarum* |

## KEYWORDS

- bacteria
- virus
- fungi
- protozoa
- helminths
- polychate
- crustacean parasites

# REFERENCES

Abdel-Salam, H. A. Assessment of Biochemical Compositions and Mineral Contents of Carapace of Some Important Commercially Crustaceans and Molluscs Organisms from Egyptian and Saudi Arabia Coasts as a New Animal Feed. *Am. J. BioSci.* **2013**, *1*, 35–44.

Abdullah, A.; Nurjanah, N.; Hidayat, T.; Gifari, A. Characterize Fatty Acid of *Babylonia spirata, Meretrix meretrix, Pholas dactylus. Int. J. Chem. Biomol. Sci.* **2016**, *2*, 38–42.

Achuthankutty, C. T.; Sreekumaran Nair, S. R.; Madhupratap, M. Pearls of the Window-pane Oyster, *Placuna placenta. Mahasagar* **1979**, *12*, 187–189.

Ademolu, K. O.; Akintola, M. Y.; Olalonye, A. O.; Adelabu, B. A. Traditional Utilization and Biochemical Composition of Six Mollusc Shells in Nigeria. *Rev. Biol. Tropic.* **2015**, *63*, 1.

Afiati, N. Hermaphriditism in *Anadara granosa* (L.) and *Anadara antiquata* (L.) (Bivalvia: Arcidae) from Central Java. *J. Coast. Dev.* **2007**, *10*, 171–179.

Ajithkumar, P.; Jeganathan, N. S.; Balamurugan, K.; Manvalan, R.; Radha, K. Evaluation of Anti-ulcer Activity of *Villorita cyprinoides* Extract (Black Water Clams) against Immobilization Stress Induced Ulcer in Albino Rats. *J. Pharm. Res. Opin.* **2012**, *2*, 55–57.

Alagarswami, K. On Pearl Formation in the Venerid Bivalve, *Gafrarium tumidum* Roding. *J. Mar. Biol. Assoc. India* **1965**, *7*, 345–347.

Al-Sayed, 2010. http://www.qu.edu.qa/qulss/10/components/hashim.pdf.

Annamalai, N.; Anburaj, R.; Jayalakshmi, S.; Thavasi, R. Antibacterial Activities of Green Mussel (*Perna viridis*) and Edible Oyster (*Crassostrea madrasensis*). *Res. J. Microbiol.* **2007**, *2*, 978–982.

Anon. Infection with *Perkinsus olseni*. In *Manual of Diagnostic Tests for Aquatic Animals*; 2009; pp 354–365.

Anon. nda. http://www.dfo-mpo.gc.ca/science/aah-saa/diseases-maladies/madoy-eng.html.

Anon. ndb. http://www.pcas.org/documents/BoxtandMillerweb.pdf.

Anon. ndc. http://www.seafoodhealthfacts.org/description-top-commercial-seafood-items/oysters.

Anon. 1989. *Yield and Nutritional Value of the Commercially More Important Fish Species*; FAO 1989; p 187.

Anon. *Fungal Infections Fungi of Bivalvia Linn 1758*, 2014. http://www.rrnursingschool.biz/invertebrate-pathology/fungal-infections-fungi-of-bivalvia-linn-1758.html.

Anon. ndd. http://shodhganga.inflibnet.ac.in/bitstream/10603/64373/16/16_chapter%2011.pdf.

Anon. nde. http://www.dfo-mpo.gc.ca/science/aah-saa/diseases-maladies/rcomu-eng.html.

Anon. ndf. http://www.dfo-mpo.gc.ca/science/aah-saa/diseases-maladies/gilldpoy-eng.html.

Anon. ndg. http://www.dfo-mpo.gc.ca/science/aah-saa/diseases-maladies/mbranoy-eng.html.

Anon. ndh. http://www.dfo-mpo.gc.ca/science/aah-saa/diseases-maladies/hcoy-eng.html.

Anon. ndi. http://www.dfo-mpo.gc.ca/science/aah-saa/diseases-maladies/nemparoy-eng.html.

Anon. ndj. http://www.epharmacognosy.com/2012/06/sea-clam-shell-haigeqiao-cyclina.html.

Anon. ndk. http://www.epharmacognosy.com/2012/06/sea-clam-shell-haigeqiao-cyclina. html.

Anon. ndl. http://www.et97.com/view/92153.htm.

Anon. ndm. http://www.fooducate.com/app#!page=product&id=1E79F85E-ACEE-11E2-9B11-1231381A4CEA.

Anon. ndn. http://www.manandmollusc.net/beginners_uses/3.html.

Anon. ndo. http://www.nefsc.noaa.gov/publications/classics/galtsoff1964/chap17.pdf.

Anon. ndp. http://www.seafoodsource.com/seafoodhandbook/shellfish/clam-softshell.

Anon. ndq. http://www.seafoodsource.com/seafoodhandbook/shellfish/clam-surf.

Anon. ndr. http://www.technology-x.net/A61K/201010505519.html.

Anon. nds. http://www.topicsinresearch.com/wiki/Nutritious_composition_analysis_of_solen_gouldii_muscle.

Anon. ndt. https://researcharchive.lincoln.ac.nz/bitstream/handle/10182/2863/Shi_MAppl Sci.pdf;jsessionid=38DD43C59874CDBB2A130629500B4DE1?sequence=5.

Anon. ndu. https://en.wikipedia.org/wiki/Spondylus.

Anon. *Saucer Scallop, Amusium balloti, Amusium pleuronectes*; FRDC, 2014. http://www.fishfiles.com.au/knowing/species/molluscs/scallops/Pages/Saucer-Scallop.aspx.

Anon. ndv. http://www.fao.org/fishery/culturedspecies/Venerupis_pullastra/en.

Anon. What is a bivalve mollusk? https://oceanservice.noaa.gov/facts/bivalve.html

Argente, F. A. T.; Ilano, A. S. Susceptibility of Some Pathogenic Microbes to Soft Tissue Extract of the Mud Clam, *Polymesoda expansa* (Bivalvia: Corbiculidae). *Experiment* **2015**, *30*, 1984–1990.

Asha, K. K.; Anandan, R.; Mathew, S.; Lakshmanan, P. T. Biochemical Profile of Oyster *Crassostrea madrasensis* and Its Nutritional Attributes. *Egypt. J. Aquat. Res.* **2014**, *40*, 35–41.

Babar, A. G.; Pande, A.; Kulkarni, B. G. Antifungal Activity and Investigation of Bioactive Compounds of Marine Intertidal Bivalve *Gafrarium divaricatum* from West Coast of India. *Int. J. Pure App. Biosci.* **2016**, *4*, 211–217.

Babu, A.; Venkatesan, V.; Rajagopal, S. Biochemical Composition of Different Body Parts of *Gafrarium tumidum* (Roding, 1798) from Mandapam, South East Coast of India. *Afr. J. Biotechnol.* **2012**, *11*, 1700–1704.

Baez, O.; Isabel, R. *Aspects of the Biology of Modiolus capax (Conrad, 1837), in the Bay of La Paz, BC Sur, Mexico*, 1987. http://www.repositoriodigital.ipn.mx/handle/123456789/15397.

Baptista, M.; Repolho, T.; Maulvault, A. L.; Lopes, V. M.; Narciso, L.; Marques, A.; Bandarra, N.; Rosa, R. Temporal Dynamics of Amino and Fatty Acid Composition in the Razor Clam *Ensis siliqua* (Mollusca: Bivalvia). *Helgol. Mar. Res.* **2014**, *68*, 465–482.

Barber, B. J. Neoplastic Diseases of Commercially Important Marine Bivalves. *Aquat. Liv. Res.* **2004**, *17*, 449–466.

Bayat, Z.; Hassanshahian, M.; Hesni, M. A. Study the Symbiotic Crude Oil-Degrading Bacteria in the Mussel *Mactra stultorum* Collected from the Persian Gulf. *Mar. Pollut. Bull.* **2016**, *105*, 120–124.

Bedford, A. J.; Williams, G.; Bellamy, A. R. Virus Accumulation by Rock Oyster *Crassostrea glomerata*. *Appl. Environ. Microbiol.* **1978**, *35*, 1012–1018.

Berthou, P.; Poutiers, J. M.; Goulletquer, P.; Dao, J. C. Shelled Molluscs. *Encyclopedia of Life Support Systems (EOLSS)*. nd. http://archimer.ifremer.fr/doc/2001/publication-529. pdf, 24p).

Bieler, R.; Kappner, I.; Mikkelsen, P. M. *Periglypta listeri* (J. E. Gray, 1838) (Bivalvia: Veneridae) in the Western Atlantic: Taxonomy, Anatomy, Life Habits, and Distribution. *Malacologia* **2004**, *46*, 427–458.

Boehs, G.; Villalba, A.; Ceuta, L. O.; Luz, J. R. Parasites of Three Commercially Exploited Bivalve Mollusc Species of the Estuarine Region of the Cachoeira River (Ilhéus, Bahia, Brazil). *J. Invertebr. Pathol.* **2010**, *103*, 43–47.

Bologna, P. A. X.; Heck Jr., K. L. Relationship between Peacrab (*Pinnotheres maculates*) Parasitism and Gonad Mass of the Bay Scallop (*Argopecten irradians*). *Gulf Caribbean Res.* **2000**, *12* (Abstracts).

Borquaye, L. S.; Darko, G.; Ocansey, E.; Ankomah, E. Antimicrobial and Antioxidant Properties of the Crude Peptide Extracts of *Galatea paradoxa* and *Patella rustica*. *SpringerPlus* **2015**, *4*, 500. DOI:10.1186/s40064-015-1266-2.

Botana, L.M. ( ed.) Seafood and Freshwater Toxins ( III Edn), CRC Press, 2014.

Bower, S. M.; Meyer, G. R. *Mikrocytosis (Denman Island Disease of Oysters)*; 2004. http://afs-fhs.org/perch/resources/14069253075.2.5microcytosis2014.pdf.

Bozbaş, S. K.; Boz, Y. Low-Cost Biosorbent: *Anadara inaequivalvis* Shells for Removal of Pb(II) and Cu(II) from Aqueous Solution. *IChem. E, J.* **2016**, *103* (Part A), 144–152.

Brien, S.; Prescott, P.; Coghlan, B.; Bashir, N.; Lewith, G. Systematic Review of the Nutritional Supplement *Perna canaliculus* (Green-Lipped Mussel) in the Treatment of Osteoarthritis. *QJM: Int. J. Med.* **2008**, *101*, 167–179.

Buranon, T. R. A Comparison of Macronutrient Levels in Green Mussel (*Perna viridis*) and Brown Mussel (*Modiolus metcalfei* Hanley). *J. Sci. Soc. Thailand* **1980**, *6*, 191–197.

Burzio, L. A.; Saéz, C.; Pardo, J.; Waite, J. H.; Burzio, L. O. The Adhesive Protein of *Choromytilus chorus* (Molina, 1782) and *Aulacomya ater* (Molina, 1782): A Proline-Rich and a Glycine-Rich Polyphenolic Protein. *Biochim. Biophys. Acta* **2000**, *15*, 315–320.

Caro, A.; Gros, O.; Got, P.; Wit, R. D.; Troussellier, M. Characterization of the Population of the Sulfur-Oxidizing Symbiont of *Codakia orbicularis* (Bivalvia, Lucinidae) by Single-Cell Analyses. *Appl. Environ. Microbiol.* **2007**, *73*, 2101–2109.

Cáceres-Martínez, C.; Chávez-Villalba, J.; Garduño-Méndez, L. First Record of *Pseudomyicola spinosus* in *Argopecten ventricosus* in Baja California, Mexico. *J. Invertebr. Pathol.* **2005**, *89*, 95–100.

Cake, Jr., E. W. Larval Cestode Parasites of Edible Molluscs of the Northeastern Gulf of Mexico. *Gulf Res. Rep.* **1977**, *6*, 1–8.

Caughey, D. E.; Grigor, R. R.; Caughey, E. B.; Young, P.; Gow, P. J.; Stewart, A. W. *Perna canaliculus* in the Treatment of Rheumatoid Arthritis. *Eur. J Rheumatol. Inflamm.* **1983**, *6*, 197–200.

Chakraborty, K.; Vijayagopal, P.; Asokan, P. K.; Vijayan, K. K. Green Mussel (*Perna viridis L.) as Healthy Food, and as a Nutraceutical Supplement*. CMFRI Digital Repository, 2011, pp 89–94. http://eprints.cmfri.org.in/9763/1/Kajal_Chakraborty_Green_Mussel_Farming.pdf.

Changgui, D. Experimental Studies on Hypoglycemia and Hypolipid Effects of Hydrolysate of *Arca subcrenata*. *Chin. J. Mar. Drugs* **1996**, *15*, 13–15.

Chaohua, L.; Wenzao, X.; Guojun, Z. Studies on *Crassostrea rivularis* as a Biological Indicator of Cadmium Pollution. *J. Fish. Sci. China* **1998**, *5*, 79–83.

Chatterji, A.; Ansari, Z. A.; Baban, S. Ingole, B. S.; Bichurina, M. A.; Sovetova, M.; Boikov, Y. A. Indian Marine Bivalves: Potential Source of Antiviral Drugs. *Curr. Sci.* **2002**, *82*, 1279–1282.

Chen, H.; Chu, X.; Yan, C.; Chen, X.; Sun, M.; Wang, Y.; Wang, C.; Yu, W.. Polypeptide from *Chlamys farreri* Attenuates Murine Thymocytes Damage Induced by Ultraviolet B. *Acta Pharmacol. Sin.* **2007**, *28*, 1665–1670.

Chen, X.; Han, Y.; Zhan, S.; Wang, C.; Chen, S. *Tegillarca granosa* Extract Haishengsu Induces Apoptosis in Human Hepatocellular Carcinoma Cell Line BEL-7402 via Fas-Signaling Pathways. *Cell Biochem. Biophys.* **2015**, *71*, 837–844.

Chen, L.; Song, L.; Li, T.; Zhu, J.; Xu, J.; Zheng, Q.; Yu, R. A New Antiproliferative and Antioxidant Peptide Isolated from *Arca subcrenata. Mar. Drugs* **2013**, *11*, 1800–1814.

Cheng, L. S.; Tao, L. D.; Long, G. J. Yong, H. B.; Hua, Z. C.; Ming, H. J.; Li, Z. Lipid components of *Ostrea rivularis, Paphia undulata* and *Pinctada martensii. J. Fish. China* **2009**, *33*, 666–671.

Cheong, S. H.; Hwang, J. W.; Lee, S. H.; Kim, Y. S.; Sim, E. J.; Kim. E. K.; You, B. I.; Lee, S. H.; Park, D. J.; Ahn, C. B.; Jeon, B. T.; Moon, S. H.; Park, P. J. Anti-inflammatory Effect of Short Neck Clam (*Tapes philippinarum*) Water Extract Containing Taurine in Zebrafish Model. *Adv. Exp. Med. Biol.* **2015**, *803*, 819–831.

Chiarelli, R.; Roccheri, M. C. Marine Invertebrates as Bioindicators of Heavy Metal Pollution. *Open J. Met.* **2014**, *4*, 93–106.

Chikalovets, I. V.; Chernikova, O. V.; Pivkina, M. V.; Molchanovaa, V. I.; Litovchenkob, A. P.; Lic, W.; Lukyanova, P. A. A Lectin with Antifungal Activity from the Mussel *Crenomytilus grayanus. Fish Shellfish Immunol.* **2015**, *42*, 503–507.

Cova, A. W.; Junior, M. S.; Boehs, G.; Souza, J. M. Parasites in the Mangrove Oyster *Crassostrea rhizophorae* Cultivated in the Estuary of the Graciosa River in Taperoá. *Bahia Rev. Bras. Parasitol.* **2015**, *24*, 21–27.

Cui, Q.; Wang, H.; Yuan, C. The Preliminary Study on the Antithrombotic Mechanism of Glycosaminoglycan from *Mactra veneriformis. Blood Coagul Fibrinol.* **2014**, *25*, 16–19.

Dagorn, F.; Couzinet-Mossion, A.; Kendel, M.; Beninger, P. G.; Rabesaotra, V.; Barnathan, G.; Wielgosz-Collin, G. Exploitable Lipids and Fatty Acids in the Invasive Oyster *Crassostrea gigas* on the French Atlantic Coast. *Mar. Drugs* **2016**, *14*, 104.

Dang, V. T.; Benkendorff, K.; Green, T.; Speck, P. Marine Snails and Slugs: A Great Place to Look for Antiviral Drugs. *J. Virol.* **2015**, *89*, 8114–8118.

Davies, I. C.; Jamabo, N. A. Proximate Composition of Edible Parts of Shellfishes from Okpoka Creeks in Rivers State, Nigeria. *Int. J. Life Sci. Res.* **2016**, *4*, 247–252.

de León-Espinosa, A.; de León-González, J. A. Pycnogonids Associated with the Giant Lion's-Paw Scallop *Nodipecten subnodosus* (Sowerby) in Ojo de Liebre Bay, Guerrero Negro, Baja California Sur, Mexico. *Zookeys* **2015**, *530*, 129–149.

Díaz, S.; Renault, T.; Antonio Villalba, A.; Carballal, M. J. Disseminated Neoplasia in Cockles *Cerastoderma edule*: Ultrastructural Characterisation and Effects on Haemolymph Cell Parameters. *Inter-Res. Dis. Aquat. Organ.* **2011**, *96*, 157–167.

Dietrich, C. P.; de Paiva, J. F.; Moraes, C. T.; Takashi, H. K.; Porcionatto, M. A.; Nader, H. B. Isolation and Characterization of a Heparin with High Anticoagulant Activity from *Anomalocardia brasiliana. Biochim. Biophys. Acta* **1985**, *843*, 1–7.

Diez, M. E.; Vázquez, N.; Urteaga, D.; Cremonte, F. Species Associations and Environmental Factors Influence Activity of Borers on *Ostrea puelchana* in Northern Patagonia. *J. Mollus. Stud.* **2014**, *80*, 430–434.

Dincer, T. Differences of Turkish Clam (*Ruditapes decussatus*) and Manila Clam (*Ruditapes philippinarum*) According to their Proximate Composition and Heavy Metal Contents. *J. Shellfish Res.* **2006**, *25*, 455–459.

Do, V.; Budha, P. B.; Daniel, B. A. *Polymesoda bengalensis. The IUCN Red List of Threatened Species*, 2012: e.T166774A1141971. http://dx.doi.org/10.2305/IUCN.UK.2012-1.RLTS.T166774A1141971.en.

Dungan, C. F.; Reece, K. S.; Moss, J. A.; Hamilton, R. M.; Diggles, B. K. *Perkinsus olseni* In Vitro Isolates from the New Zealand Clam *Austrovenus stutchburyi. J. Eukaryot. Microbiol.* **2007**, *54*, 263–270.

Elston, R. A.; Kent, M. L.; Drum, A. S. Transmission of Hemic Neoplasia in the Bay Mussel, *Mytilus edulis*, Using Whole Cells and Cell Homogenate. *Dev. Compar. Immunol.* **1988**, *12*, 719–727.

Erazo-Pagador, G. A Parasitological Survey of Slipper-Cupped Oysters (*Crassostrea iredalei*, Faustino, 1932) in the Philippines. *J. Shellfish Res.* **2010**, *29*, 177–179.

Estradaa, N.; Ascencioa, F.; Liora Shoshani, L.; Contreras, R. G. Apoptosis of Hemocytes from Lions-Paw Scallop *Nodipecten subnodosus* Induced with Paralyzing Shellfish Poison from *Gymnodinium catenatum. Immunobiology* **2014**, *219*, 964–997.

Eswar, A.; Nanda, R. K.; Ramamoorthy, K.; Isha, Z.; Gokulakrishnan, S. Biochemical Composition and Preliminary Qualitative Analysis of Marine Clam *Gafrarium divaricatum* (Gmelin) from Mumbai, West Coast of India. *Asian J. Biomed. Pharm. Sci.* **2016**, *6*, 1–6.

FAO. *The State of World Fisheries and Aquaculture*, 2014. http://www.fao.org/3/a-i3720e/i3720e01.pdf.

Fengör, G. F.; Gün, H.; Kalafatoglu, H. Determination of the Amino Acid and Chemical Composition of Canned Smoked Mussels (*Mytilus galloprovincialis*, L.) *Turk. J. Vet. Anim. Sci.* **2008**, *32*, 1–5.

Fernández-Reiriz, M. J.; Labarta, U.; Albentosa, M.; Pérez-Camacho, A. Lipid Profile and Growth of the Clam Spat, *Ruditapes decussatus* (L), Fed with Microalgal Diets and Cornstarch. *Compar. Biochem. Physiol., B* **1999**, *124*, 309–318.

FISHEAT. nd. http://www.fisheat.it/donax-clam-donax-trunculus/Nutritionalfacts.

Frazier, J.; Margaritoulis, D.; Muldoon, K.; Potters, W.; Rosewater, J.; Ruckdesche, C.; Salas, S. Epizoan Communities on Marine Turtles I. Bivalve and Gastropod Molluscs. *P. S. Z. N. I: Mar. Ecol.* **1985**, *6*, 127–140.

Freites, L.; Fernández-Reir, M. J.; Labarta, U. Fatty Acid Profiles of *Mytilus galloprovincialis* (Lmk) Mussel of Subtidal and Rocky Shore Origin. *Compar. Biochem. Physiol., B: Biochem. Mol. Biol.* **2002**, *132*, 453–461.

Friedman, C. S.; Beaman, B. L.; Chun, J.; Goodfellow, M.; Gee, A.; Hedrick, R. P. *Nocardia crassostreae* sp. Nov.: The Causal Agent of Nocardiosis in Pacific Oysters. *Int. J. Syst. Bacteriol.* **1998**, *48*, 237–246.

Galimany, E.; Sunila, I. Several Cases of Disseminated Neoplasia in Mussels *Mytilus edulis* (L.) in Western Long Island Sound. *J. Shellfish Res.* **2008**, *27*, 1201–1207.

García-Morales, H.; Gutiérrez-Millán, L. E.; Valdez, M. A.; Burgos-Hernández, A.; Gollas-Galván, T.; Burboa, M. G. Antiproliferative Activity of Protein Extracts from the Black Clam (*Chione fluctifraga*) on Human Cervical and Breast Cancer Cell Lines. *Afr. J. Biotechnol.* **2016**, *15*, 214–220.

Giftson, H.; Patterson, J. Antibacterial Activity of the Shell Extracts of Marine Mollusc *Donax faba* against Pathogens. *Int. J. Microbiol. Res.* **2014**, *5*, 140–143.

Gnanambal; et al.; nd. http://nopr.niscair.res.in/handle/123456789/1694.

Goggin, C. L.; Lester, R. J. G. Occurrence of *Perkinsus* Species (Protozoa, Apicomplexa) in Bivalves from the Great Barner Reef. *Dis. Aquat. Org.* **1987**, *3*, 113–117.

Gomes, A. M.; Kozlowskia, E. O.; Pominb, V. H.; de Barrosa, C. M.; Zaganeli, J. L.; Mauro S. G.; Pavãoa, M. S. G. Unique Extracellular Matrix Heparan Sulfate from the Bivalve *Nodipecten nodosus* (Linnaeus, 1758) Safely Inhibits Arterial Thrombosis after Photo Chemically Induced Endothelial Lesion. *J. Biol. Chem.* **2010**, *285*, 7312–7323.

Góngora-Gómez, A. M.; Muñoz-Sevilla, N. P.; Hernández-Sepúlveda, J. A.; García-UlloaII, M.; García-Ulloa, M. Association between the Pen Shell *Atrina tuberculosa* and the Shrimp *Pontonia margarita*. *Symbiosis* **2015**, *66*, 107–110.

Gopalsamy, I.; Arumugam, M.; Kumaresan, S.; Thangavel, B. Nutritional Value of Marine Bivalve, *Donax cuneatus* (Linnaeus, 1758) from Cuddalore Coastal Waters, Southeast Coast of India. *Invent. Impact: Life Style* **2014**, *1*, 15–19.

Gopeechund, A.; Bhagooli, R.; Sharadha, V.; Bhujun, N.; Bahoun, T. Antioxidant Activities of Edible Marine Molluscs from a Tropical Indian Ocean Island. nd. http://symposium.wiomsa.org/wp-content/uploads/2015/10/A.-GOPEECHUND.pdf.

Gosling, E. *Bivalve Molluscs: Biology, Ecology and Culture*; John Wiley and Sons: Hoboken, NJ, 2003; 456 p.

Green, T. J.; Raftos, D.; Speck, P.; Montagnani, C. Antiviral Immunity in Marine Molluscs. *J. Gen. Virol.* **2015**, *96*, 2471–2482.

Gunawarman, Affi, J.; Ilhamdi Gundini, R.; Ahli, A. Characterization of Bioceramic Powder from Clamshell (*Anadara antiquata*) Prepared By Mechanical and Heat Treatments for Medical Application. In: *Proceeding Seminar Nasional Tahunan Teknik Mesin* XIV (SNTTM XIV), 2015, material 9; 6 p.

Himmelman, J. H.; Guderley, H. E.; Duncan, P. F. Responses of the Saucer Scallop *Amusium balloti* to Potential Predators. *J. Exp. Mar. Biol. Ecol.* **2009**, *378*, 58–61.

Hine, P. M.; Thorne, T. A Survey of Some Parasites and Diseases of Several Species of Bivalve Mollusc in Northern Western Australia. *Dis. Aquat. Organ.* **2000**, *40*, 67–78.

Ho,J.W.Y.;Zhou,C.*NaturalPearlsReportedlyfromaSpondylusSpecies("Thorny"Oyster)*. nd. http://www.gia.edu/gems-gemology/fall-2014-labnotes-natural-pearls-spondylus-species.

Holm, G. P. nd. Pearl Found in *Solen Sicarius*. http://www.bily.com/pnwsc/web-content/Articles/Pearl%20found%20in%20Solen%20sicarius.pdf.

Hornell, J.; Southwell, T. Description of a New Species of *Pinnotheres* from *Placuna placenta*, with a Note on the Genus. *Report to Government of Baroda on Marine Zoology of Olchamandal in Kathiawar*, Part I, London, 1903; pp 99–103.

Hu, X.; Song, L.; Huang, L.; Zheng, Q.; Yu, R. Antitumor Effect of a Polypeptide Fraction from *Arca subcrenata* In Vitro and In Vivo. *Mar. Drugs* **2012**, *10*, 2782–2794.

Hubert, F.; Knaap, W.; Noël, T.; Roch, P. Cytotoxic and Antibacterial Properties of *Mytilus galloprovincialis*, *Ostrea edulis* and *Crassostrea gigas* (bivalve molluscs) Hemolyrnph. *Aquat. Living Resour.* **1996**, *9*, 115–124.

Hwang, J. Reproductive Cycles of the Pearl Oysters, *Pinctada fucata* (gould) and *Pinctada margaritifera* (Linnaeus) (Bivalvia: Pteriidae) in Southwestern Taiwan Waters. *J. Mar. Sci. Technol.* **2007**, *15*, 67–75.

Hyun, D. W.; Shin, N. R.; Kim, M. S.; Oh, S. J.; Kim, P. S.; Whon, T. W. *Endozoicomonas atrinae* sp. nov., Isolated from the Intestine of a Comb Pen Shell *Atrina pectinata*. *Int. J. Syst. Evol. Microbiol.* **2014**, *64*, 2312–2318.

Jaramillo, E.; Navarro, J.; Winter, J. The Association between *Mytilus chilensis* Hupe (Bivalvia, Mytilidae) and *Edotea magellanica* Cunningham (Isopoda, Valvifera) in Southern Chile. *Biol. Bull.* **1981**, *160*, 107–113.

Jasim, U. M.; Zulfigar, Y.; Munawar, K.; Shau-Hwai, A. T. Parasites of Blood Cockle *Anadara granosa* (Linnaeus, 1758) from the Straits of Malacca. *J. Shellfish Res.* **2010**, *30*, 875–880.

Jiang, C.; Xiong, Q.; Gan, D.; Jiao, Y.; Liu, J.; Ma, L.; Zeng, X. Antioxidant Activity and Potential Hepatoprotective Effect of Polysaccharides from *Cyclina sinensis*. *Carbohydr. Polym.* **2013**, *91*, 262–268.

Jiménez, M.; Pineda, F.; Sánchez, I.; Orozco, I.; Senent, C. Allergy Due to *Ensis macha*. *Allergy (Eur. J. Allergy Clin. Immunol.)* **2005**, *60*, 1090–1091.

Jing-jing, S.; Hui-Hui, L.; Shi-Quan, Z.; Xin-Chao, W.; Mei-Hua, F.; Wang, S.; Zzhi, L. A Novel Antimicrobial Peptide Identified from *Mytilus coruscus*. *Acta Hydrobiol. Sin.* **2014**, *38*, 563–570.

Joy, M.; Chakraborty, K.; Pananghat, V. Comparative Bioactive Properties of Bivalve Clams against Different Disease Molecular Targets. *J. Food Biochem.* **2016**, *40*, 593–602.

Jung, W.; Kim, S-J. Isolation and Characterisation of an Anticoagulant Oligopeptide from Blue Mussel, *Mytilus edulis*. *Food Chem.* **2009**, *117*, 687–692.

Jung, W.; Rajapakse, N.; Kim, S. Antioxidative Activity of a Low Molecular Weight Peptide Derived from the Sauce of Fermented Blue Mussel, *Mytilus edulis*. *Eur. Food Res. Technol.* **2005**, *220*, 535–539.

Karnjanapratum, S.; Benjakul, S.; Kishimura, H.; Tsai, Y. Chemical compositions and nutritional value of Asian hard clam (Meretrix lusoria) from the coast of Andaman Sea. *Food Chem.* **2013**, *141*, 4138–4145.

Karunasagar, I. *Bivalve mollusc Production, Trade and Codex Guidelines*. FAO. nd. http://www.eurofish.dk/~efweb/images/stories/files/Albania/01.pdf.

Kern, F. Dermo Disease of Oysters Caused by *Perkinsus marinus*. In *ICES* 2011, Leaflet No. 30; 5 p.

Khiri, M. Z. A.; Matori, K. A.; Zainuddin, N.; Abdullah, C. A. C.; Alassan, Z. N.; Baharuddin, N. F.; Zaid, M. H. M. The Usability of Ark Clam Shell (*Anadara granosa*) as Calcium Precursor to Produce Hydroxyapatite Nanoparticle via Wet Chemical Precipitate Method in Various Sintering Temperature. *SpringerPlus* **2016**, *5*, 1206.

Kim, S. Marine Medicinal Foods: Implications and Applications—Animals and Microbes. *Food and Nutr. Res.* **2012**, *65*, 523.

Kim, S. *Seafood Science: Advances in Chemistry, Technology and Applications*; CRC Press: Boca Raton, FL, 2014; p 606.

Kim, E.; Kim, Y.; Hwang, J.; Park, P. Purification and Characterization of a Novel Anti-cancer Peptide Derived from *Ruditapes philippinarum*. *Process Biochem.* **2013**, *48*, 1086–1090.

Kiran, N.; Siddiqui, G.; Khan, A. N.; Ibrar, K.; Tushar, P. Extraction and Screening of Bioactive Compounds with Antimicrobial Properties from Selected Species of Mollusc and Crustaceans. *Clin. Cell. Immunol.* **2014**, *5*, 5.

Ko, R. C. *Echinocephalus sinensis* n. sp. (Nematoda: Gnathostomatidae) from the Ray (*Aetabatus flagellum*) in Hong Kong, Southern China. *Can. J. Zool.* **2011**, *53*, 490–500.

Koyama, T.; Chounan, R.; Uemura, D.; Yamaguchi, K.; Yazawa, K. Hepatoprotective Effect of a Hot-Water Extract from the Edible Thorny Oyster *Spondylus varius* on Carbon Tetra-chloride-Induced Liver Injury in Mice. *Biosci., Biotechnol. Biochem.* **2006**, *70*, 729–731.

Krzynowek, J.; Murphy, J. *Proximate Composition, Energy, Fatty Acid, Sodium, and Cholesterol Content of Finfish, Shellfish, and their Products*; NOAA Technical Report NMFS 55, 1987; 53 p.

Laxmilatha, P. Proximate Composition of the Surf Clam *Mactra violacea* (Gmelin 1791). *Indian J. Fish.* **2009**, *56*, 147–150.

Leslie et al. nd. http://edis.ifas.ufl.edu/fe568.

Li, Y.; Sadiq, F. A.; Fu, L.; Zhu, H.; Zhong, M.; Sohail, M. Identification of Angiotensin I-Converting Enzyme Inhibitory Peptides Derived from Enzymatic Hydrolysates of Razor Clam *Sinonovacula constricta*. *Mar. Drugs* **2016**, *14*, 110.

Li, L.; Li, H.; Qian, J.; He, Y.; Zheng, J.; Lu, Z.; Xu, Z.; Shi, J. Structural and Immuno-logical Activity Characterization of a Polysaccharide Isolated from *Meretrix meretrix* Linnaeus. *Mar. Drugs* **2016**, *14*, 6.

Liao, J. H.; et al. A Multivalent Marine Lectin from *Crenomytilus grayanus* Possesses Anti-cancer Activity through Recognizing Globotriose Gb3. *J. Am. Chem. Soc.* **2016**, *138*, 4787–4795.

Lin, H.; Yashiro, H.; Yamada, T.; Oshim, Y. Purification and Characterization of Paralytic Shellfish Toxin Transforming Enzyme from *Mactra chinensis*. *Toxicon* **2004**, *44*, 657–668.

Liu, R.; Zhu, Y.; Chen, J.; Wu, H.; Shi, L.; Wang, X.; Wang, L. Characterization of ACE Inhibitory Peptides from *Mactra veneriformis* Hydrolysate by Nano-Liquid Chromatog-raphy Electrospray Ionization Mass Spectrometry (Nano-LC-ESI-MS) and Molecular Docking. *Mar. Drugs* **2014**, *12*, 3917–3928.

Liu, W.; Li, Q.; Kong, L. Estradiol-17β and Testosterone Levels in the Cockle *Fulvia mutica* during the Annual Reproductive Cycle. *New Zeal. J. Mar. Freshwater Res.* **2008**, *42*, 417–424.

Luna-González, A.; Maeda-Martínez, A.; Campa-Córdova, A.; Orduña-Rojas, J. Antibac-terial Activity in the Hemolymph of the Catarina Scallop *Argopecten ventricosus*. *Hidro-biológica* **2007**, *17*, 87–89.

Luppi, E.; Cesaretti, M.; Volpi, N. Purification and Characterization of Heparin from the Italian Clam *Callista chione*. *Biomacromolecules*, **2005**, *6*, 1672–1678.

Machkevsky, V. K. Endosymbionts of Mangrove Oyster in Nature and under Cultivation. In: Ross, J. D., Guerra, A., Eds.; *Ecology of Marine Molluscs*; Scientia Marina, CSIC: Barcelona, 1997; pp 99–107.

Madhu, V. N.; Sivaperumal, P.; Kamala, K.; Ambekar, A. A.; Kulkarni, B. G. Antibacte-rial and Antioxidant Activities of the Tissue Extract of *Perna viridis* Linnaeus, 1758

(Mollusca: Bivalvia) from Versova Coast, Mumbai. *Int. J. Pharmacy Pharm. Sci.* **2014**, *6*, 704–707.

Maeno, Y.; Yurimoto, T.; Nasu, H.; Ito, S. Virus-Like Particles Associated with Mass Mortalities of the Pen Shell *Atrina pectinata* in Japan. *Dis. Aquat. Org.* **2006**, *71*, 169–173.

Malhotra, P.; Dasaradhi, P. V. N.; Mohammed, A.; Mukherji, S.; Manivel, V.; Rao, K. V. S.; Mishra, G. C.; Parameswaran, P. S.; Chatterj, A. Use of Phosphono Derivatives of Selected Aliphatic Acids for Anti-malarial Activity. Indian Patent 0268NF, 2003.

Maoka, T.; Fujiwara, Y.; Hashimoto, K.; Akimoto, N. Carotenoids in Three Species of *Corbicula* Clams, *Corbicula japonica, Corbicula sandai,* and *Corbicula* sp. (Chinese Freshwater Corbicula Clam). *J. Agric. Food Chem.* **2005**, *53*, 8357–8364.

Marichamy, G.; Shanker, S.; Saradha, A.; Nazar, A. R.; Haq, M. A. B. Proximate Composition and Bioaccumulation of Metals in Some Finfishes and Shellfishes of Vellar Estuary (South East Coast of India). *Eur. J. Exp. Biol.* **2011**, *1*, 47–55.

Marie-Agnès, T.; Katharine, M.; Ana, R.; Friedman, C. S. Bacterial Diseases in Marine Bivalves. *J. Invertebr. Pathol.* **2015**, *131*, 11–31.

Martínez-Córdova, L. R.; López-Elías, J. A.; Martínez-Porchas, M.; Bernal-Jaspeado, T.; Miranda-Baeza, A. Studies on the Bioremediation Capacity of the Adult Black Clam, *Chione fluctifraga*, of Shrimp Culture Effluents. *Rev. Biol. Mar. Oceanogr.* **2011**, *46*, 105–113.

Matsuno, T.; Sakaguchi, S.; Ookubo, M.; Maoka, T. Isolation and Identification of Amarouciaxanthin A from the Bivalve *Paphia euglypta. Bul. Jpn. Soc. Sci. Fish.* **1985**, *51*, 1999.

McNabb, P. S.; Taylor, D. I.; Ogilvie, S. C.; Wilkinson, L.; Anderson, A.; Hamon, D.; Wood, S. A.; Peake, B. M. First Detection of Tetrodotoxin in the Bivalve *Paphies australis* by Liquid Chromatography Coupled to Triple Quadrupole Mass Spectrometry with and without Precolumn Reaction. *JAOAC Int.* **2014**, *97*, 325–333.

Meng, J.; Zhang, G.; Huang, Z. Structure and Identification of Solenin: A Novel Fibrous Protein from Bivalve *Solen grandis* Ligament. *J. Nanomater.* **2014**, Article ID 241975, 7 p.

Miller and Boxt. nd. http://www.pcas.org/documents/BoxtandMillerweb.pdf.

Miranda, A.; Voltolina, D.; Fierro, G. I.; López, I. S.; Sandoval, I. Removal of Suspended Solids from the Effluents of a Shrimp Farm by the Blood Ark *Anadara tuberculosa* (Sowerby, 1833). *Hidrobiológica* **2009**, *19*, 173–176.

Mirshahi, M.; Mirshahi, P.; Negro, S.; Soria, J.; Sreekumar, P. K.; Kotnala, S.; Therwath, A.; Pertanika, A. C. J. Extract of Indian Green Mussel, *Perna viridis* (L.) Shows Inhibition of Blood Capillary Formation In Vitro. *Tropic. Agric. Sci.* **2009**, *32*, 35–42.

Mohammad, S. H.; Yusuf, M. S. Proximate Evaluation of Some Economical Seafood as a Human Diet and as an Alternative Prospective Valuable of Fish Meal. *J. Fish. Aquat. Sci.* **2016**, *11*, 12–27.

Mohanraj, T.; Prabhu, K.; Lakshmanasenthil, S. Antimicrobial Activity of *Pteria penguin* against Human Pathogens from the Southeast Coast of India. *Int. J. Pharma BioSci.* **2012**, *3*, 65–70.

Moss, J. A.; Burreson, E. M.; Cordes, J. F.; Dungan, C. F.; Brown, G. D.; Wang, A.; Wu, X.; Reece, K. S. Pathogens of *Crassostrea ariakensis* and Other Asian Oyster Species: Implications for Non-native Oyster Introduction to Chesapeake Bay. *Dis. Aquat. Org.* **2007**, *77*, 207–223.

Moyer, M.; Blake, N.; Arnold, W. An Ascetosporan Disease Causing Mass Mortality in the Atlantic Calico Scallop, *Argopecten gibbus* (Linnaeus, 1758). *J. Shellfish Res.* **1993**, *12*, 305–310.

Murphy, K. J.; Mann, N. J.; Sinclair, A. J. Fatty Acid and Sterol Composition of Frozen and Freeze-Dried New Zealand Green Lipped Mussel (*Perna canaliculus*) from Three Sites in New Zealand. *Proc. Nutr. Soc. Austr.* **2002**, *26*, S305.

Nagvenkar, S. S.; Jagtap, T. G. Spatio-temporal Variations in Biochemical Composition, Condition Index and Percentage Edibility of the Clam, *Paphia malabarica* (Chemnitz) from Estuarine Region of Goa. *Indian J. Mar. Sci.* **2013**, *42*, 786–793.

Nair, D. G.; Weiskirchen, R.; Al-Musharafi, S. K. The Use of Marine-Derived Bioactive Compounds as Potential Hepatoprotective Agents. *Acta Pharmacol. Sin.* **2015**, *36*, 158–170.

Nazeer, R. A.; Prabha, K. R.; Kumar, N. S.; Ganesh, R. J. Isolation of Antioxidant Peptides from Clam, *Meretrix casta* (Chemnitz). *J. Food Sci. Technol.* **2013**, *50*, 777–783.

Nicholaus, R.; Zheng, Z. The Effects of Bioturbation by the Venus Clam *Cyclina sinensis* on the Fluxes of Nutrients across the Sediment–Water Interface in Aquaculture Ponds. *Aquacult. Int.* **2014**, *22*, 913–924.

Ninan, L.; Stroshine, R. L.; Wilker, J. J.; Shi, R. Adhesive Strength and Curing Rate of Marine Mussel Protein Extracts on Porcine Small Intestinal Submucosa. *Acta Biomater.* **2007**, *3*, 687–694.

Ning, X.; Zhao, J.; Zhang, Y.; Cao, S.; Liu, M.; Ling, P.; Lin, X. A Novel Anti-tumor Protein Extracted from *Meretrix meretrix* Linnaeus Induces Cell Death by Increasing Cell Permeability and Inhibiting Tubulin Polymerization. *Int. J. Oncol.* **2009**, *35*, 805–812.

Normah, I.; Noorasma, M. Physicochemical Properties of Mud Clam (*Polymesoda erosa*) Hydrolysates Obtained Using Different Microbial Enzymes. *Int. Food Res. J.* **2015**, *22*, 1103–1111.

Ohnoki, S.; Mitomi, Y.; Hata, R.; Satake, K. Heterogeneity of Hemoglobin from *Arca* (*Anadara satowi*) Molecular Weights and Oxygen Equilibria of Arca Hb I and II. *J. Biochem.* **1973**, *73*, 717–725.

Paez-Osuna, F.; Zazueta-Padilla, H. M.; Osuna-López, J. Biochemical Composition of the Oysters *Crassostrea iridescens* Hanley and *Crassostrea corteziensis* Hertlein in the Northwest Coast of Mexico: Seasonal Changes. *J. Exp. Mar. Biol. Ecol.* **1993**, *170*, 1–9.

Pan, M. H.; Huang, Y. T.; Ho, C. T.; Chang, C. I.; Hsu, P. C.; Sun, Pan, B. Induction of Apoptosis by *Meretrix lusoria* through Reactive Oxygen Species Production, Glutathione Depletion, and Caspase Activation in Human Leukemia Cells. *Life Sci.* **2006**, *79*, 1140–1152.

Pandian, V.; Thirugnanasambandan, S. Glycosaminoglycans (GAG) from Backwater Clam *Marcia opima*. *Iran. J. Pharmacol. Therap.* **2008**, *7*, 147–151.

Pawar, R. T.; Nagvenkar, S. S.; Jagtap, T. G.; Dergisi, T. B. Protective Role of Edible Clam *Paphia malabarica* (Chemnitz) against Lipid Peroxidation and Free Radicals. *Turk. J. Biochem.* **2013**, *38*, 138–144.

Pejler, G.; Danielsson, A.; Bjork, I.; Lindahl, U.; Nader, H. B.; Dietrich, C. P. Structure and Antithrombin-Binding Properties of Heparin Isolated from the Clams *Anomalocardia brasiliana* and *Tivela mactroides*. *J. Biol. Chem.* **1987**, *262*, 11413–11421.

Periyasamy, N.; Murugan, S.; Bharadhirajan, P. Isolation and Characterization of Anticoagulant Compound from Marine Mollusc *Donax faba* (Gmelin, 1791) from Thazhanguda, Southeast Coast of India. *Afr. J. Biotechnol.* **2013**, *12*, 5968–5974.

Periyasamy, N.; Murugan, S.; Bharadhirajan, P. Biochemical Composition of Marine Bivalve *Donax incarnatus* (Gmelin, 1791) from Cuddalore Southeast Coast of India. *Int. J. Adv. iPharmacy, Biol. Chem.* **2014**, *3*, 575–582.

Perkins, F. O. *Dermocystidium marinum* Infection in Oysters. *Mar. Fish. Rev.* **1976**, http://spo.nmfs.noaa.gov/mfr3810/mfr38107.pdf.

Pogoda, B.; Buck, B. H.; Saborowski, R.; Hagen, W. Biochemical and Elemental Composition of the Offshore-Cultivated Oysters *Ostrea edulis* and *Crassostrea gigas*. *Aquaculture* **2013**, *400–401*, 53–60.

Pradit, S.; Shazili, N. A.; Towatana, P.; Saengmanee, W. Accumulation of Trace Metals in *Anadara granosa* and *Anadara inaequivalvis* from Pattani Bay and the Setiu Wetlands. *Bull. Environ. Contam. Toxicol.* **2016**, *96*, 472–477.

Qian, Z.; Jung, W.; Byun, H. G.; Kim, S. Protective Effect of an Antioxidative Peptide Purified from Gastrointestinal Digests of Oyster, *Crassostrea gigas* against Free Radical Induced DNA Damage. *Bioresour. Technol.* **2008**, *99*, 3365–3371.

Queiroga, F. R.; Vianna, R. T.; Vieira, C. B.; Farias, N. D. Parasites Infecting the Cultured Oyster *Crassostrea gasar* (Adanson, 1757) in Northeast Brazil. *Parasitology* **2015**, *142*, 756–766.

Quezada, A. T. Age Determination and Growth of *Chione californiensis* (Broderip, 1835 (Bivalvia: Veneridae) in the Bahia de La Paz, Gulf of California, México. In *45th Annual Meeting of the Western Society of Malacologists and International Workshop on Opisthobranchs*, Santa Cruz, CA, 2012; p 62.

Radić, I. D.; Carić, M.; Najdek, M.; Jasprica, N.; Bolotin, J.; Peharda, M.; Cetinić, A. B. Biochemical and Fatty Acid Composition of *Arca noae* (Bivalvia: Arcidae) from the Mali Ston Bay, Adriatic Sea. *Medit. Mar. Sci.* **2014**, *15*, 520–531.

Rahman, M. A.; Parvej, M. R.; Rashid, M. H.; Hoq, M. E. Availability of Pearl Producing Marine Bivalves in South-Eastern Coast of Bangladesh and Culture Potentialities. *J. Fish.* **2015**, *3*, 293–296.

RaLonde, R. Paralytic Shellfish Poisoning: The Alaska Problem. *Alaska's Mar. Resour.* **1996**, *8*, 7.

Ramasamy, M.; Balasubramanian, U. Identification of Bioactive Compounds and Antimicrobial Activity of Marine Clam *Anadara granosa* (Linn.) *Int. J. Sci. Nat.* **2016**, *3*, 263–266.

Ramesh, S.; Sankar, V.; Santhanam, R. *Marine Pharmaceutical Compounds*; Lambert Academic Publishing 2012; 249 p.

Rao, K. V. S.; Wani, M. R.; Manivel, V.; Parameswaran, P. S.; Singh, V. K.; Anand, R. V.; Desa, E.; Mishra, G. C.; Chatterji, A. *Novel Molecules to Develop Drug for the Treatment of Osteoporosis*. Indian Patent #0412NF2003, US, 30/12/2003, 10/747, 671.

Renault, T. Appearance and Spread of Diseases among Bivalve Molluscs in the Northern Hemisphere in Relation to International Trade. *Rev. Sci. Tech. Off. Int. Epiz.* **1996**, *15*, 551–561.

Renault, T.; Chollet, B.; Cochennec, N.; Gerard, A. Shell Disease in Eastern Oysters, *Crassostrea virginica*, Reared in France. *J. Invertebr. Pathol.* **2002**, *80*, 1–6.

Rice, M. A. *Environmental Effects of Shellfish Aquaculture in the Northeast*. NRAC Publication No. 105, 2008; p 6.

Romanenkoa, L. A.; Uchinob, M.; Kalinovskayaa, N. I.; Mikhailova, V. V. Isolation, Phylogenetic Analysis and Screening of Marine Mollusc-Associated Bacteria for Antimicrobial, Hemolytic and Surface Activities. *Microbiol. Res.* **2008**, *163*, 633–644.

Salaskar, G. M.; Nayak, V. N. Nutritional Quality of Bivalves, *Crassostrea madrasensis* and *Perna viridis* in the Kali Estuary, Karnataka, India. *Rec. Res. Sci. Technol.* **2011**, *3*, 6–11.

Sani, N. K.; Vijayan, K. K. nd. *Diseases and Parasites of Bivalves*. http://eprints.cmfri.org.in/9764/1/N_K_Sanil_Green_Mussel_Farming.pdf.

Saritha, K.; Mary, D.; Patterson, J. Nutritional Status of Green Mussel *Perna viridis* at Tamil Nadu, Southwest Coast of India. *J. Nutr. Food Sci.* **2015**, *S14*, 003. DOI:10.4172/2155-9600.S14-003.

Sarizan, N. M. B. *Identification of Anti-atherosclerotic Compounds from Marine Molluscs of Bidong Archipelago.* M.Sc. Thesis of Universiti Malaysi, Teregganu, 2013.

Sasaki, J.; Wang, M.; Liu, J.; Wang, J.; Uchisawa, H.; Lu, C. Fired Shell Powder of Bivalve *Corbicula Japonica* Improves Mal-Function of Liver—Possible Development of Multi-Functional Calcium. *J. US–China Med. Sci.* **2011**, *8*, 449–457.

Sasaki, S.; Ohta, T. Seasonal Variations of Chemical Components in Surfcalm (*Spisula sachalinensis*). *Food Sci. Technol.* **1999**, *5*, 311–315.

Sathyan, N.; Philip, R.; Chaithanya, E. R.; Anil Kumar, P. R. Identification and Molecular Characterization of Molluscin, a Histone-H2A-Derived Antimicrobial Peptide from Molluscs. *ISRN Mol. Biol.* **2012**, *2012*, Article ID 219656, 6 p.

Saxena, A. *Text Book of Biochemistry*; Discovery Publishing House 2006; 604 p.

Scarratt, K.; Pearce, C.; Johnson, P. A Note on a Pearl Attached to the Interior of *Crassostrea virginica* (Gmelin, 1791) (an Edible Oyster, Common Names, American or Eastern Oyster). *J. Gemm.* **2006**, *30*, 43–50.

Schmeer, M. R.; Beery, G. Mercenene, Growth-Inhibitor Extracted from the Clam *Mercenaria campechensis*. A Preliminary Investigation of In Vivo and in In Vitro Activity. *Life Sci.* **1965**, *4*, 2157–2163.

Scotti, P. D.; Dearing, S. C.; Greenwood, D. R.; Newcomb, R. D. Pernin: A Novel, Self-Aggregating Haemolymph Protein from the New Zealand Greenlipped Mussel, *Perna canaliculus* (Bivalvia: Mytilidae). *Compar. Biochem. Physiol. B* **2001**, *128*, 767–779.

Seraspe, E. B.; Abaracoso, M. Screening of Extracts of Diwal (*Pholas orientalis*) for Antimicrobial Activities. *J. Aquacult. Mar. Biol.* **2014**, *1*, 00002.

Sharma, S.; Anil Chatterji, A.; Das, P. Pertanika Effect of Different Extraction Procedures on Antimicrobial Activity of Marine Bivalves: A Comparison. *J. Trop. Agric. Sci.* **2009**, *32*, 77–83.

Shinn, A. P.; Pratoomyot, J.; Bron, E.; Paladini, G.; Brooker, E. E.; Brooker, A. J. Economic Costs of Protistan and Metazoan Parasites to Global Mariculture. *Parasitology* **2015**, *142*, 196–270.

Silva, B. A.; Araujo, E. F.; Lorenzo, L. H.; Ferreira, G. R. S.; Lake, A. R.; Steps, M.; Silva, N. S. Minerals and Amino Acids Profile of Oysters (*Crassostrea gasar*) Grown in New Olinda. nd. http://iufost.org.br/sites/iufost.org.br/files/anais/11162.pdf.

Sindermann, C. J.; Rosenfield, A. Principal Diseases of Commercially Important Marine Bivalve Mollusca and Crustacean. *Fish. Bull.* **1967**, *66*, 335–385.

Singh, Y. T.; Krishnamoorthy, M.; Thippeswamy, S. Seasonal Changes in the Biochemical Composition of Wedge Clam, *Donax scortum* from the Padukere Beach, Karnataka. *Rec. Res. Sci. Technol.* **2012**, *4*, 12–17.

Sirbu, R.; Bechir, A.; Negreanu-Pîrjol, T.; Sava, C.; Negreanu-Pîrjol, B.; Zaharia, T.; Ursache, C.; Stoicescu, R. M. Studies on the Nutrient Content of Species *Mytilus galloprovincialis* of the Black Sea. *Sci. Study Research, Chem. Chem. Eng., Biotechnol., Food Ind.* **2011**, *12*, 221–228.

Sreejamole, K. L.; Radhakrishnan, C. K. Antioxidant and Cytotoxic Activities of Ethyl Acetate Extract of the Indian Green Mussel *Perna viridis. Asian J. Pharm. Clin. Res.* **2013**, *6*, 197–201.

Sri Kantha, S. Carotenoids of Edible Molluscs: A Review. *J. Food Biochem.* **1989**, *13*, 429–442.

Srilatha, G.; Chamundeeswari, K.; Ramamoorthy, K.; Sankar, G.; Varadharajan, D. Proximate, Amino Acid, Fatty Acid and Mineral Analysis of Clam, *Meretrix casta* (Chemnitz) from Cuddalore and Parangipettai Coast, South East Coast of India. *J. Mar. Biol. Oceanogr.* **2013**, *2*, 2. DOI:http://dx.doi.org/10.4172/2324-8661.1000111.

Sturmer, L. N.; Morgan, K. L.; Degner, R. L. *Nutritional Composition and Marketable Shelf-Life of Blood Ark Clams and Ponderous Ark Clams.* Publication #FE568, *University of Florida, IFAS Extension.* nd. http://edis.ifas.ufl.edu/fe568.

Subhapradha, N.; Suman, S.; Ramasamy, P.; Saravanan, R.; Shanmugam, V.; Srinivasan, A.; Shanmugam, A. Anticoagulant and Antioxidant Activity of Sulfated Chitosan from the Shell of Donacid Clam *Donax scortum* (Linnaeus, 1758). *Int. J. Nutr. Pharmacol. Neurol. Dis.* **2013**, *3*, 39–45.

Sujit, S.; Deshmukh, V. D.; Raje, S. G. Population Eruption of Sunset Shell *Siliqua radiata* (Linnaeus, 1758) along Versova Beach in Mumbai. *J. Mar. Biol. Assoc. India* **2010**, *52*, 99–101.

Takarina, N. D.; Bengen, D. G.; Sanusi, H. S.; Riani, E. Bioconcentration Factor of Copper, Lead, and Zinc in Anadara indica Related to the Water Quality in Coastal Areas. *Makara Seri Sains* **2013**, *17*, 23–28.

Tamari, M.; Kandatsu, M. Occurrence of Ceramide Aminoethylphosphonate in Edible Shellfish, AGEMAKI, *Sinonovacula constricta. Agric. Biol. Chem.* **1986**, *50*, 1495–1501.

Tan, Y.; Xing, J.; Zhan, W. Agglutination Activities of Haemolymph and Tissue Extracts in Scallop *Chlamys farreri* and Purification of Mannan-Binding Lectin from Haemolymph. *Aquaculture* **2013**, *400–401*, 148–152.

Teshima, S.; Kanazawa, A.; Koshio, S.; Mukai, H.; Yamasaki, S.; Hirata, H. Fatty Acid Details for Bivalves, *Tapes philippinarum* and *Corbicula japonica*, and Marine Types of Algae, *Nannochloropsis* sp. and *Chlorella* sp. *Mem. Fac. Fish. Kagoshima Univ.* **1990**, *39*, 137–149.

Tinoco-Orta, G. D.; Caceres Martinez, J. Infestation of the Clam *Chione fluctifraga* by the Burrowing Worm *Polydora* sp. Nov. in Laboratory Conditions. *J. Invertebr. Pathol.* **2003**, *83*, 196–205.

Trottier, O.; Jeffs, A. G. Biological Characteristics of Parasitic *Nepinnotheres novaezelandiae* within a *Perna canaliculus* Farm. *Dis Aquat Organ.* **2012**, *101*, 61–68.

Tun, K. L.; Shimizu, Y.; Yamanoi, H.; Yoshinaga, T.; Ogawa, K. Seasonality in the Infection and Invasion of *Marteilioides chungmuensis* in the Pacific Oyster *Crassostrea gigas. Dis. Aquat. Organ.* **2008**, *80*, 157–165.

Umayaparvathi, S.;. Arumugam, M.; Meenakshi, S.; Dräger, G.; Kirschning, A.; Balasu-bramanian, T. Purification and Characterization of Antioxidant Peptides from Oyster (*Saccostrea cucullata*) Hydrolysate and the Anticancer Activity of Hydrolysate on Human Colon Cancer Cell Lines. *Int. J. Pept. Res. Therap.* **2014**, *20*, 231–243.

Venkatesan, V. *Marine Ornamental Molluscs.* National Training Programme on Ornamental Fish Culture, 2010. http://eprints.cmfri.org.in/8671/1/Marine_Ornamental_Fish_Culture.pdf.

Vyncke, W.; Hillewaert, H.; Guns, M.; van Hoeyweghen, P. Trace Metals in Cut Trough Shell (*Spisula subtruncata*) from Belgian Coastal Waters. *Food Addit. Contam.* **1999**, *16*, 1–8.

Waite, J. H. Mussel Glue from *Mytilus californianus* Conrad: A Comparative Study. *J. Comp. Physiol. B* **1986**, *156*, 491–496.

Wang, Y.; He, H.; Wang, G.; Wu, H.; Zhou, B.; Xiu-Lan Chen, X.;, Zhang, Y. Oyster (*Crassostrea gigas*) Hydrolysates Produced on a Plant Scale Have Antitumor Activity and Immunostimulating Effects in Balb/c Mice. *Mar. Drugs* **2010**, *8*, 255–268.

Wang, L.; Wu, H.; Chang, N.; Zhang, K. Anti-hyperglycemic Effect of the Polysaccharide Fraction Isolated from *Mactra veneriformis*. *Front. Chem. Sci. Eng.* **2011**, *5*, 238–244.

WILD Fact Sheets. nd. http://www.wildsingapore.com/wildfacts/mollusca/bivalvia/spondylidae/spondylidae.htm.

Wu, X.; Zhao, H.; Che, J.; Feng, R.; Li, C.; Zhang, Z.; Zhang, C.; Li, G.; Zhao, Y. The History and Outlook of Animal Drugs Treating Asthma, Chronic Bronchitis, and Haze Episode-Induced Respiratory Diseases. *Int. J. Biotechnol. Wellness Ind.* **2014**, *3*, 69–78.

Xu, J.; Chen, Z.; Song, L.; Chen, L.; Zhu, J.; Lv, S.; Yu, R. A New in Vitro Anti-Tumor Polypeptide Isolated from *Arca inflata*. *Mar. Drugs* **2013**, *11*, 4773–4787.

Yadzir, Z. H. M.; Misnan, R.; Bakhtiar, F.; Abdullah, N.; Murad, S. Tropomyosin and Actin Identified as Major Allergens of the Carpet Clam (*Paphia textile*) and the Effect of Cooking on Their Allergenicity. *BioMed. Res. Int.* **2015**, Article ID 254152, 6.

Yang, H.; Sturmer, L. N.; Baker, S. *Molluscan Shellfish Aquaculture and Production.* nd. http://edis.ifas.ufl.edu/fa191.

Yuan, Q.; Zhao, L.; Cha, Q.; Sun, Y.; Ye, H.; Zeng, X. Structural Characterization and Immunostimulatory Activity of a Homogeneous Polysaccharide from *Sinonovacula constricta*. *J. Agric. Food Chem.* **2015**, *63*, 7986–7994.

Zawadzki, M.; Janosch, C.; Szechinski, J. *Perna canaliculus* Lipid Complex PCSO-524 Demonstrated Pain Relief for Osteoarthritis Patients Benchmarked against Fish Oil, a Randomized Trial, without Placebo Control. *Mar. Drugs* **2013**, *11*, 1920–1935.

Zeidan; et al. nd. http://dx.doi.org/10.1590/S1984-29612012000400009.

# INDEX

Milton Keynes UK
Ingram Content Group UK Ltd.
UKHW022042141024
449569UK00022B/791

9 781774 630648